Advances of Heat Transfer in Porous Media

Advances of Heat Transfer in Porous Media

Editors

Moghtada Mobedi
Kamel Hooman

MDPI • Basel • Beijing • Wuhan • Barcelona • Belgrade • Manchester • Tokyo • Cluj • Tianjin

Editors
Moghtada Mobedi
Shizuoka University
Japan

Kamel Hooman
Technische Universiteit Delft
The Netherlands

Editorial Office
MDPI
St. Alban-Anlage 66
4052 Basel, Switzerland

This is a reprint of articles from the Special Issue published online in the open access journal *Energies* (ISSN 1996-1073) (available at: https://www.mdpi.com/journal/energies/special_issues/heat_transfer_in_porous_media).

For citation purposes, cite each article independently as indicated on the article page online and as indicated below:

LastName, A.A.; LastName, B.B.; LastName, C.C. Article Title. *Journal Name* **Year**, *Volume Number*, Page Range.

ISBN 978-3-0365-6710-5 (Hbk)
ISBN 978-3-0365-6711-2 (PDF)

© 2023 by the authors. Articles in this book are Open Access and distributed under the Creative Commons Attribution (CC BY) license, which allows users to download, copy and build upon published articles, as long as the author and publisher are properly credited, which ensures maximum dissemination and a wider impact of our publications.

The book as a whole is distributed by MDPI under the terms and conditions of the Creative Commons license CC BY-NC-ND.

Contents

About the Editors vii

Preface to "Advances of Heat Transfer in Porous Media" ix

Jacqueline F. B. Diniz, João M. P. Q. Delgado, Anderson F. Vilela, Ricardo S. Gomez, Arianne D. Viana, Maria J. Figueiredo, and et al.
Drying of Sisal Fiber: A Numerical Analysis by Finite-Volumes
Reprinted from: *Energies* **2021**, *14*, 2514, doi:10.3390/en14092514 1

João C. S. Melo, João M. P. Q. Delgado, Wilton P. Silva, Antonio Gilson B. Lima, Ricardo S. Gomez, Josivanda P. Gomes, and et al.
Non-Equilibrium Thermodynamics-Based Convective Drying Model Applied to Oblate Spheroidal Porous Bodies: A Finite-Volume Analysis
Reprinted from: *Energies* **2021**, *14*, 3405, doi:10.3390/en14123405 27

Elisiane S. Lima, João M. P. Q. Delgado, Ana S. Guimarães, Wanderson M. P. B. Lima, Ivonete B. Santos, Josivanda P. Gomes, and et al.
Drying and Heating Processes in Arbitrarily Shaped Clay Materials Using Lumped Phenomenological Modeling
Reprinted from: *Energies* **2021**, *14*, 4294, doi:10.3390/en14144294 49

Nihad Dukhan
Equivalent Parallel Strands Modeling of Highly-Porous Media for Two-Dimensional Heat Transfer: Application to Metal Foam
Reprinted from: *Energies* **2021**, *14*, 6308, doi:10.3390/en14196308 75

Trilok G, N Gnanasekaran and Moghtada Mobedi
Various Trade-Off Scenarios in Thermo-Hydrodynamic Performance of Metal Foams Due to Variations in Their Thickness and Structural Conditions
Reprinted from: *Energies* **2021**, *14*, 8343, doi:10.3390/en14248343 93

Kazuhisa Yuki, Risako Kibushi, Ryohei Kubota, Noriyuki Unno, Shigeru Tanaka and Kazuyuki Hokamoto
Heat Transfer Potential of Unidirectional Porous Tubes for Gas Cooling under High Heat Flux Conditions
Reprinted from: *Energies* **2022**, *15*, 1042, doi:10.3390/en15031042 117

Jian Zhang, Haochun Zhang, Yiyi Li, Qi Wang and Wenbo Sun
Thermal Cloaking in Nanoscale Porous Silicon Structure by Molecular Dynamics
Reprinted from: *Energies* **2022**, *15*, 1827, doi:10.3390/en15051827 127

Trilok G, Kurma Eshwar Sai Srinivas, Devika Harikrishnan, Gnanasekaran N and Moghtada Mobedi
Correlations and Numerical Modeling of Stacked Woven Wire-Mesh Porous Media for Heat Exchange Applications
Reprinted from: *Energies* **2022**, *15*, 2371, doi:10.3390/en15072371 141

Stephane K. B. M. Silva, Carlos J. Araújo, João M. P. Q. Delgado, Ricardo S. Gomez, Hortência L. F. Magalhães, Maria J. Figueredo, and et al.
Heat and Mass Transfer in Structural Ceramic Blocks: An Analytical and Phenomenological Approach
Reprinted from: *Energies* **2022**, *15*, 7150, doi:10.3390/en15197150 167

Rawal Diganjit, N. Gnanasekaran and Moghtada Mobedi
Numerical Study for Enhancement of Heat Transfer Using Discrete Metal Foam with Varying Thickness and Porosity in Solar Air Heater by LTNE Method
Reprinted from: *Energies* **2022**, *15*, 8952, doi:10.3390/en15238952 . . . . **183**

About the Editors

Moghtada Mobedi

Moghtada Mobedi is a professor of heat transfer, and he works in the Mechanical Engineering Department, Shizuoka University, in Japan. He received his PhD from Middle East Technical University, Turkey in 1994. After working in a HVAC company as a project manager, he worked in the Mechanical Engineering Department of Izmir Institute of Technology in Turkey between 2003 and 2015. Since 2015, he has been working in Shizuoka University and continues his research in Japan. He has taught many bachelor, master and PhD courses such as Heat Transfer, Computational Fluid Dynamics, Convective Heat Transfer, Numerical Methods in Heat Transfer in Turkey, Japan and European countries. His research interests include heat transfer enhancement in solid/liquid phase change, heat and mass transfer in porous media, adsorption heat pump and computational fluid dynamics. He published more than 70 papers in international journals as well as 100 papers in national and international conferences and 5 book chapters on the various applications of heat transfer. He edited a book called *Solid/liquid thermal energy storage: modeling and applications*, published in 2022. He has supervised many master and PhD students both in Turkey and Japan. He received fellowships from the Japan Society for the Promotion of Science, European Union, and Cracow University of Technology to visit laboratories of different universities in Japan, Poland, Italy, Sweden and Austria. He has led many projects funded by "State Planning Department of Turkey", "Scientific and Technological Research Council of Turkey", "Japan Society for the Promotion of Science" and "Suzuki Foundation" aimed at discovering innovative methods for heat transfer enhancement for single-convection heat transfer, adsorbent beds as well as for solid/liquid phase change thermal storage.

Kamel Hooman

Kamel Hooman is a professor of heat transformation technology at Delft University of Technology. He received his PhD from The University of Queensland in 2009 where he has worked for almost two decades. He is working closely with industry in the field of thermo-fluids engineering. He was named Australia's Research Field Leader in Thermal Sciences in 2019. His books *Convective Heat Transfer in Porous Media* and *Solid/liquid thermal energy storage: modeling and applications*, published in 2019 and 2022, respectively, (CRC press) are devised to help both undergraduate and postgraduate students who work on porous media flows and thermal energy storage. He has given numerous national and international invited lectures, keynote addresses, and presentations. He has been awarded fellowships/awards from Emerald, Australian Research Council, National Science Foundation China, Australian Academy of Sciences, and Chinese Academy of Sciences with visiting professor/researcher positions at different universities across the globe. He is the associate editor for the *International Journal of Heat and Mass Transfer*, *Heat Transfer Engineering*, and *Journal of Porous Media* while serving on the editorial/advisory board of some international journals and conferences in the field of energy storage, conversion, and management. As an editor for *Heat Exchanger Design Handbook* (Begell House), he relies on his practical experience to ensure the latest development in the field of heat exchangers is kept up to date and shared with the practicing engineers. He has been the organizer and chair of the International Conference on Cooling Tower and Heat Exchanger sponsored by IAHR. He has carried out various sponsored research projects through companies, governmental funding agencies, and national labs. He has also consulted for various companies and governments, in Australia, and overseas.

Preface to "Advances of Heat Transfer in Porous Media"

Porous media can be defined as solid materials containing voids. The voids can be connected to each other, and fluid can flow through the porous media, or they can be isolated and porous media does not permit fluid flowing, but heat can flow. Practically, the size of the voids can be nanoscale such as membranes or larger such as rocks. The heat and fluid flow in porous media has wide applications from sweating (flowing of water through skin) and air flow in lungs to the flowing of water in soil or a liquid flow in woods or in concretes. Recently, cities or forests have been accepted as porous media, and heat and fluid flow through them are modeled by the governing equations of porous media.

This wide application of heat and fluid flow through porous media motivated authors to apply for a Special Issue named "Advances of Heat Transfer in Porous Media" under the support of the journal *Energies*. Valuable studies were submitted to this Special Issue and published in *Energies* by expert researchers in this field. The present book is a reprint of the published papers as a book. The first three chapters of the book are about the modelling of the drying process. Numerical models based on finite volume methods and lumped analysis are developed for the drying process and applied for different porous media such as clay and lentil grain. The fourth chapter relates to the development of a new modeling technique for highly porous metal foam or similar complex porous structures. The suggested model can be used for macroscopically two-dimensional heat transfer such as metal foam between two parallel plates. Recently, studies on the use of metal foams for the enhancement of heat transfer have increased in the literature, and the fifth chapter is about the enhancement of heat transfer in a channel by using metal foam. It investigates the thickness and structural parameters of the metal foam layer on heat transfer enhancement. The sixth chapter studies unidirectional porous tubes for gas cooling under high heat flux conditions with the application of helium gas cooling in a nuclear fusion divertor. Thermal cloaking in a nanoscale porous silicon structure by molecular dynamics is studied in chapter 7. The study calculates the thermal conductivity of porous structures and crystalline silicon films, and a rectangular nano-cloak porous structure was built and investigated in this study. Wire mesh can be a serious alternative for metal foams (whose manufacturing is difficult and expensive) for the enhancement of heat transfer. The seventh chapter investigates the use of stacked woven wire-mesh porous media for heat exchange applications. Among the various stages of production of ceramic materials, drying is one of the most energy-consuming processes. Heat and mass transfer in structural ceramic blocks is analyzed by an analytical and phenomenological approach in chapter 9. The use of clean energy sources such as solar energy is suggested to avoid the harmful effects of fossil fuels. By using solar air collectors, solar energy can be converted into thermal energy for many practical applications such as the preheating of air for a drying process. A numerical study for the enhancement of heat transfer using discrete metal foam with varying thickness and porosity for use in solar air heaters has been performed and reported in chapter 10.

The studies reported in 10 chapters of this book are samples of the recent advanced studies in the field of heat transfer in porous media. The methods, models and comments that are explained in the chapters will be helpful particularly for young researchers in this field. The authors would like to thank Ms. Gillian Yang, who is the section managing editor of *Energies*, for her great assistance.

Moghtada Mobedi and Kamel Hooman
Editors

Article

Drying of Sisal Fiber: A Numerical Analysis by Finite-Volumes

Jacqueline F. B. Diniz [1], João M. P. Q. Delgado [2,*], Anderson F. Vilela [3], Ricardo S. Gomez [4], Arianne D. Viana [3], Maria J. Figueiredo [3], Diego D. S. Diniz [5], Isis S. Rodrigues [6], Fagno D. Rolim [7], Ivonete B. Santos [8], João E. F. Carmo [4] and Antonio G. B. Lima [4]

[1] Department of Mathematics, Federal University of Campina Grande, Campina Grande 58429-900, PB, Brazil; jacqueline@mat.ufcg.edu.br

[2] CONSTRUCT-LFC, Civil Engineering Department, Faculty of Engineering, University of Porto, 4200-465 Porto, Portugal

[3] Department of Agro-industrial Management and Technology, Federal University of Paraíba, Bananeiras 58220-000, PB, Brazil; prof.ufpb.anderson@gmail.com (A.F.V.); arianneviana@hotmail.com (A.D.V.); mariaufp@gmail.com (M.J.F.)

[4] Department of Mechanical Engineering, Federal University of Campina Grande, Campina Grande 58429-900, PB, Brazil; ricardosoaresgomez@gmail.com (R.S.G.); jevan.franco@gmail.com (J.E.F.C.); antonio.gilson@ufcg.edu.br (A.G.B.L.)

[5] Department of Engineering, Rural Federal University of Semi-Arid, Caraúbas 59780-000, RN, Brazil; diego.diniz@ufersa.edu.br

[6] Department of Chemical Engineering, Federal University of Campina Grande, Campina Grande 58429-900, PB, Brazil; isisrodrigues07@gmail.com

[7] Teacher Training Center, Federal University of Campina Grande, Cajazeiras 58900-000, PB, Brazil; dallino@hotmail.com

[8] Department of Physics, State University of Paraíba, Campina Grande 58429-500, PB, Brazil; ivoneetebs@gmail.com

* Correspondence: jdelgado@fe.up.pt; Tel.: +351-225081404

Abstract: Vegetable fibers have inspired studies in academia and industry, because of their good characteristics appropriated for many technological applications. Sisal fibers (Agave sisalana variety), when extracted from the leaf, are wet and must be dried to reduce moisture content, minimizing deterioration and degradation for long time. The control of the drying process plays an important role to guarantee maximum quality of the fibers related to mechanical strength and color. In this sense, this research aims to evaluate the drying of sisal fibers in an oven with mechanical air circulation. For this purpose, a transient and 3D mathematical model has been developed to predict moisture removal and heating of a fiber porous bed, and drying experiments were carried out at different drying conditions. The advanced model considers bed porosity, fiber and bed moisture, simultaneous heat and mass transfer, and heat transport due to conduction, convection and evaporation. Simulated drying and heating curves and the hygroscopic equilibrium moisture content of the sisal fibers are presented and compared with the experimental data, and good concordance was obtained. Results of moisture content and temperature distribution within the fiber porous bed are presented and discussed in details. It was observed that the moisture removal and temperature kinetics of the sisal fibers were affected by the temperature and relative humidity of the drying air, being more accentuated at higher temperature and lower relative humidity, and the drying process occurred in a falling rate period.

Keywords: mass; heat; sisal fiber; experimental; simulation

1. Introduction

Vegetable fibers originate from plants and, basically, are composed of a solid skeleton and pores filled with fluid. The major components of these materials are cellulose, hemicellulose and lignin, with minor percentages of pectin, proteins, wax, inorganic salts and other water-soluble substances. Based on the cultivation region, soil type and climatic conditions, the chemical composition of vegetable fibers present some differences in values [1]. In Brazil, there are large varieties of vegetable fibers, for example: caroá, macambira, curauá, pineapple leaf, jute, cotton, and sisal.

Sisal (Agave sisalana perrine) is a plant cultivated in semiarid regions, resistant to aridity and strong solar irradiation. The leaf of sisal contains about 4% fiber, 1% film (cuticle), 8% dry matter and 87% water. After the fiber is extracted, the rest of the materials are processing waste, frequently used for different purposes, for example, as organic fertilizer, animal feed and by the pharmaceutical industry [2,3].

The sisal fiber when extracted from the leaf is moist. Thus, these biological materials must be dried, in order to avoid deterioration and to be used for most purposes. Green sisal fibers have a color that varies from white to light yellow, but it soon becomes discolored, turning a creamy-white or dark yellow, when submitted to intense heating for long time. However, the sisal fiber processing is responsible for more than half a million direct or indirect jobs, such as, activities of crop maintenance, harvesting, leaf cutting, fiber shredding and processing [2]. After processing, the sisal fibers are commonly used in different industrial sectors, such as in the automotive industry, replacing synthetic fibers, and as reinforcement in polymer composite materials. The vast application field of these fibers is due to their various advantages, such as low cost, low density, high specific strength and stiffness, low abrasiveness to process equipment, biodegradability, nontoxicity and nonpollution, significant reduction of environmental problems. Mainly they originate from renewable sources and are found in many regions in the world [4–7]. Despite the large advantages, these fibers are very susceptible to moisture and temperature actions, which strongly affect their mechanical properties, mainly when used as reinforcement in polymer composite.

Drying is a complex and coupled phenomenon of heat and mass transfer, momentum and sometimes dimensional variations [8]. The drying process control of sisal fibers plays an important role in obtaining the optimized drying conditions. The idea is to reduce product losses and increase energy saving.

According to literature, sisal fiber drying is realized currently by putting the fibers into the oven; however, the monitoring of the moisture and temperature transient behavior along the process is a secondary goal. However, severe changes in the temperature and moisture inside the fibers generate hydric-thermo-mechanical stresses which can cause fiber deterioration, rupture and weakness in the material. Thus, to adequately control drying process is important.

Unfortunately, few works about drying of sisal fibers have been reported in the literature. These works, the majority is focused in experiments [9–12] while others are dedicated to theoretical analysis by using analytical and numerical procedures [13–20].

Ferreira et al. [9] reported that the cellulose polysaccharide chains are more tightly arranged after moisture removal during drying and, consequently, the microfibrils come together in the dry state as a result of increased packaging. Besides, according to the authors, fiber voids are significantly reduced as consequence of the drying process, and cannot be completely reopened with rehumidification process. The minor water absorption rate minimizes the fiber dimensional changes, resulting in larger fiber dimensional stability.

Santos et al. [10] evaluated the effects of convective drying (in oven) on the mechanical properties of sisal fibers. In this research, they studied the moisture loss and heating processes, and the effect of the drying air conditions on the mechanical properties of these materials. According to the authors, air temperature ranging from 60 to 105 °C significantly affected not the tensile mechanical properties, elongation at rupture and modulus of elasticity of the sisal fibers, but caused discoloration.

Work by Ghosh [12] indicates that artificial drying of sisal fiber, after decortication and squeezing to remove free water, currently is realized in dryers at air temperature ranging from 100–110 °C. However, according to the author, sisal fiber can be dried at temperature of around 180 °C, and inevitable problem such as discoloration can be minimized by adequately controlling drying time and temperature. In this way, it is possible to obtain products with an acceptable quality level.

Diniz et al. [13], in their research related to the drying of sisal fibers, proposed lumped mathematical models to predict the transient behavior of the average moisture content and surface temperature, and also, the equilibrium moisture content of vegetable fiber as a function of the drying air temperature and water vapor concentration in the fiber porous bed. According to the authors, predicted results were in good fit with experimental data; for all drying conditions used correlation coefficients greater than 0.99 were obtained, and thus, was shown the potential of the proposed models to appropriately predict the drying process of sisal fiber, and other related materials.

Nordon and David [17] studied the coupled diffusion of heat and moisture in hygroscopic textile materials. The authors presented a 1D-numerical solution of the governing equations by finite differences, based on the "double scan" method for the nonlinear differential equations. The study was applied to wool, as an example of hygroscopic material and showed that the moisture transfers from the air to the wool and from the wool to the air are not symmetrical processes. According to the authors, the method does not predict the physical problem adequately.

Haghi [18] studied the simultaneous heat and moisture transfer in porous systems. For this purpose and based on the model reported by Nordon and David (1967) [17], the authors developed a one-dimensional mathematical model and its numerical solution using the finite difference method. The effects of operational parameters, such as temperature and moisture in the dryer, the initial moisture content of the porous material and the mass and heat transfer coefficients are examined using this model. The aim of this study was to describe adequately the heat and mass transfer during convective drying. According to the authors the proposed model can be used to predict transient variations in temperature distribution and moisture content in fabrics with reasonable accuracy.

Xiao et al. [19] and Xiao et al. [20] reported studies about mass transfer (imbibition process) in fibrous porous media using the fractal theory including surface roughness. The global model is based on the capillary theory (Hagen–Poiseuille model and Kozeny-Carman equation), assuming that the porous media is formed by a numbers of capillaries tubes. According to the authors, the proposed model predicts adequately the physical mechanisms of fluid transport inside fibrous porous media, and it contains no empirical constant, as commonly required for different models.

As already mentioned in previous comments, different distributed models have been reported in the literature applied to heat and/or mass transfer in fibrous porous media; however, restricted to one-dimensional approaches, no works used the finite volume technique for numerical solution of the governing equations. Then, in complement to these studies, this work aims to study the drying of sisal fibers in an oven using a 3D advanced mathematical modeling that considers the coupled phenomena of heat and mass transport inside the material, including fiber bed porosity and sorption heat. Here, the finite-volumes numerical method has been used to solve the governing equations. The idea is to assist engineers, industrials and academics in a rigorous understanding of the dominant mechanisms involved in the diffusion process occurring in the drying process.

2. Methodology

2.1. Mathematical Modeling

2.1.1. Geometry and the Physical Problem

In this paper, the physical problem consists in predicting the moisture and heat transfer within a wet fibrous medium in the shape of a rectangular prism, as illustrated in the Figure 1. The fibrous medium is formed by a series of rigid and nonreactive fibers. In the drying process, hot air flows around the porous bed supplying heat to the wet fibers. The physical effect is to evaporate water inside the fiber, to heat the fibers, and to remove the water in the vapor phase. Water molecules in the vapor phase are free to move between the voids through the fiber bed, and to be absorbed or desorbed by the fibers. Volume variations provoked by water removal and heating of the fiber are not considered.

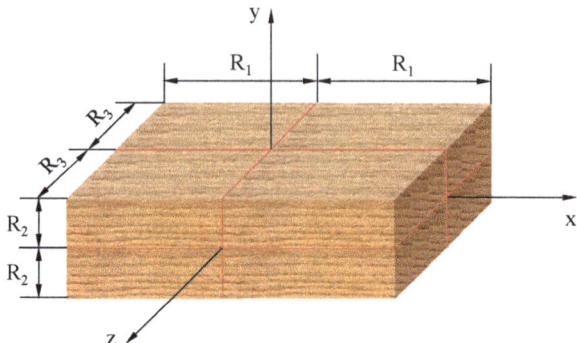

Figure 1. Geometry considered in this research.

2.1.2. The Governing Equations

Vapor Diffusion Equation

To predict the water vapor concentration behavior inside the fiber bed along the drying process, the following equation was considered:

$$[\varepsilon + (1-\varepsilon)\rho_s \sigma]\frac{\partial C}{\partial t} = \left[\frac{\partial}{\partial x}\left(D\frac{\partial C}{\partial x}\right) + \frac{\partial}{\partial y}\left(D\frac{\partial C}{\partial y}\right) + \frac{\partial}{\partial z}\left(D\frac{\partial C}{\partial z}\right)\right] + (1-\varepsilon)\rho_s \beta \frac{\partial T}{\partial t}, \quad (1)$$

where σ and β are constants from the equilibrium equation; ε is the porosity; ρ_s is the fiber density; C is the water vapor concentration (interfibers); T is the temperature, and t is the time. The parameter $D = \varepsilon D\prime$ and $D\prime$ represents the vapor diffusion coefficient within the fibrous medium. In Equation (1), it is clear the effect of heat transfer in the mass transfer phenomenon.

In the model, the following initial and boundary conditions were adopted:

- Initial condition:

$$C(x, y, z, t = 0) = C_o, \quad (2)$$

- Boundary conditions:

$$-D\frac{\partial C}{\partial x}\bigg|_{x=R_1} = h_{mx}(C - C_{eq}), \quad \forall (x = R_1, y, z, t > 0); \quad (3)$$

$$-D\frac{\partial C}{\partial y}\bigg|_{y=R_2} = h_{my}(C - C_{eq}), \quad \forall (x, y = R_2, z, t > 0); \quad (4)$$

$$-D\frac{\partial C}{\partial z}\bigg|_{z=R_3} = h_{mz}(C - C_{eq}), \quad \forall (x, y, z = R_3, t > 0). \quad (5)$$

Heat Conduction Equation

To predict the temperature of the fibers bed along the drying process, the following energy equation was used:

$$\rho(c_P + h_s\beta)\frac{\partial T}{\partial t} = \left[\frac{\partial}{\partial x}\left(K\frac{\partial T}{\partial x}\right) + \frac{\partial}{\partial y}\left(K\frac{\partial T}{\partial y}\right) + \frac{\partial}{\partial z}\left(K\frac{\partial T}{\partial z}\right)\right] + h_s \rho \sigma \frac{\partial C}{\partial t}, \quad (6)$$

where c_p is the specific heat; K is the thermal conductivity; h_s is the enthalpy and, T is the temperature. In Equation (6), it is considered that the energy of the wet air inside the fiber bed is negligible compared to the energy of the fibers. In Equation (6), the effect of mass transfer in the heat transfer phenomenon is clear.

- Initial condition:

$$T(x, y, z, t = 0) = T_o. \tag{7}$$

- Boundary conditions for heat transfer:

$$-K \frac{\partial T}{\partial x}\bigg|_{x=R_1} = h_{cx}(T - T_{eq}), \forall (x = R_1, y, z, t > 0); \tag{8}$$

$$-K \frac{\partial T}{\partial y}\bigg|_{y=R_2} = h_{cy}(T - T_{eq}), \forall (x, y = R_2, z, t > 0); \tag{9}$$

$$-K \frac{\partial T}{\partial z}\bigg|_{z=R_3} = h_{cz}(T - T_{eq}), \forall (x, y, z = R_3, t > 0). \tag{10}$$

∧ Equilibrium Equation

According to Crank [21], the following equilibrium equation was considered:

$$M = \alpha + \sigma C - \beta T, \tag{11}$$

where M is the moisture absorbed per unit mass of fiber ($kg_{vapor}/kg_{dry\ fiber}$), and α, σ and β are constants, that can be obtained by fitting to experimental data.

From Equation (11), it is possible to obtain the following derivative:

$$\frac{\partial M}{\partial t} = \sigma \frac{\partial C}{\partial t} - \beta \frac{\partial T}{\partial t} \tag{12}$$

The average value of the potential of interest can be determined as follows:

$$\overline{\Phi} = \frac{1}{V} \int_V \Phi dV, \tag{13}$$

where Φ can be C, T or M (Equation (11)) and V is the volume of the fibrous medium.

2.2. Numerical Procedure

Equations (1) and (6) are highly nonlinear differential equations. Thus, exact solutions of these equation isn't possible. Thus, in this paper, these equations are solved using the finite-volume method [22–24]. In the discretization process, the following assumptions were adopted:

(a) The solid is homogeneous and isotropic;
(b) The only mechanism of water transport within the solid is diffusion;
(c) The thermophysical parameters are constant along the drying process;
(d) The diffusion coefficient is constant along the drying;
(e) Convective mass and heat transfer coefficients are constant along the drying.

Due to the geometric shape (rectangular prism), it is possible to numerically solve the physical problem treated here using only one symmetrical part of the domain, as illustrated in Figure 2. In Figure 3 is shown the computational domain utilized in the discretization process of the governing equation.

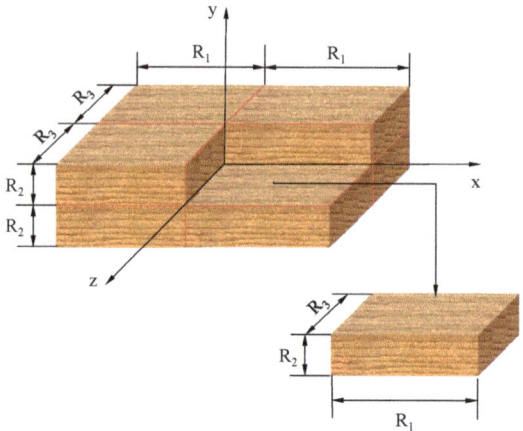

Figure 2. Illustration of the symmetric domain used in the numerical solution.

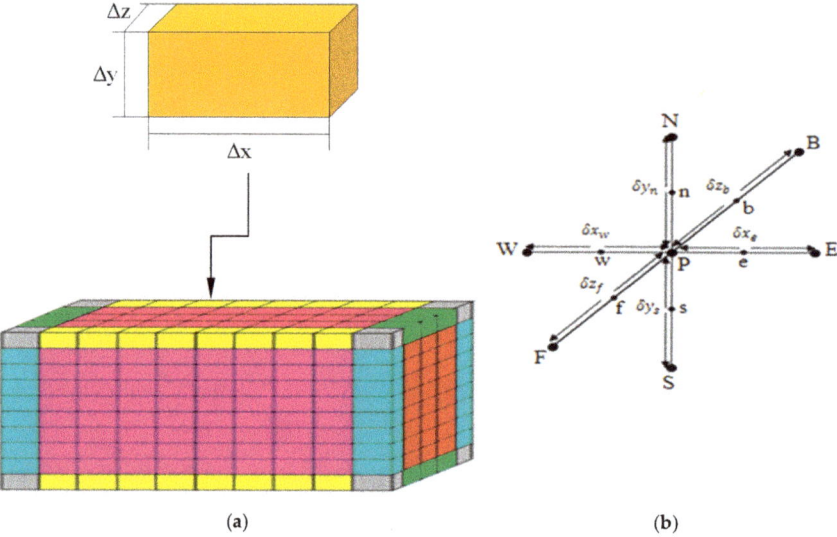

Figure 3. Illustration of the computational domain used in the numerical solution: (**a**) Identifications of 27 types of control volumes and (**b**) Control volume parameters.

Based on the Figure 3, and integrating Equation (1) into volume and time, the following linear algebraic equation is obtained for the internal control volumes P:

$$A_P C_P = A_e C_E + A_w C_W + A_n C_N + A_s C_S + A_f C_F + A_b C_B + B, \tag{14}$$

where,

$$A_P = [\varepsilon + (1-\varepsilon)\rho_s\,\sigma]\frac{\Delta x \Delta y \Delta z}{\Delta t} + D_e^C \frac{\Delta y \Delta z}{\delta x_e} + D_w^C \frac{\Delta y \Delta z}{\delta x_w} + D_n^C \frac{\Delta x \Delta z}{\delta y_n} + D_s^C \frac{\Delta x \Delta z}{\delta y_s} + D_f^C \frac{\Delta x \Delta y}{\delta z_f} + D_b^C \frac{\Delta x \Delta y}{\delta z_b}, \tag{15}$$

$$A_e = D_e^C \frac{\Delta y \Delta z}{\delta x_e}, \tag{16}$$

$$A_w = D_w^C \frac{\Delta y \Delta z}{\delta x_w}, \qquad (17)$$

$$A_n = D_n^C \frac{\Delta x \Delta z}{\delta y_n}, \qquad (18)$$

$$A_s = D_s^C \frac{\Delta x \Delta z}{\delta y_s}, \qquad (19)$$

$$A_f = D_f^C \frac{\Delta x \Delta y}{\delta z_f}, \qquad (20)$$

$$A_b = D_b^C \frac{\Delta x \Delta y}{\delta z_b}, \qquad (21)$$

$$B = [\varepsilon + (1-\varepsilon)\rho_s \, \sigma] \frac{\Delta x \Delta y \Delta z}{\Delta t} C_P^0 + S^C \, \Delta x \Delta y \Delta z. \qquad (22)$$

For boundary volumes, the integration process of the governing equation is similar, as described for the internal points. However, must be considered the existing boundary conditions. For example, considering the symmetry condition, the mass fluxes in the west, back and south faces are null, i.e., $C_w'' = 0$, $C_b'' = 0$ and $C_s'' = 0$, as illustrated in Figure 4.

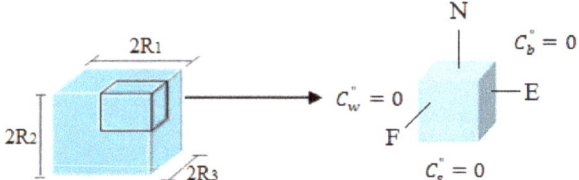

Figure 4. Illustration of the symmetry condition at the internal faces of the domain.

For the heat transfer equation after integration procedure, the following discretized linear algebraic equation is obtained:

$$A_P T_P = A_e T_E + A_w T_W + A_n T_N + A_s T_S + A_f T_F + A_b T_B + B, \qquad (23)$$

where,

$$A_P = \frac{\rho(c_P + h_s \beta) \Delta x \Delta y \Delta z}{\Delta t} + K_e^T \frac{\Delta y \Delta z}{\delta x_e} + K_w^T \frac{\Delta y \Delta z}{\delta x_w} + K_n^T \frac{\Delta x \Delta z}{\delta y_n} + K_b^T \frac{\Delta x \Delta z}{\delta y_s} + K_f^T \frac{\Delta x \Delta y}{\delta z_f} + K_b^T \frac{\Delta x \Delta y}{\delta z_b}, \qquad (24)$$

$$A_e = K_e^T \frac{\Delta y \Delta z}{\delta x_e}, \qquad (25)$$

$$A_w = K_w^T \frac{\Delta y \Delta z}{\delta x_w}, \qquad (26)$$

$$A_n = K_n^T \frac{\Delta x \Delta z}{\delta y_n}, \qquad (27)$$

$$A_s = K_s^T \frac{\Delta x \Delta z}{\delta y_s}, \qquad (28)$$

$$A_f = K_f^T \frac{\Delta x \Delta y}{\delta z_f}, \qquad (29)$$

$$A_b = K_b^T \frac{\Delta x \Delta y}{\delta z_b}, \qquad (30)$$

$$B = \frac{\rho(c_P + h_s \beta) \, \Delta x \Delta y \Delta z}{\Delta t} T_P^0 + S^T \, \Delta x \Delta y \Delta z. \qquad (31)$$

In the discretized form, Equation (6) applied to vapor concentration assumes the form:

$$\overline{C} = \frac{1}{V} \sum_{i=2}^{npx-1} \sum_{j=2}^{npy-1} \sum_{k=2}^{npz-1} C_{ijk} \Delta V'_{ijk}, \qquad (32)$$

where,

$$V = \sum_{i=2}^{npx-1} \sum_{j=2}^{npy-1} \sum_{k=2}^{npz-1} \Delta V'_{ijk}, \qquad (33)$$

where i, j and k define the position nodal point, $\Delta V'_{ijk} = \Delta x \Delta y \Delta z$ is the volume of the elemental volume, and npx-2, npy-2, and npz-2 are the number of control volumes in the x, y, and z directions, respectively.

To solve iteratively (Gauss–Seidel method) the systems of algebraic equations originated from the Equations (14) and (23) as applied for all control-volume, a computational code using Mathematica® software was developed. In the numerical treatment, the following convergence criterion was used at each nodal point and process time:

$$\left| \Phi^{n+1} - \Phi^n \right| \leq 10^{-8}, \qquad (34)$$

where Φ can be C or T, and n represents the nth iteration at each time point. Simulations were performed using a numerical mesh of 20 × 20 × 20 nodal points and Δt = 20s, which were obtained after a rigorous mesh and time step refining study.

2.3. Experimental Study

In this work, sisal fibers (Agave sisalana variety) were used in the drying experiments. The wet fibers were obtained in a farm located close to the town of Pocinhos, State of Paraiba, Brazil. The fibers were submitted to drying in oven with forced air circulation at different operating conditions. In Figure 5 is shown the sisal fibers bed used in the experiment. Table 1 summarizes information about the fiber and drying air.

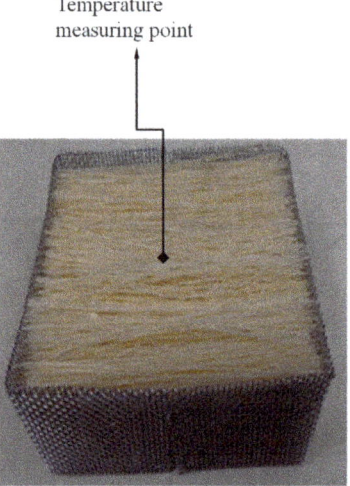

Figure 5. Sisal fibers bed used in the drying tests.

Table 1. Experimental conditions of air and sisal fiber.

Air			Fibrous Medium							Process Time
RH (%)	T (°C)	v (m/s)	$2R_1$ (m)	$2R_2$ (m)	$2R_3$ (m)	M_o (d.b.)	M_{eq} (d.b.)	T_o (°C)	T_f (°C)	t (h)
17.52	50	0.05	0.1	0.05	0.1	0.11327	0.03837	29.8	46.3	7.7
11.04	60	0.06	0.1	0.05	0.1	0.11118	0.02606	29.8	56.6	6.7
6.89	70	0.07	0.1	0.05	0.1	0.11148	0.02015	31.5	67.3	5.7
4.19	80	0.08	0.1	0.05	0.1	0.11030	0.01390	29.6	76.3	5.0
3.28	90	0.09	0.1	0.05	0.1	0.11342	0.00525	30.4	87.4	4.7

In each drying experiments, the sample were withdraw from the oven periodically. In each step, the mass of was evaluated using a digital electronic device (0.001 g precision), and the surface temperature was measured using an infrared thermometer. These measurements were taken every 5 min (about 30 min), after 10, 15, 20, 25 and 30 min time intervals. After this period, the measurements were taken every 60 min until that constant mass was reached. Following that, the sample was dried for 24 h at the same drying temperature, in order to obtain the sample mass at the equilibrium condition, and then for a further 24 h at 105 °C to obtain the dried sample mass. Furthermore, at every measurement instant, temperature and relative humidity of the air outside of the oven were measured. This information was used to calculate air relative humidity inside the oven. In each drying experiments, air velocity inside the oven was measured using a hot wire anemometer. Figure 5 illustrates the point where surface temperatures were taking.

2.4. Parameters Estimation

2.4.1. Transport Coefficients

In this research, the estimation of the mass diffusion coefficient and convective heat and mass transfer coefficients was performed using the least square error technique. Deviations between experimental and predicted values and variance were determined by [25]:

$$\text{ERMQ} = \sum_{i=1}^{n}\left(\Phi_{i,\,\text{Num}} - \Phi_{i,\,\text{Exp}}\right)^2, \qquad (35)$$

$$\overline{S}^2 = \frac{\text{ERMQ}}{(n - \hat{n})}, \qquad (36)$$

where n is the number of experimental points, n̂ is the number of fitted parameters, and Φ can be $\overline{M}$ or $\overline{T}$.

The initial guess of the convective heat transfer coefficient was determined by common correlations for the average Nusselt, Reynolds and Prandtl numbers [26], applied to air flowing over a plane plate by:

$$h'_{cj} = \frac{\overline{Nu_j} \times k}{R_j}, \qquad (37)$$

where, $\overline{Nu_j} = 0.664 \times Re_j^{\frac{1}{2}} \times Pr^{\frac{1}{3}}$, $Re_j = \frac{\rho \times v \times R_j}{\mu}$, k is the air thermal conductivity and j = 1,2 and 3 (Figure 2), valid in the intervals $5 \times 10^5 < Re \leq 1 \times 10^8$.

The initial guess of the convective mass transfer coefficient applied to the air was determined by common correlations for average Sherwood numbers and Schmidt, by [26]:

$$h'_{mj} = \frac{\overline{Sh_j} \times D_{AB}}{R_j}, \qquad (38)$$

where, $\overline{Sh_j} = 0.664 \times Re_j^{\frac{1}{2}} \times Sc^{\frac{1}{3}}$ and D_{AB} is the diffusivity of water vapor in the air.

2.4.2. Equilibrium Equation

Considers the Equation (11) re-written in the form:

$$M = a_1 + a_2 \times T, \quad (39)$$

where $a_1 = \alpha + \sigma \times C$ and $a_2 = -\beta$.

With the values of T_{eq} and M_{eq} for each experimental condition, a linear fit of Equation (39) to the experimental data was performed by the quasi-Newton method using the Statistica® Software (convergence criterion 9.9×10^{-5}). After this statistical procedure the values of the parameters a_1 and a_2 were obtained.

On the other hand, considering the drying air as an ideal gas, the air density inside the fibrous bed was calculated by:

$$\rho_{air} = \frac{P}{\overline{R} \times T_{abs}}, \quad (40)$$

where P is the atmospheric pressure, ρ_{air} is the air density, $\overline{R}$ is the particular gas constant (atmospheric air), and T_{abs} is the absolute temperature of the drying-air.

Then, from Equation (40), the equilibrium water vapor concentration between the fibers, can be determined by:

$$C_{eq} = \rho_{air} \times AH, \quad (41)$$

where AH is the air absolute humidity on the voids of the porous bed.

2.4.3. Estimation of the Fiber and Sample Parameters

The apparent density of the fiber bed was calculated by:

$$\rho_{sample} = \frac{m_{fiber}}{V_{sample}}, \quad (42)$$

where $V = 2R_1 \times 2R_2 \times 2R_3$ is the volume of the fiber bed.

To obtain the bed porosity, the following equation was used:

$$\varepsilon = 1 - \frac{V_{fiber}}{V_{sample}}, \quad (43)$$

where $V_{fiber} = N \times L \times \frac{\pi d^2}{4}$ is the total volume of fibers, L and d, are the length and diameter of the sisal fiber, respectively, and N represents the number of fibers (Figure 6).

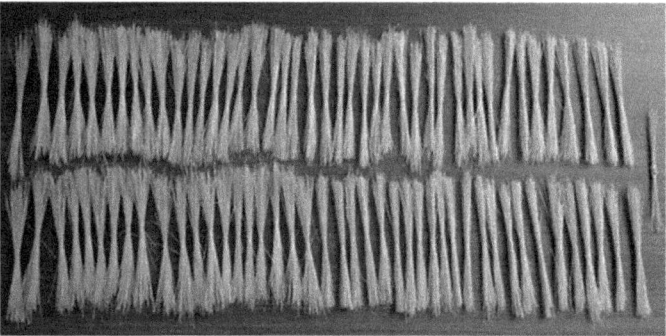

Figure 6. Amount of fibers in a sample used in the drying experiment.

The parameters thermal conductivity and specific heat of the fiber bed were determined as follows:

$$k_{sample} = (1-\varepsilon) \times k_{fiber} + \varepsilon \times k_{air}, \qquad (44)$$

and

$$c_{psample} = (1-\varepsilon) \times c_{pfiber} + \varepsilon \times c_{pair} \qquad (45)$$

Table 2 summarizes the geometrical data of the fiber and samples before drying [11,13–16].

Table 2. Geometrical parameters of the fibers and porous bed before drying for different drying conditions.

T (°C)	Fiber Length (m)	* Fiber Diameter (μm)	** Fiber Volume (nm³)	Numbers of Fibers/Sample (—)	V_{fiber} (m³)	V_{sample} (m³)	ε (—)
50	0.1	247	4.789	8539	40.895×10^{-6}	5.00×10^{-4}	0.918
60	0.1	247	4.789	8190	39.224×10^{-6}	5.00×10^{-4}	0.922
70	0.1	247	4.789	8490	40.660×10^{-6}	5.00×10^{-4}	0.919
80	0.1	247	4.789	8627	41.316×10^{-6}	5.00×10^{-4}	0.917
90	0.1	247	4.789	8869	42.475×10^{-6}	5.00×10^{-4}	0.915

* Santos [15], ** Calculated by equation.

2.5. Thermophysical Properties of the Materials

2.5.1. For Drying Air

Tables 3 and 4 present some drying air parameters at atmospheric pressure for each experiments.

Table 3. Air parameters at atmospheric pressure for different drying conditions.

T (°C)	Parameter				
	ρ_{air} (kg/m³)	k_{air} (W/m·K)	μ_{air} (N·s/m²)	AH_{air} (kg$_{vapor}$/kg$_{dry\,air}$)	$AH_{air,\,sat}$ (kg$_{vapor}$/kg$_{dry\,air}$)
50	1.0835	0.0269	19.054×10^{-6}	0.01363	0.0863
60	1.0508	0.0276	19.509×10^{-6}	0.01387	0.1524
70	1.0204	0.0283	19.966×10^{-6}	0.01354	0.2765
80	0.9921	0.0290	20.427×10^{-6}	0.01247	0.5464
90	0.9637	0.0296	20.847×10^{-6}	0.01449	1.3990
	D_{AB} (m²/s)	c_p (kJ/kg K)	h_s (kJ/kg)	Pr (—)	Sc (—)
50	28.06×10^{-6}	1.0331	2382.90	0.73210	0.62671
60	29.60×10^{-6}	1.0335	2358.34	0.73123	0.62726
70	31.17×10^{-6}	1.0335	2333.26	0.73016	0.62774
80	32.78×10^{-6}	1.0328	2307.62	0.72884	0.62818
90	34.42×10^{-6}	1.0348	2281.39	0.72877	0.62855

Table 4. Air dimensionless parameters at atmospheric pressure for different drying conditions.

Dimensionless Parameter	T (°C)				
	50	60	70	80	90
Re_x	142.1625	134.6619	127.7713	121.4225	115.5627
Re_y	71.0812	67.3309	63.8856	60.7112	57.7813
Re_z	142.1625	134.6619	127.7713	121.4225	115.5627
$\overline{Nu}_x$	7.13539	6.94186	6.75860	6.58460	6.42355
$\overline{Nu}_y$	5.04548	4.90864	4.77905	4.65601	4.54213
$\overline{Nu}_z$	7.13539	6.94186	6.75860	6.58460	6.42355
$\overline{Sh}_x$	6.77511	6.59588	6.42655	6.26630	6.11445
$\overline{Sh}_y$	4.79072	4.66399	4.54425	4.43094	4.32357
$\overline{Sh}_z$	6.77511	6.59588	6.42655	6.26630	6.11445

2.5.2. For Water and Fiber

Table 5 presents some fibrous medium parameters for different drying experiments.

Table 5. Sisal fiber parameters and sample for different drying conditions [11,13–16].

T (°C)	Parameter			
	C_o (kg_{vapor}/m^3)	C_{eq} (kg_{vapor}/m^3)	ρ_{sample} (kg/m^3)	ρ_{fiber} (kg/m^3)
50	0.04146	0.01482	83.088	1450.00
60	0.04052	0.01458	74.960	1450.00
70	0.04125	0.01381	81.956	1450.00
80	0.04005	0.01439	85.138	1450.00
90	0.4174	0.01403	90.782	1450.00
	$c_{p\,sample}$ (J/kg·K)	$c_{p\,fiber}$ (J/kg·K)	k_{sample} (W/m·K)	k_{fiber} (W/m·K)
50	960.8454	149.65	0.03017	0.067
60	964.2007	149.65	0.03067	0.067
70	961.6180	149.65	0.03141	0.067
80	959.8464	149.65	0.03209	0.067
90	959.6211	149.65	0.03278	0.067

3. Results and Discussion

3.1. Fiber Morphology

The scanning electron micrographs (SEM) of untreated sisal fibers in moist condition are illustrated in Figure 7. Upon analyzing this figure, it can be verified that the arrangement of this particular fiber is similar to other natural fibers. It can be seen a spongier aspect, voids, rougher surface and a thin and compacted cellular arrangement. Furthermore, it can be observed that parenchyma cells are widely distributed along the fiber. The fibers presented diameters almost constant along the length and, minor differences between the fibers mean diameter were observed.

3.2. Analysis of Equilibrium Equation

$$M = 0.044702 + 2.217337 \times C - 0.000784 \times T. \tag{46}$$

Equation (46) predict the linear dependence of the moisture content with the temperature and water vapor concentration, fundamental in predicting the coupling between the Equations (1) and (6). A correlation coefficient R = 0.99 was obtained after the performed linear regression. Most details about the statistical procedure can found in the literature [11,13–16].

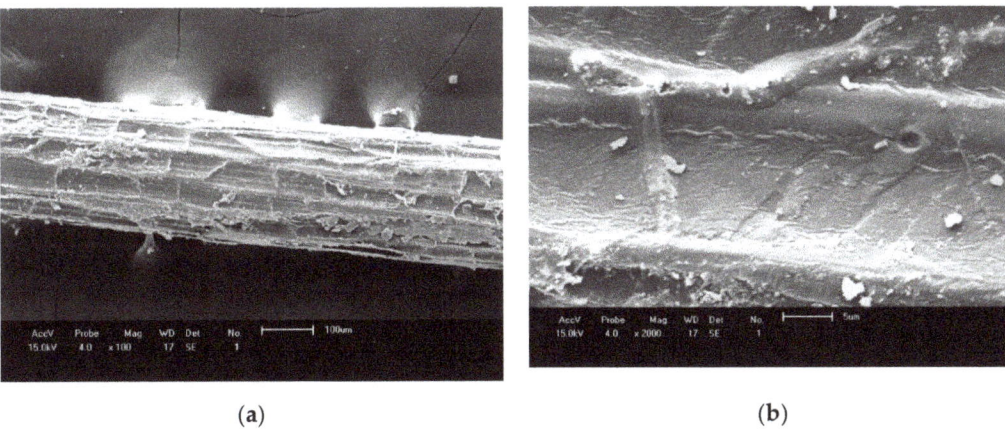

Figure 7. SEM micrographs of untreated sisal fiber: (**a**) 100× magnification and (**b**) 2000× magnification.

3.3. Moisture Removal and Heat Transfer Analysis

3.3.1. Drying and Heating Kinetics

Figures 8 and 9 illustrate the predicted and experimental average moisture content (dry basis) of the fiber sample as a function of time for the drying air temperature of 50 °C and 80 °C. Upon analyzing these figures, it can be seen that there is a good concordance between the simulated and experimental of the average moisture content along the process for every condition. Besides, it was verified that the average moisture content decreases with time until to reach the hygroscopic equilibrium condition. Furthermore, the drying rate increased and equilibrium moisture content decreased when higher drying air temperature and lower air relative humidity were used in the experiment. Therefore, this physical situation resulted in a short drying time.

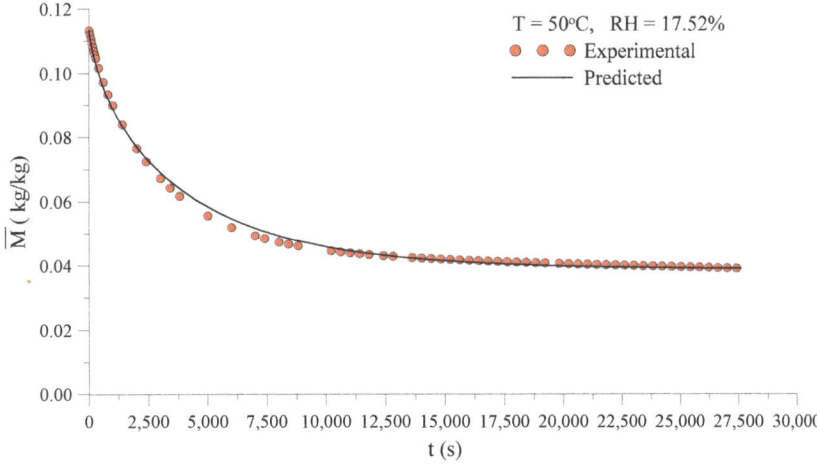

Figure 8. Transient behavior of the predicted and experimental average moisture content of the fiber bed (T = 50 °C).

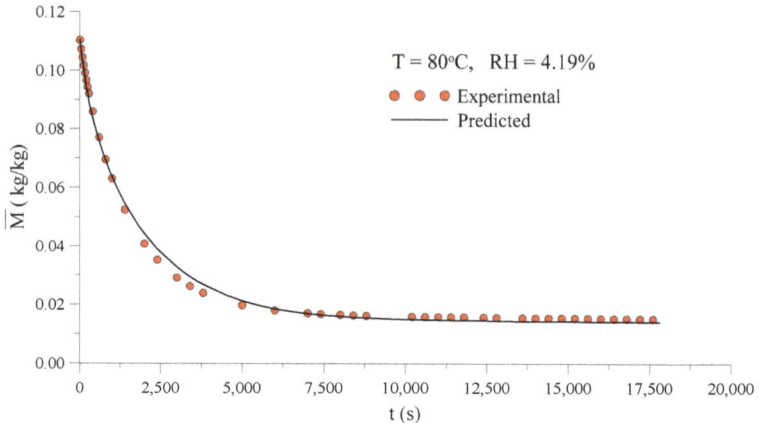

Figure 9. Transient behavior of the predicted and experimental average moisture content of the fiber bed (T = 80 °C).

In Figures 10 and 11 are shown the comparison between the predicted and experimental transient values of the temperature at the sample surface at 50 °C and 80 °C. Upon analyzing these figures, it can be noted that there is good concordance between surface temperature values in the first 4000 s of process followed by minor discrepancies for long time. For example, in the drying at 50 °C and 27,400 s elapsed time difference, between the predicted and experimental surface temperature data there is 2.8 °C difference, while at 80 °C and 17,800 s elapsed time, the difference in these values is 4.5 °C. These discrepancies may be attributed to minor errors in the temperature measurement procedure in each the experiments or even by the accuracy in the device itself. Besides the differences verified for long time may also be associated with the consideration of constant convective heat coefficient and null dimensional variations along the process as established in the mathematical modeling.

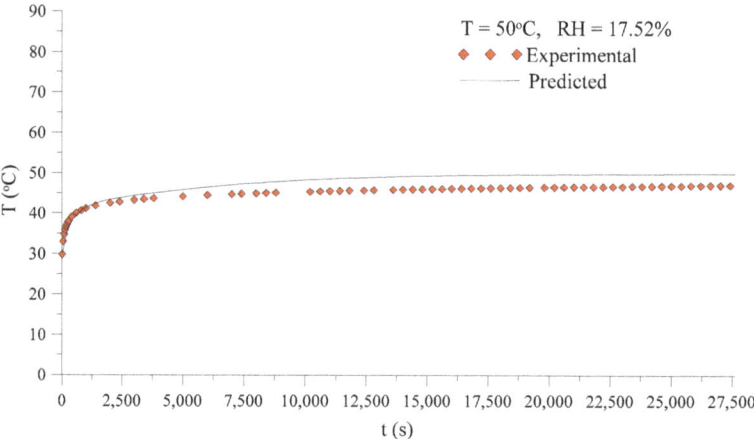

Figure 10. Transient behavior of the predicted and experimental temperature at the surface of the fiber sample (T = 50 °C).

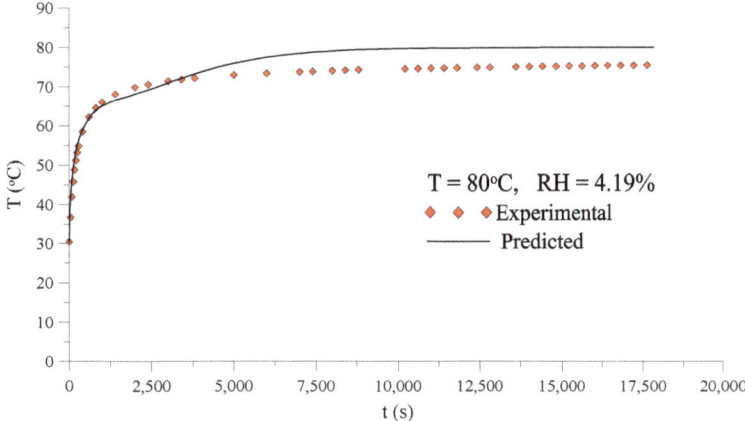

Figure 11. Transient behavior of the predicted and experimental temperature at the surface of the fiber sample (T = 80 °C).

Upon analyzing the Figures 7 and 8, it can be seen that the temperature at the surface of the fibrous bed changed along the drying, reaching the thermal equilibrium condition faster at the highest air temperature. The increases in the surface temperature of the fiber with progress of drying indicates the moisture removal occurred on the falling drying rate period, i.e., the water flux inside the fiber to its surface is less than the water flux removed by the heated air at the surface. Furthermore, it was observed that the fiber temperature increased as the drying rate decreased until the equilibrium condition. These phenomena were cited in the literature [10,11,13–16].

3.3.2. Water Vapor Concentration and Temperature Distributions

In Figures 12 and 13 is illustrated the water vapor concentration field inside the fiber bed, at the plans x = 0.025 m ($R_1/2$) and y = 0.0125 m ($R_2/2$), at the moments 200 s, 700 s, 5000 s and 8000 s of the drying process, respectively.

By analyzing the Figures 12 and 13, it is possible to see that water vapor concentration presented the highest values in the center of the fiber bed at any time. Thus, moisture flows from the center to surface. Besides, vapor concentration decreased at any position and time, tending towards its equilibrium condition for long drying time. The regions close to the edge dry faster than the others regions inside the porous bed, especially in the vertex region.

Figures 14 and 15 show the distribution of temperature inside the fiber bed, analyzed in the plans x = 0.025 m ($R_1/2$) and y = 0.0125 m ($R_2/2$), at the moments 200 s, 700 s, 5000 s and 8000 s of the drying process, respectively.

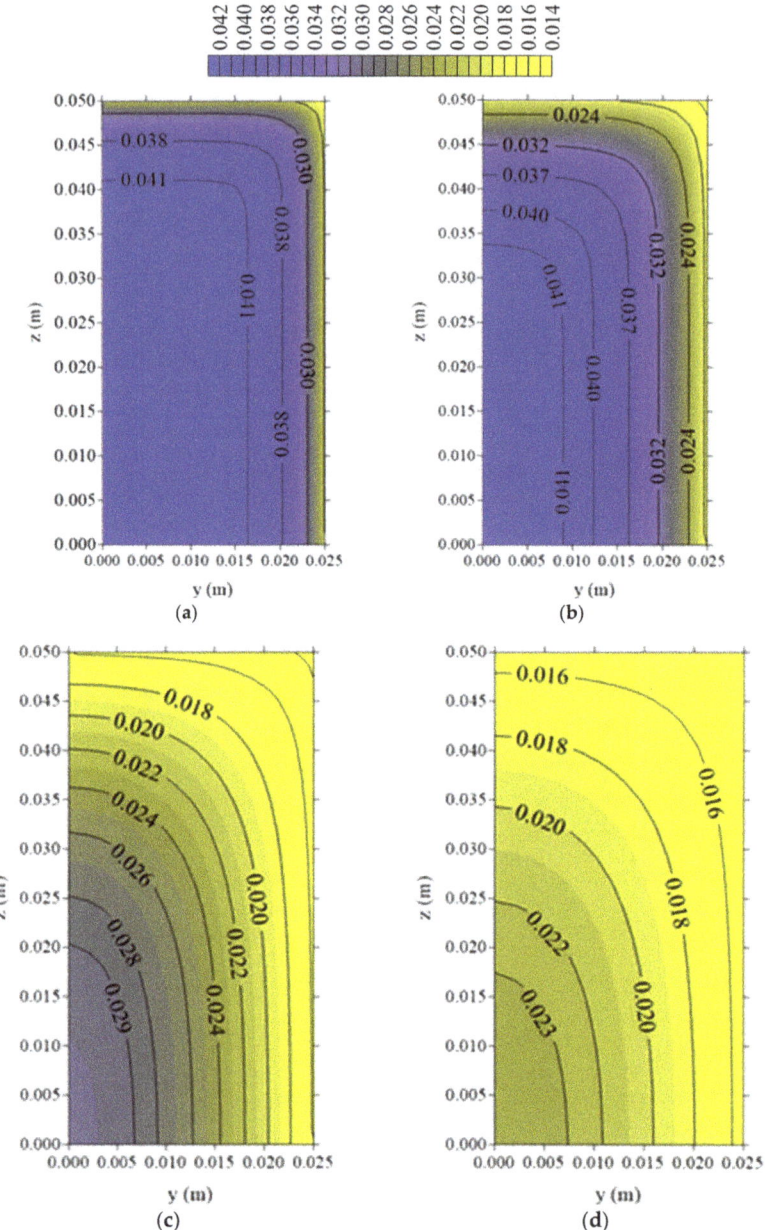

Figure 12. Water vapor concentration field (kg$_{vapor}$/m^3) in the zy plan at x = 0.025 m (R$_1$/2) (T= 50°C). (**a**) t = 200 s, (**b**) t = 700 s, (**c**) t = 5000 s and (**d**) t = 8000 s.

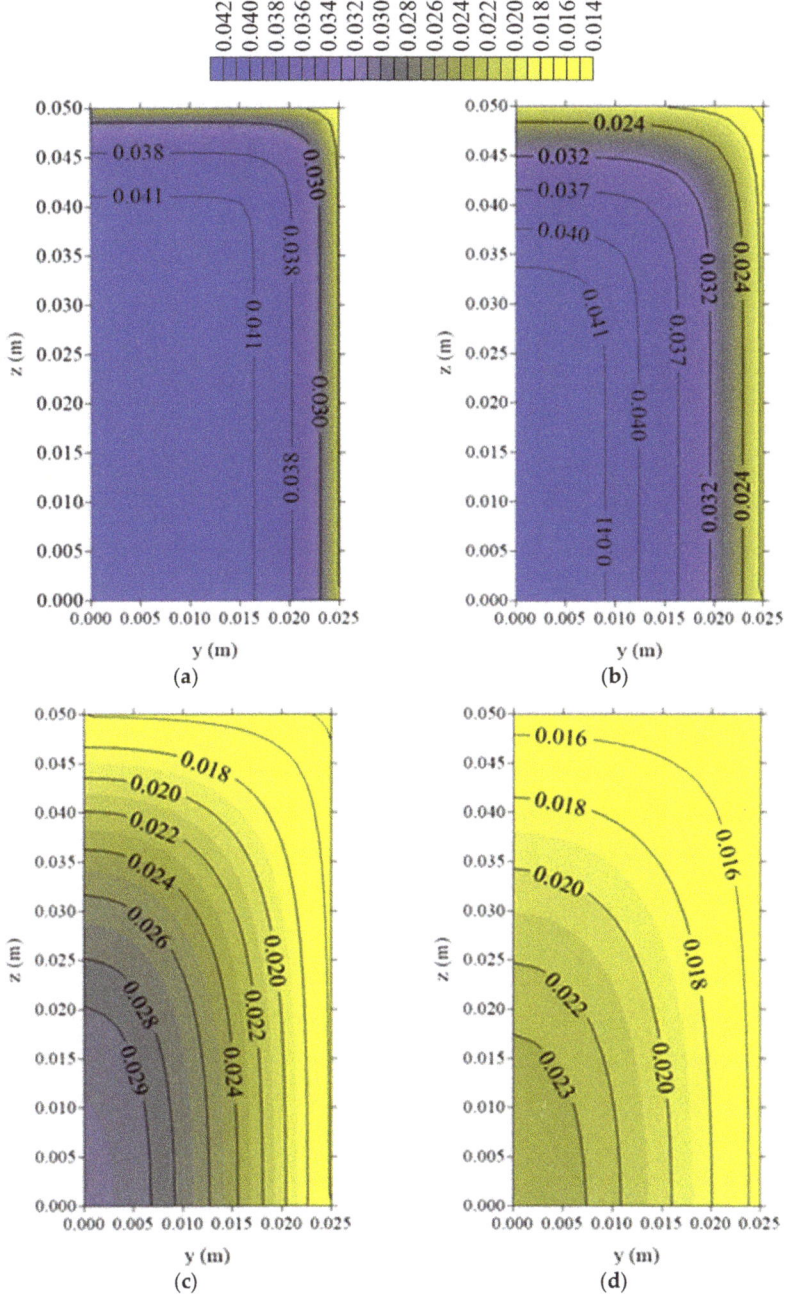

Figure 13. Water vapor concentration field (kg_{vapor}/m^3) in the zx plan at y = 0.0125 m ($R_2/2$) (T = 50°C). (**a**) t = 200 s, (**b**) t = 700 s, (**c**) t = 5000 s and (**d**) t = 8000 s.

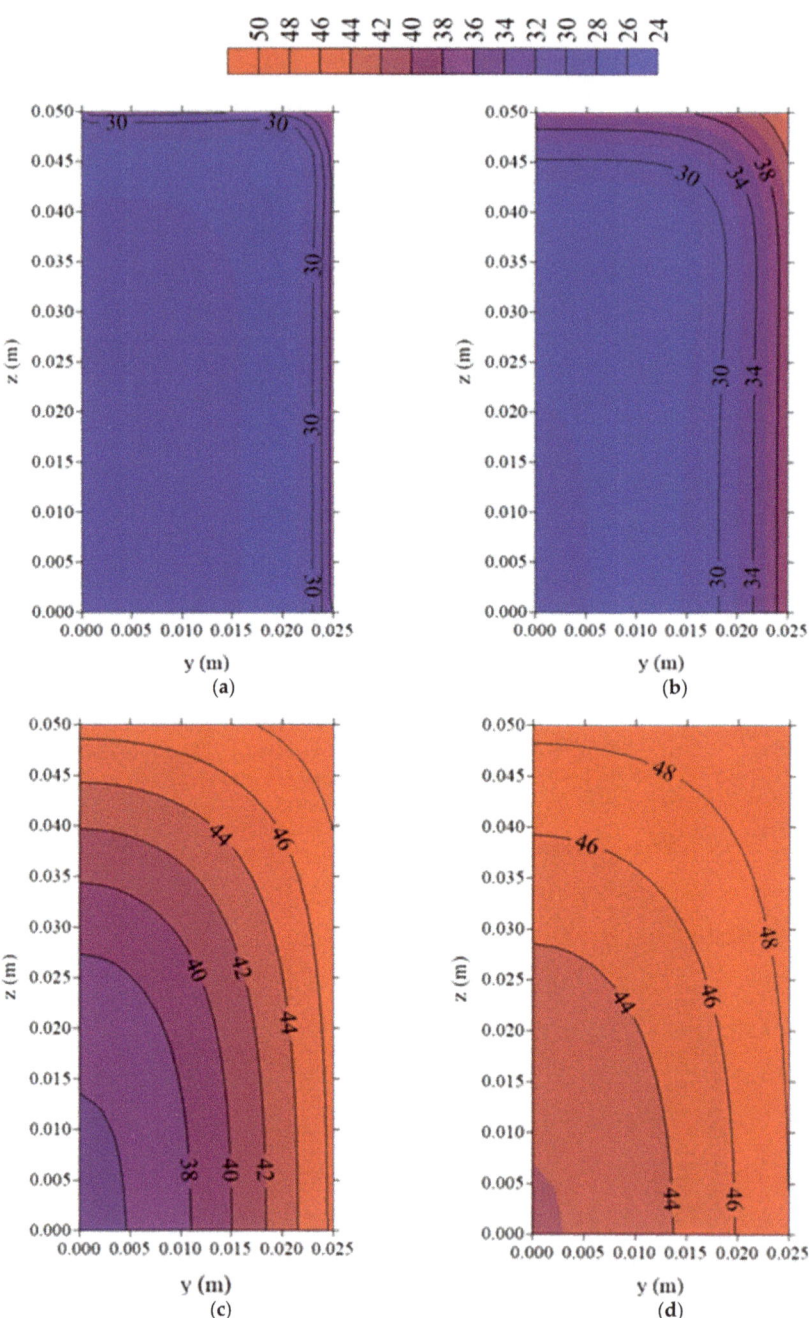

Figure 14. Temperature field (°C) in the zy plan at x = 0.025 m ($R_1/2$) (T = 50°C). (**a**) t = 200 s, (**b**) t=700 s, (**c**) t = 5000 s and (**d**) t = 8000 s.

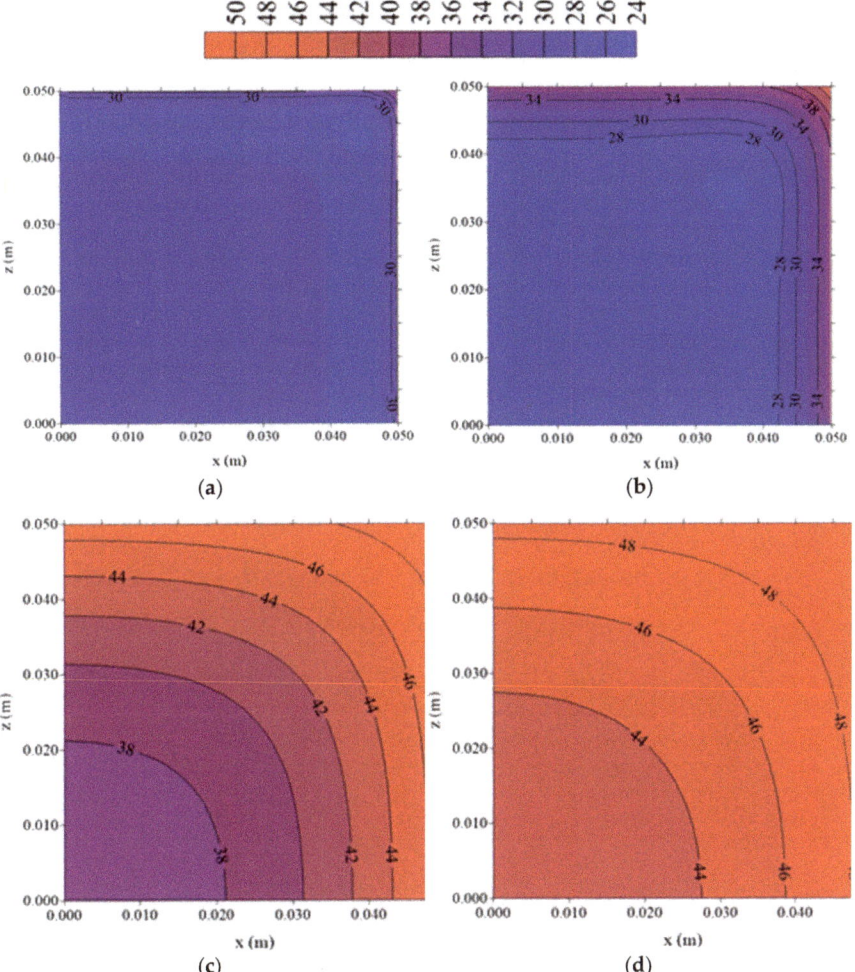

Figure 15. Temperature field (°C) in the plane zx at y = 0.0125 m ($R_2/2$) (T = 50°C). (a) t = 200 s, (b) t=700 s, (c) t = 5000 s and (d) t = 8000 s.

Upon analyzing these Fs, it can be seen that the temperature has the lowest results in the center of the fibrous medium along the process. The temperature also increased in any position, tending to the thermal equilibrium condition for long drying time. This behavior proved that heat flux occurs from the surface to the center of the porous sample. Furthermore, it is observed that the maximum temperature occurred close to the edge, especially in the vertex of the sample.

From already mentioned results, the regions close to the vertex dry and heat faster. Therefore, these regions are most susceptible to higher moisture and temperature gradients, which originate hydric and thermal stresses. Based on the intensity of these stresses, the material can achieve unfavorable deformation and rupture levels, and poor quality for a specific application. Under the industrial aspect these results are important to assist engineers in making decision about how the material must be dried and the fiber distributed in the bed, in order to minimize these effects.

3.3.3. Estimation of Transport Coefficients (D, h_m and h_c)

During drying process, at the surface of the fiber sample different phenomena occur simultaneously: heat and mass convection due to hot air flowing over the sample, and heat and mass transfer due to the water evaporation. Analyzing the effects of convective heat and mass transfer at the surface of the porous sample, boundary conditions of third kind were considered in this research. Thus, two phenomena occur, simultaneously: moisture flux at the surface that is proportional to the difference between the moisture content at the surface of the fiber sample and the equilibrium moisture content on the air temperature, and heat flux at the surface that is proportional to the difference between the temperature at the surface of the fiber sample and the drying air temperature. For these situations, the convective heat transfer coefficient (h_c) and convective mass transfer coefficient (h_m) were obtained using the least square error technique. In this technique, the predicted and experimental data of fiber average moisture content and surface temperature during drying at different conditions are compared. The best values of these parameters correspond to the lowest value of the variance statistical parameter.

As already described earlier, from these initial values of the convective heat and mass transfer coefficients, an iterative mathematical procedure was performed to determine the optimized values of these coefficients, including the mass diffusion coefficient. Tables 6 and 7 summarize the values of the parameters estimated, relative error and variance for all experimental tests obtained after the statistical procedure.

Table 6. Estimated mass transport coefficients for different drying air conditions.

T (°C)	Parameter					
	h_{mx} (m/s)	h_{my} (m/s)	h_{mz} (m/s)	D' (m^2/s)	$ERMQ_M$ $(kg/kg)^2$	$\bar{S}^2_M$ $(kg/kg)^2$
50	3.8022×10^{-4}	5.3771×10^{-4}	3.8022×10^{-4}	0.612×10^{-6}	0.0001576	1.1939×10^{-6}
60	4.0436×10^{-4}	20.2160×10^{-4}	4.0436×10^{-4}	1.012×10^{-6}	0.0004562	4.0015×10^{-6}
70	5.0622×10^{-4}	21.6564×10^{-4}	5.0622×10^{-4}	1.612×10^{-6}	0.0000722	7.5257×10^{-7}
80	7.6108×10^{-4}	7.7809×10^{-4}	7.6108×10^{-4}	2.212×10^{-6}	0.0002573	3.0629×10^{-6}
90	4.2089×10^{-4}	5.9523×10^{-4}	4.2089×10^{-4}	2.312×10^{-6}	0.0003805	4.8164×10^{-6}

Table 7. Estimated heat transport coefficients for different drying air conditions.

T (°C)	Parameters					
	h_{cx} ($W/m^2 K$)	h_{cy} ($W/m^2 \cdot K$)	h_{cz} ($W/m^2 \cdot K$)	α (m^2/s)	$ERMQ_T$ $(°C)^2$	$\bar{S}^2_T$ $(°C)^2$
50	4.2208	5.9692	4.2208	3.7789×10^{-7}	989.6026	7.4406
60	4.9768	7.0382	4.9768	4.2430×10^{-7}	1516.8398	13.1899
70	5.5389	7.8332	5.5389	3.9856×10^{-7}	967.74872	9.9768
80	5.1463	7.2780	5.1463	3.9270×10^{-7}	1324.2250	15.5791
90	6.6552	9.4119	6.6552	3.7627×10^{-7}	1273.6510	15.9206

After analysis of the Tables 5 and 6, small errors and variances can be seen, proving that the all mathematical procedures used in estimating the transport coefficients are appropriate. Further, we verified that both convective heat and mass transfer coefficient have small values typical of natural convections, due to small values of the air velocity inside the oven, and tend to increase when air drying temperature is increased. Some oscillatory behavior was verified in the convective mass transfer coefficient that can be attributed, for example, to the fact that the effects of dimensional variations due to heating (dilation) and moisture removal (shrinkage) were not considered, and the assumptions of constant fiber bed porosity and equilibrium equation has linear behavior during the drying process.

Particularly, the mass diffusion coefficient (D) was obtained by multiplying the vapor diffusion coefficient within the fibrous bed by the bed porosity, as already mentioned before, i.e. $D = \varepsilon D'$. Upon analyzing of the Table 6, we verified that major mass diffusion coefficients are found when higher air temperature was used. Furthermore, since the porosity of the medium presented minor variation in the experiments ($0.915 < \varepsilon < 0.922$), the drying of the fibrous bed presented similar behavior than the drying of individual fibers, which can be confirmed from the value of the parameter D' which is less than the parameter D_{AB} (Table 3) for all experimental tests, at least an order of magnitude. Therefore, we state that the water vapor flux within the fibrous bed in less than in the air outside the porous sample [27,28].

Finally, it is well known that the drying process requires high energy consumption and is responsible for high pollutant emissions in the atmosphere due to the use of different fuels for drying-air heating. Thus, it is important to control drying systems in laboratory or industrial scales, minimizing both energy consumption and environmental impact.

In this research, we develop macroscopic and advanced heat and mass transfer equations applicable for capillary-porous bodies with parallelepiped configuration. Both the model and the finite-volume method used herein demonstrated great potential, being accurate and efficient to be applied in many practical physical problems such as:

a. Diffusion processes of wetting, drying, heating and cooling, whether coupled or separate;
b. Coupled diffusion processes of liquid, vapor and heat in different porous bodies;
c. It is possible to study different diffusion problems assuming variable or constant thermophysical properties, and equilibrium or convective boundary conditions;
d. Under numerical aspects it is possible to use uniform or nonuniform grids, and to consider changes in the dimensions (dilation and shrinkage) of the bodies in the transient simulations.
e. The finite volume method is unconditionally stable even when applied to solving nonlinear partial differential equations;
f. No restriction about the nature of the porous body bed (fruits, grains, vegetables, textiles, etc.) is required when it is considered as continuum media;
g. Good estimation of process time to dry the product, contributing to reduction in energy consumption and increase in productivity of dried porous bodies;
h. Rigorous checkup and understanding of the effect of process variables on product quality during drying.

Despite the advantages presented above, some limitations of the modeling can be cited:

(a) Chemical transformation, thermal and structure inside the porous body that, in general, are responsible to provoke no uniform distribution of void are not considered;
(b) Inadequacy to be applied in drying processes where high air-drying temperature and effects of pressure are important;
(c) Mass diffusion is the only moisture migration mechanism, and gravity, capillarity, and filtration effects, for example, are not considered;
(d) It is necessary to use estimated parameters after fitting nonlinear regression or other techniques for this purpose.

4. Conclusions

From the predicted results, it can be concluded that:

(a) The proposed mathematical model proved to be useful for describing the drying process of sisal fibers, considering the good agreement between the predicted and experimental data of average moisture content and surface temperature of the sisal fibers obtained in all drying conditions;
(b) Despite the highly nonlinear character of the governing equations that represent the mathematical model, the finite-volume method proved to be useful to solve the cited equations and to assist in predicting the phenomenon of heat and mass transfer within the porous sample;

(c) Drying of sisal fibers occurred at a falling drying rate period;
(d) The largest water vapor concentration and temperature gradients are located in the regions near the vertex of the fiber porous bed. Then, these regions are more suitable to deformation and hydric and thermal effects, which are responsible for reducing product quality after drying process.

Finally, in continuation of this research, the authors strongly recommend studying drying problems occurring with dimensional variations of the bed and intermittent drying that is a technique useful in minimizing energy consumption.

Author Contributions: All the authors contributed to the development, analysis, writing, and revision of the paper: con-ceptualization, J.F.B.D. and A.F.V.; methodology, J.F.B.D., F.D.R. and R.S.G.; software, J.F.B.D. and A.G.B.L.; validation, J.F.B.D., D.D.S.D. and I.S.R.; formal analysis, J.F.B.D., J.M.P.Q.D. and A.G.B.L.; investigation, J.F.B.D., I.B.S. and J.M.P.Q.D.; resources, A.D.V. and M.J.F.; writ-ing—original draft preparation, J.F.B.D. and J.M.P.Q.D.; writing—review and editing, J.F.B.D., J.M.P.Q.D. and J.E.F.C.; visualization, R.S.G., J.E.F.C. and I.B.S.; supervision, A.G.B.L. and J.M.P.Q.D. All authors have read and agreed to the published version of the manuscript.

Funding: This work was supported by base funding (UIDB/04708/2020) and programmatic funding (UIDP/04708/2020) of the CONSTRUCT—Instituto de I & D em Estruturas e Construções, which is funded by national funds through the Fundação para a Ciência e a Tecnologia (FCT/MCTES), Central Government Investment and Expenditure Program (PIDDAC) and CNPq, CAPES, and FINEP (Brazilian Research Agencies).

Institutional Review Board Statement: Not applicable.

Informed Consent Statement: Not applicable.

Data Availability Statement: The data that support the findings of this study are available upon request from the authors.

Acknowledgments: The authors are grateful to the Federal University of Campina Grande (Brazil) for the research infrastructure and the references cited in the manuscript.

Conflicts of Interest: The authors declare no conflict of interest.

Abbreviations

A_b	Coefficient of the discretized linear equation	—
A_e	Coefficient of the discretized linear equation	—
A_f	Coefficient of the discretized linear equation	—
A_n	Coefficient of the discretized linear equation	—
A_p	Coefficient of the discretized linear equation	—
A_s	Coefficient of the discretized linear equation	—
A_w	Coefficient of the discretized linear equation	—
AH_{air}	Air absolute humidity	$kg_{vapor}/kg_{dry\,air}$
$AH_{air,\,sat}$	Air absolute humidity at the saturation	$kg_{vapor}/kg_{dry\,air}$
B	Source term	—
C	Water vapor concentration	kg_{vapor}/m^3
C''	Specific water vapor concentration	$kg_{vapor}/m^3/m^2$
C_{eq}	Equilibrium water vapor concentration	kg_{vapor}/m^3
C_o	Initial water vapor concentration	kg_{vapor}/m^3
$\overline{C}$	Average water vapor concentration	kg_{vapor}/m^3
C_B	Water vapor concentration in the nodal point B	kg_{vapor}/m^3
C_E	Water vapor concentration in the nodal point E	kg_{vapor}/m^3
C_F	Water vapor concentration in the nodal point F	kg_{vapor}/m^3
C_N	Water vapor concentration in the nodal point N	kg_{vapor}/m^3
C_P	Water vapor concentration in the nodal point P	kg_{vapor}/m^3
c_p	Specific heat;	$J/kg \cdot K$
C_P^0	Old water vapor concentration in the nodal point P	kg_{vapor}/m^3
C_S	Water vapor concentration in the nodal point S	kg_{vapor}/m^3

C_W	Water vapor concentration in the nodal point W	kg_{vapor}/m^3
D_{AB}	Diffusivity of water vapor in the air	m^2/s
D_e^C	Water vapor diffusion coefficient in the x-direction	m^2/s
D_f^C	Water vapor diffusion coefficient in the z-direction	m^2/s
D_n^C	Water vapor diffusion coefficient in the y-direction	m^2/s
D_b^C	Water vapor diffusion coefficient in the z-direction	m^2/s
D_s^C	Water vapor diffusion coefficient in the y-direction	m^2/s
D_w^C	Water vapor diffusion coefficient in the x-direction	m^2/s
D	Water vapor diffusion coefficient	m^2/s
d	Fiber diameter	m
ERMQ	Deviations between experimental and predicted values	—
h_s	Enthalpy	J/kg
h'_{cj}	Convective heat transfer coefficient	$W/m^2 \cdot K$
h'_{mj}	Convective mass transfer coefficient applied to the air	m/s
h_{cx}	Convective heat transfer coefficient	$W/m^2 \cdot K$
h_{cy}	Convective heat transfer coefficient	$W/m^2 \cdot K$
h_{cz}	Convective heat transfer coefficient	$W/m^2 \cdot K$
h_{mx}	Convective mass transfer coefficient	m/s
h_{my}	Convective mass transfer coefficient	m/s
h_{mz}	Convective mass transfer coefficient	m/s
K	Thermal conductivity;	$W/m \cdot K$
k	Air thermal conductivity	$W/m \cdot K$
K_b^T	Thermal conductivity in the z-direction	$W/m \cdot K$
K_b^T	Thermal conductivity in the y-direction	$W/m \cdot K$
K_e^T	Thermal conductivity in the x-direction	$W/m \cdot K$
K_f^T	Thermal conductivity in the z-direction	$W/m \cdot K$
K_n^T	Thermal conductivity in the y-direction	$W/m \cdot K$
K_w^T	Thermal conductivity in the x-direction	$W/m \cdot K$
L	Fiber length	m
M	Moisture content	$kg_{vapor}/kg_{dry\ fiber}$
M_{eq}	Equilibrium moisture content	$kg_{vapor}/kg_{dry\ fiber}$
n	Number of experimental points	—
$\hat{n}$	Number of fitted parameters	—
$\overline{Nu}_j$	Average Nusselt number	—
P	Atmospheric pressure	Pa
Pr	Average Prandtl number	—
R_1	Half-length of the bed in the x-direction	m
R_2	Half-length of the bed in the y-direction	m
R_3	Half-length of the bed in the z-direction	m
RH	Relative humidity	—
$\overline{R}$	Particular gas constant (atmospheric air)	J/mol/K
Re_j	Average Reynolds number	—
Sc	Average Schmidt number	—
$\overline{S}^2$	Variance	—
$\overline{Sh}_j$	Average Sherwood number	—
S^C	Source term	—
S^T	Source term	—
t	Time.	s
T	Temperature,	°C
T_B	Temperature in the nodal point B	°C
	Temperature in the nodal point E	°C
	Temperature in the nodal point F	°C
T_N	Temperature in the nodal point N	°C
	Temperature in the nodal point P	°C
	Old temperature in the nodal point P	°C
T_S	Temperature in the nodal point S	°C

T_W	Temperature in the nodal point W	°C
T_{abs}	Absolute temperature of the drying-air	°C
T_{eq}	Equilibrium temperature	°C
T_f	Final temperature	°C
T_o	Initial temperature	°C
T_o	Initial temperature	°C
V	Volume of the fiber bed	m^3
v	Air velocity	m/s
x	Cartesian coordinate	m
w, e, n, s, f, t	Faces of the control-volume	—
W, E, N, S, F, T, P	Nodal points	—
y	Cartesian coordinate	m
z	Cartesian coordinate	m
α	Constant of the equilibrium equation	$kg_{vapor}/kg_{dry\ fiber}$
β	Constant of the equilibrium equation	$kg_{vapor}/kg_{dry\ fiber}/°C$
δx_w	Distance between nodal points in the x-direction	m
δy_n	Distance between nodal points in the y-direction	m
δy_s	Distance between nodal points in the y-direction	m
δz_b	Distance between nodal points in the z-direction	m
δz_f	Distance between nodal points in the z-direction	m
δx_e	Distance between nodal points in the x-direction	m
Δt	Time step	s
$\Delta V'_{ijk}$	Volume of the elemental volume,	m^3
Δx	Length of the control-volume in the x-direction	m
Δy	Length of the control-volume in the y-direction	m
Δz	Length of the control-volume in the z-direction	m
ε	Bed porosity	—
μ	Air viscosity	$N·s/m^2$
ρ_s	Dry solid density	kg/m^3
σ	Constant of the equilibrium equation	$m^3/kg_{dry\ fiber}$
Φ	Potential of interest	—
$\overline{\Phi}$	Average value of the potential of interest	—

References

1. Silva, R.V. Polyurethane Resin Composite Derived from Castor Oil and Vegetable Fibers. Ph.D. Thesis, University of São Paulo, São Carlos, Brazil, 2003. Available online: https://www.teses.usp.br/teses/disponiveis/88/88131/tde-29082003-105440/pt-br.php (accessed on 10 April 2020). (In Portuguese).
2. Martin, A.R.; Martins, M.A.; Mattoso, L.H.C.; Silva, O.R.R.F. Chemical and structural characterization of sisal fibers from Agave Sisalana variety. *Polímeros Ciência e. Tecnol.* **2009**, *19*, 40–46. (In Portuguese) [CrossRef]
3. Wei, J.; Meyer, C. Improving degradation resistance of sisal fiber in concrete through fiber surface treatment. *Appl. Surface Sci.* **2014**, *289*, 511–523. [CrossRef]
4. Cruz, V.C.A.; Nóbrega, M.M.S.; Silva, W.P.; Carvalho, L.H.; Lima, A.G.B. An experimental study of water absorption in polyester composites reinforced with macambira natural fiber. *Mater. Werkst.* **2011**, *42*, 979–984. [CrossRef]
5. Melo Filho, J.A.; Silva, F.A.; Toledo Filho, R.D. Degradation kinetics and aging mechanisms on sisal fiber cement composite systems. *Cem. Concr. Compos.* **2013**, *40*, 30–39. [CrossRef]
6. Nóbrega, M.M.S.; Cavalcanti, W.S.; Carvalho, L.H.; Lima, A.G.B. Water absorption in unsaturated polyester composites reinforced with caroá fiber fabrics: Modeling and simulation. *Mater. Werkst.* **2010**, *41*, 300–305. [CrossRef]
7. Zhou, F.; Cheng, G.; Jiang, B. Effect of silane treatment on microstructure of sisal fibers. *Appl. Surf. Sci.* **2014**, *292*, 806–812. [CrossRef]
8. Lima, A.G.B.; Silva, J.B.; Almeida, G.S.; Nascimento, J.J.S.; Tavares, F.V.S.; Silva, V.S. Clay products convective drying: Foundations, modeling and applications. In *Drying and Energy Technologies. Advanced Structured Materials*; Delgado, J.M.P.Q., Lima, A.G.B., Eds.; Springer: Berlin/Heidelberg, Germany, 2016; Volume 63, pp. 43–70. [CrossRef]
9. Ferreira, S.R.; Lima, P.R.L.; Silva, F.A.; Toledo Filho, R.D. Effect of sisal fiber humidification on the adhesion with portland cement matrices. *Matéria* **2012**, *17*, 1024–1034. (In Portuguese) [CrossRef]
10. Santos, D.G.; Lima, A.G.B.; Costa, P.S. The Effect of the Drying Temperature on the Moisture Removal and Mechanical Properties of Sisal Fibers. *Defect Diffus. Forum* **2017**, *380*, 66–71. [CrossRef]
11. Diniz, J.F.B.; Lima, E.S.; Magalhães, H.L.F.; Lima, W.M.P.B.; Porto, T.R.N.; Gomez, R.S.; Moreira, G.; Lima, A.G.B. Drying of sisal fibers in oven with forced air circulation: An experimental study. *Res. Soc. Dev.* **2020**, *9*, e8639109342. [CrossRef]

12. Ghosh, B.N. Studies on the High Temperature Drying and Discoloration of Sisal Fibre. *J. Agric. Engng. Res.* **1966**, *11*, 69–75. [CrossRef]
13. Diniz, J.F.B.; Magalhães, H.L.F.; Lima, E.S.; Gomez, R.S.; Porto, T.R.N.; Moreira, G.; Lima, W.M.P.B.; Lima, A.G.B. Drying of sisal fibers in fixed bed: A predictive analysis using lumped models. *Res. Soc. Dev.* **2020**, *9*. (In Portuguese) [CrossRef]
14. Diniz, J.F.B.; Lima, H.G.G.M.; Lima, A.G.B.; Nascimento, J.J.S.; Ramos, A.D.O. Drying of Sisal Fiber: A Theoretical and Experimental Investigation. *Defect Diffus. Forum* **2019**, *391*, 36–41. [CrossRef]
15. Santos, D.G. Thermo-Hydric Study and Mechanical Characterization of Polymeric Matrix Composites Reinforced with Vegetable Fiber: 3D Simulation and Experimentation. Ph.D. Thesis, Federal University of Campina Grande, Campina Grande, Brazil, 2017. Available online: http://dspace.sti.ufcg.edu.br:8080/jspui/handle/riufcg/946 (accessed on 15 April 2020). (In Portuguese).
16. Diniz, J.F.B.; Lima, A.R.C.; Oliveira, I.R.; Farias, R.P.; Batista, F.A.; Lima, A.G.B.; Andrade, R.O. Vegetable Fiber Drying: Theory, Advanced Modeling and Application. In *Transport Process and Separation Technologies. Advanced Structured Materials*, 1st ed.; Delgado, J.M.P.Q., Lima, A.G.B., Eds.; Springer: Cham, Switzerland, 2021; Volume 133, pp. 31–60. [CrossRef]
17. Nordon, P.; David, H.G. Coupled diffusion of moisture and heat in hygroscopic textile materials. *Int. J. Heat Mass Transf.* **1967**, *10*, 853–866. [CrossRef]
18. Haghi, A.K. Simultaneous moisture and heat transfer in porous systems. *J. Comput. Appl. Mech.* **2001**, *2*, 195–204.
19. Xiao, B.; Huang, Q.; Chen, H.; Chen, X.; Long, G. A fractal model for capillary flow through a single tortuous capillary with roughened surfaces in fibrous porous media. *Fractals* **2021**, *29*, 2150017. [CrossRef]
20. Xiao, B.; Zhang, Y.; Wang, Y.; Jiang, G.; Liang, M.; Chen, X.; Long, G. A fractal model for kozeny–carman constant and dimensionless permeability of fibrous porous media with roughened surfaces. *Fractals* **2019**, *27*, 1950116. [CrossRef]
21. Crank, J. *The Mathematics of Diffusion*, 2nd ed.; Oxford University Press: London, UK, 1975.
22. Maliska, C.R. *Computational Heat Transfer and Fluid Mechanics*, 2nd ed.; LTC: Rio de Janeiro, Brazil, 2004.
23. Patankar, S.V. *Numerical Heat Transfer and Fluid Flow*; CRC Press: Boca Raton, FL, USA, 1980.
24. Versteeg, H.K.; Malalasekera, W. *An Introduction to Computational Fluid Dynamics: The Finite Volume Method*, 2nd ed.; Pearson Education Limited: Harlow, UK, 2007.
25. Figliola, R.S.; Beasley, D.E. *Theory and Design for Mechanical Measurements*; John Wiley & Sons: New York, NY, USA, 1995.
26. Bergman, T.L.; Incropera, F.P.; DeWitt, D.P.; Lavine, A.S. *Fundamentals of Heat and Mass Transfer*, 7th ed.; John Wiley & Sons: Hoboken, NJ, USA, 2011.
27. Diniz, J.F.B.; Moreira, G.; Nascimento, J.J.S.; Farias, R.P.; Magalhães, H.L.F.; Barbalho, G.H.A.; Lima, A.G.B. Drying of Sisal Fiber: Three-Dimensional Mathematical Modeling and Simulation. *Defect Diffus. Forum* **2020**, *399*, 202–207. [CrossRef]
28. Diniz., J.F.B. Heat and Mass Transfer in Porous Solids with Parallelepiped Shape. Case Study: Drying of Sisal Fibers. Ph.D. Thesis, Federal University of Campina Grande, Campina Grande, Brazil, 2018. Available online: http://dspace.sti.ufcg.edu.br:8080/jspui/handle/riufcg/2424 (accessed on 21 April 2020). (In Portuguese).

Article

Non-Equilibrium Thermodynamics-Based Convective Drying Model Applied to Oblate Spheroidal Porous Bodies: A Finite-Volume Analysis

João C. S. Melo [1], João M. P. Q. Delgado [2,*], Wilton P. Silva [3], Antonio Gilson B. Lima [4], Ricardo S. Gomez [4], Josivanda P. Gomes [5], Rossana M. F. Figueirêdo [5], Alexandre J. M. Queiroz [5], Ivonete B. Santos [6], Maria C. N. Machado [7], Wanderson M. P. B. Lima [4] and João E. F. Carmo [4]

1. Federal Institute of Education, Science and Technology of Rio Grande do Norte, Caicó 59300-000, Brazil; carlos.soares@ifrn.edu.br
2. CONSTRUCT-LFC, Department of Civil Engineering, Faculty of Engineering, University of Porto, 4200-465 Porto, Portugal
3. Department of Physics, Federal University of Campina Grande, Campina Grande 58429-900, Brazil; wiltonps@uol.com.br
4. Department of Mechanical Engineering, Federal University of Campina Grande, Campina Grande 58429-900, Brazil; antonio.gilson@ufcg.edu.br (A.G.B.L.); ricardosoaresgomez@gmail.com (R.S.G.); wan_magno@hotmail.com (W.M.P.B.L.); jevan.franco@gmail.com (J.E.F.C.)
5. Department of Agricultural Engineering, Federal University of Campina Grande, Campina Grande 58429-900, Brazil; josivanda@gmail.com (J.P.G.); rossanamff@gmail.com (R.M.F.F.); alexandrejmq@gmail.com (A.J.M.Q.)
6. Department of Physics, State University of Paraíba, Campina Grande 58429-500, Brazil; ivoneetebs@gmail.com
7. Department of Chemical, State University of Paraíba, Campina Grande 58429-500, Brazil; ceicamachado3@gmail.com
* Correspondence: jdelgado@fe.up.pt; Tel.: +351-225081404

Citation: Melo, J.C.S.; Delgado, J.M.P.Q.; Silva, W.P.; B. Lima, A.G.; Gomez, R.S.; Gomes, J.P.; Figueirêdo, R.M.F.; Queiroz, A.J.M.; Santos, I.B.; Machado, M.C.N.; et al. Non-Equilibrium Thermodynamics-Based Convective Drying Model Applied to Oblate Spheroidal Porous Bodies: A Finite-Volume Analysis. *Energies* **2021**, *14*, 3405. https://doi.org/10.3390/en14123405

Academic Editor: Moghtada Mobedi

Received: 11 May 2021
Accepted: 8 June 2021
Published: 9 June 2021

Publisher's Note: MDPI stays neutral with regard to jurisdictional claims in published maps and institutional affiliations.

Copyright: © 2021 by the authors. Licensee MDPI, Basel, Switzerland. This article is an open access article distributed under the terms and conditions of the Creative Commons Attribution (CC BY) license (https://creativecommons.org/licenses/by/4.0/).

Abstract: Commonly based on the liquid diffusion theory, drying theoretical studies in porous materials has been directed to plate, cylinder, and sphere, and few works are applied to non-conventional geometries. In this sense, this work aims to study, theoretically, the drying of solids with oblate spheroidal geometry based on the thermodynamics of irreversible processes. Mathematical modeling is proposed to describe, simultaneously, the heat and mass transfer (liquid and vapor) during the drying process, considering the variability of the transport coefficients and the convective boundary conditions on the solid surface, with particular reference to convective drying of lentil grains at low temperature and moderate air relative humidity. All the governing equations were written in the oblate spheroidal coordinates system and solved numerically using the finite-volume technique and the iterative Gauss–Seidel method. Numerical results of moisture content, temperature, liquid, vapor, and heat fluxes during the drying process were obtained, analyzed, and compared with experimental data, with a suitable agreement. It was observed that the areas near the focal point of the lentil grain dry and heat up faster; consequently, these areas are more susceptible to the appearance of cracks that can compromise the quality of the product. In addition, it was found that the vapor flux was predominant during the drying process when compared to the liquid flux.

Keywords: drying; lentil grain; oblate spheroid; modeling; numerical simulation

1. Introduction

Water is the main constituent present in high concentrations in fresh foods, which considerably influences the palatability, digestibility, and physical structure of the food. The deterioration that occurs in food is practically influenced in one way or another by the concentration and mobility of water inside it [1]. The removal of water from solid foods is used as a way to reduce water activity by inhibiting microbial growth, thus avoiding its deterioration. Water removal has become of great importance in reducing the costs of energy, transport, packaging, and storage of these foods [2].

Most of the methods of food preservation are based on the reduction in water mobility by the use of humectants materials and freezing, and also by physical removal of water through osmotic dehydration, drying, evaporation, or lyophilization [3]. The main idea is to decrease the amount of water in the product to acceptable levels and maintain the physical-chemical and sensory properties of agricultural products in order to increase the shelf life of the products.

Although several technological processes of food conservation can be applied in order to increase the shelf life of agricultural products, the drying process has several advantages, as: the facility in the preservation of the product; stability of aromatic components at room temperature for long periods of time; protection against enzymatic and oxidative degradation; reducing of the weight; energy savings because it does not require refrigeration and product availability during any time of the year [4]. In literature, it is possible to find several studies that showed the use of drying processes as a method to increase the shelf life of products such as: lentils [5,6]; grape [7]; carrots [8]; banana [9]; corn [10,11]; red chilies [12]; wheat [13]; bean [14]; strawberry [15], etc.

Drying is a unitary operation of water removal of a product by evaporation or sublimation, by applying heat under controlled conditions, which has, as its purpose, to conserve food and its nutritional and organoleptic properties by reducing the activity of water inside it [16].

During the drying process, the material undergoes variations in its chemical, physical and biological characteristics, which, depending on the intensity of the effect, may cause their loss or disable them for some applicability [7,17]. Thus, the criterion of preserving the quality of the product, which depends on the final use that will be made, is the one that regulates the type of drying and storage process [2].

From the theoretical point of view, the modeling and prediction of the drying process can be approached in two ways: based on external parameters to the solid such as relative humidity, temperature, and air velocity, correlating them with the drying rate of the solid, while the other has as characteristic, the internal conditions to the porous material and the mechanisms of moisture transport in this material [18].

The complexity of drying leads several researchers to propose various theories and models to predict the moisture transfer inside the material, such as: liquid diffusion theory, capillary theory, condensation and evaporation theory, Luikov theory, Krischer theory, Berger and Pei theory, Philip and De Vries theory, and the theory of Fortes and Okos [19].

The Luikov model [20] and Fortes and Okos [19] model are examples of models based on the thermodynamics of irreversible processes. Luikov's theory takes into account the mechanisms of diffusion, effusion, and convection of water inside the porous medium. Therefore, the equations that define the Luikov model take into account that the molecular transport of water vapor, air, and liquid happen simultaneously [21]. On the other hand, Fortes [22] and Fortes and Okos [19,23] proposed that the driving force for isothermal transfer, both liquid and vapor, is the gradient of equilibrium moisture content, due to the local equilibrium hypothesis, which, in turn, is a function of temperature, relative humidity, and equilibrium moisture content. Detail about the non-equilibrium thermodynamic can be found in the references cited in the text.

Oliveira and Lima [24] used the model proposed by Fortes and Okos [19] to study the heat and mass transfer in the simulation of wheat drying and concluded that the numerical results of the average moisture content showed a suitable agreement with the experimental values. Oliveira et al. [25] and Oliveira et al. [26] observed, applying the model developed by Fortes and Okos [19] to study the phenomenon of shrinkage during a drying process of wheat grain, that the model used was accurate and effective and could be used to simulate many practical diffusion problems, such as heating, cooling, wetting and drying in prolate spheroidal solids, including spherical solids.

In previous studies, Carmo and Lima [27] reported studies of drying of oblate spheroidal porous solids using the liquid diffusion theory, with particular reference to lentil grains, considering constant thermophysical properties and equilibrium, and convective boundary conditions on the solid surface. Melo et al. [28] and Melo et al. [29] reported the transient behavior of heat and mass transfer, also in oblate spheroidal solids (lentil grains) using a mathematical model based on the thermodynamics of irreversible processes and considering equilibrium boundary condition on the surface of the solid. In this study, the authors reported vapor flux as the dominant mechanism for mass transfer during the drying process.

In this sense, due to the fact that few works studied coupled heat and mass transfer in solids with complex shape [18,24–27]; that during the drying process of agricultural products with low moisture content, the predominant mass flux is the vapor flux (not predicted by the liquid diffusion theory l), and the equilibrium boundary condition (hygroscopic and thermal balances) does not accurately reflect the physical phenomenon of heat and mass transport on the surface of the grain; this work aims to study the transfer of heat and mass in oblate spheroidal porous materials using mathematical modeling based on the non-equilibrium thermodynamics. This model considers variable thermophysical properties and convective boundary conditions on the surface of the porous solid. The application has been given for lentil grains drying. The idea is to provide subsidies for a better understanding of the drying process in solids with complex shapes in order to allow its optimization in terms of energy consumption (prediction of drying time) and product quality (evaluated by the moisture and heat transfer inside the hygroscopic porous solid).

2. Methodology

2.1. Mathematical Modeling

The model under study involves the determination of heat and mass transfer in a two-dimensional case for solids with ellipsoidal geometry. In the mathematical formulation, the following assumptions were adopted:

(a) The solid is homogeneous and isotropic;
(b) Mass transfer in the single particle occurs by diffusion of liquid and vapor, under decreasing drying rate;
(c) At the beginning of the drying process, the distributions of the moisture and temperature content are considered uniform and symmetrical around the z-axis;
(d) The thermophysical properties are variable during the drying process and dependent on the position and moisture content inside the material;
(e) Volume shrinkage negligible;
(f) No capillarity effect;
(g) Moisture transfer inside the solid by liquid and vapor diffusion, and evaporation and convection on a solid surface;
(h) Heat transfer inside the solid by conduction, evaporation, and convection on a solid surface.

Based on a mechanistic approach and considering equality between the chemical potential and the potential of water (suction), Fortes [22], Fortes and Okos [23], and Fortes and Okos [19] reported a set of partial differential equations to describe the drying process of porous hygroscopic solids. According to these authors, the expression that describes the liquid flux is given by:

$$\vec{J}_\ell = -\rho_\ell k_\ell R_v \ell n H \nabla T - \rho_\ell k_\ell \left(\frac{R_v T}{H}\right) \frac{\partial H}{\partial M} \nabla M + \rho_\ell k_\ell \vec{g} \qquad (1)$$

where ρ_l is the liquid density, k_l is the liquid conductivity, H is the relative humidity, T is the temperature, M is the moisture content, R_v is the universal gas constant, and g is the acceleration of gravity.

The expression that describes the vapor flux is given by:

$$\vec{J}_v = -k_v\left(\rho_{v0}\frac{\partial H}{\partial T} + H\frac{d\rho_{v0}}{dT}\right)\nabla T - k_v\rho_{v0}\frac{\partial H}{\partial M}\nabla M + \rho_l k_l \vec{g} \qquad (2)$$

where k_v is the vapor conductivity and ρ_{v0} is the saturation density. The heat flux is given by the following equation:

$$\begin{aligned}\vec{J}_q = & -k_T\nabla T - \left[\rho_\ell k_\ell R_v \ell n H + k_v\left(\rho_{v0}\frac{\partial H}{\partial T} + H\frac{d\rho_{v0}}{dT}\right)\right]\frac{R_v T^2}{H}\frac{\partial H}{\partial M}\nabla M + \\ & + T\left[\rho_\ell k_\ell R_v \ell n H + k_v\left(\rho_{v0}\frac{\partial H}{\partial T} + H\frac{d\rho_{v0}}{dT}\right)\right]\vec{g}\end{aligned} \qquad (3)$$

where k_T is the effective apparent thermal conductivity of the porous medium, valid for conditions that do not involve mass transport.

Considering the absence of ice and that several factors are negligible, as the air mass, the vapor mass (but not its flux) in relation to the liquid mass and shrinkage of the medium, the differential equation for the mass transfer in the vapor and liquid phases inside the material, and applied to an elementary control volume, is given as follows:

$$\rho_s\frac{\partial M}{\partial t} = -\nabla\cdot\left(\vec{J}_\ell + \vec{J}_v\right), \qquad (4)$$

where t is the time and $M = M_\ell + M_v \cong M_\ell$.

The energy balance equation could be obtained by the relation between the rate of variation of the volumetric enthalpy of the system and the adsorption heat with the divergence of the enthalpy flux (liquid and vapor phases),

$$\frac{\partial}{\partial t}(\rho_s c_b T) - \frac{\partial}{\partial t}(\rho_s h_w M) = -\nabla\cdot\vec{J}_q - \nabla\cdot\left(h_{fg}\vec{J}_v\right) - \vec{J}_\ell\cdot c_\ell\nabla T - \vec{J}_v\cdot c_v\nabla T \qquad (5)$$

where h_w is the differential specific heat of sorption, h_{fg} is the specific latent heat of water vaporization, and c_b is the specific heat of the humid medium, given by:

$$c_b = c_s + c_\ell M_\ell + c_v M_v \qquad (6)$$

where c_s is the specific heat of the dry product, c_l is the specific heat of the liquid, and c_v is the vapor specific heat.

Furthermore, considering the gravitational effects, Equations (4) and (5) could be written in a more compact form as,

$$\frac{\partial}{\partial t}(\rho_s M) = \nabla\cdot\left(\Gamma_1^\Phi\nabla M\right) + \nabla\cdot\left(\Gamma_2^\Phi\nabla T\right), \qquad (7)$$

and

$$\begin{aligned}\frac{\partial}{\partial t}(\rho_s c_b T) - \frac{\partial}{\partial t}(\rho_s h_w M) = & \nabla\cdot\left(\Gamma_3^\Phi\nabla T\right) + \nabla\cdot\left(\Gamma_4^\Phi\nabla M\right) + \nabla\cdot\left(\Gamma_5^\Phi\nabla T\right) \\ & + \nabla\cdot\left(\Gamma_6^\Phi\nabla M\right) + \Gamma_7^\Phi\nabla T\cdot\nabla T + \Gamma_8^\Phi\nabla M\cdot\nabla T,\end{aligned} \qquad (8)$$

where the parameter Γ_i^Φ is given in Appendix A.

In this work, a porous solid with oblate spheroidal geometry was considered. Figure 1 shows half of the solid and some geometric parameters of this body. In this figure, μ, φ and ω represent the elliptical coordinates, and L_1 and L_2 are the minor and major axis of the ellipse.

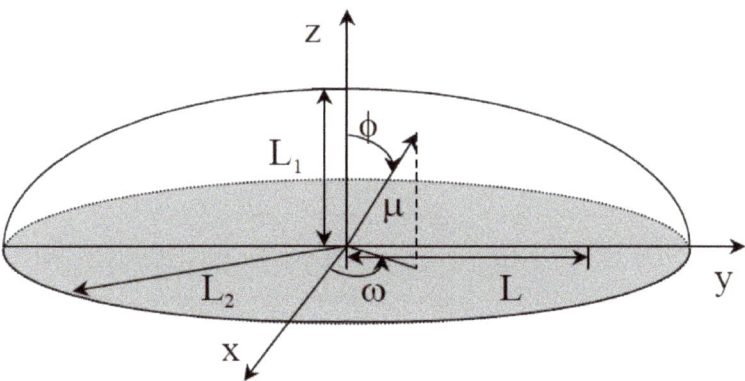

Figure 1. Features of an oblate spheroidal body.

In order to present the diffusion equation in an oblate spheroidal coordinates system, the following relationships are used [30]:

$$x = L\sqrt{(1+\xi^2)(1-\eta^2)}\,\zeta, \tag{9}$$

$$y = L\sqrt{(1+\xi^2)(1-\eta^2)}\sqrt{1-\zeta^2}, \tag{10}$$

$$z = L\xi\eta, \tag{11}$$

where $\xi = \sinh\mu$, $\eta = \cos\varphi$ and $\zeta = \cos\omega$. Finally, L is the focal length given by:

$$L = \sqrt{L_2^2 - L_1^2}. \tag{12}$$

The domain of the new spheroidal variables ξ, η and ζ (in terms of ω), related with Figure 1 is given by: $0 \leq \xi \leq L_1/L$, $0 \leq \eta \leq 1$ and $0 \leq \omega \leq 2\pi$.

Using the methodology described by Maliska [31] in order to obtain the diffusion equation in the oblate spheroidal coordinates system and, considering the symmetry of the solid, the following diffusion equations for the two-dimensional case are obtained:

$$\frac{\partial(\rho_s M)}{\partial t} = \left[\frac{1}{L^2(\xi^2+\eta^2)}\frac{\partial}{\partial \xi}\left((\xi^2+1)\Gamma_1^\Phi \frac{\partial M}{\partial \xi}\right)\right] + \left[\frac{1}{L^2(\xi^2+\eta^2)}\frac{\partial}{\partial \eta}\left((1-\eta^2)\Gamma_1^\Phi \frac{\partial M}{\partial \eta}\right)\right] \\ + \left[\frac{1}{L^2(\xi^2+\eta^2)}\frac{\partial}{\partial \xi}\left((\xi^2+1)\Gamma_2^\Phi \frac{\partial T}{\partial \xi}\right)\right] + \left[\frac{1}{L^2(\xi^2+\eta^2)}\frac{\partial}{\partial \eta}\left((1-\eta^2)\Gamma_2^\Phi \frac{\partial T}{\partial \eta}\right)\right], \tag{13}$$

and

$$\frac{\partial}{\partial t}(\rho_s c_b T) - \frac{\partial}{\partial t}(\rho_s h_w M) = \frac{1}{L^2(\xi^2+\eta^2)}\left\{\frac{\partial}{\partial \xi}\left[(\xi^2+1)\Gamma_3^\Phi \frac{\partial T}{\partial \xi}\right] + \frac{\partial}{\partial \eta}\left[(1-\eta^2)\Gamma_3^\Phi \frac{\partial T}{\partial \eta}\right]\right\} \\ + \frac{1}{L^2(\xi^2+\eta^2)}\left\{\frac{\partial}{\partial \xi}\left[(\xi^2+1)\Gamma_5^\Phi \frac{\partial T}{\partial \xi}\right] + \frac{\partial}{\partial \eta}\left[(1-\eta^2)\Gamma_5^\Phi \frac{\partial T}{\partial \eta}\right]\right\} + \frac{\Gamma_7^\Phi}{L^2(\xi^2+\eta^2)} \\ \left[(\xi^2+1)\left(\frac{\partial T}{\partial \xi}\right)^2 + (1-\eta^2)\left(\frac{\partial T}{\partial \eta}\right)^2\right] + \frac{\Gamma_8^\Phi}{L^2(\xi^2+\eta^2)}\left[(\xi^2+1)\frac{\partial M}{\partial \xi}\frac{\partial T}{\partial \xi} + (1-\eta^2)\frac{\partial M}{\partial \eta}\frac{\partial T}{\partial \eta}\right] + \\ + \frac{1}{L^2(\xi^2+\eta^2)}\left\{\frac{\partial}{\partial \xi}\left[(\xi^2+1)\Gamma_4^\Phi \frac{\partial M}{\partial \xi}\right] + \frac{\partial}{\partial \eta}\left[(1-\eta^2)\Gamma_4^\Phi \frac{\partial M}{\partial \eta}\right]\right\} + \\ + \frac{1}{L^2(\xi^2+\eta^2)}\left\{\frac{\partial}{\partial \xi}\left[(\xi^2+1)\Gamma_6^\Phi \frac{\partial M}{\partial \xi}\right] + \frac{\partial}{\partial \eta}\left[(1-\eta^2)\Gamma_6^\Phi \frac{\partial M}{\partial \eta}\right]\right\} \tag{14}$$

Equations (13) and (14) can be rewritten as:

$$\frac{\partial}{\partial t}\left(\frac{\lambda_1 \Phi_1}{J}\right) = \frac{\partial}{\partial \xi}\left(\alpha_{11} J \Gamma_1^\Phi \frac{\partial \Phi_1}{\partial \xi}\right) + \frac{\partial}{\partial \eta}\left(\alpha_{22} J \Gamma_1^\Phi \frac{\partial \Phi_1}{\partial \eta}\right) + S_1^\Phi \qquad (15)$$

and

$$\frac{\partial}{\partial t}\left(\frac{\lambda_2 \Phi_2}{J}\right) = \frac{\partial}{\partial \xi}\left(\alpha_{11} J \Gamma_3^\Phi \frac{\partial \Phi_2}{\partial \xi}\right) + \frac{\partial}{\partial \eta}\left(\alpha_{22} J \Gamma_3^\Phi \frac{\partial \Phi_2}{\partial \eta}\right) + \frac{\partial}{\partial \xi}\left(\alpha_{11} J \Gamma_5^\Phi \frac{\partial \Phi_2}{\partial \xi}\right) +$$
$$+ \frac{\partial}{\partial \eta}\left(\alpha_{22} J \Gamma_5^\Phi \frac{\partial \Phi_2}{\partial \eta}\right) + S_2^\Phi \qquad (16)$$

where the source terms S_1^Φ and S_2^Φ are given in Appendix A.

For a well-posed formulation, the initial boundary and symmetry conditions for the proposed model are the following:

(a) Mass
- Initial:
$$M(\xi, \eta, t = 0) = M_o. \qquad (17)$$

- Symmetry planes: In mass transfer the angular and radial gradients of the moisture content are equal to zero in the symmetry planes.

$$\frac{\partial M(\xi, 1, t)}{\partial \eta} = 0 \qquad (18)$$

$$\frac{\partial M(\xi, 0, t)}{\partial \eta} = 0 \qquad (19)$$

$$\frac{\partial M(0, \eta, t)}{\partial \xi} = 0 \qquad (20)$$

- Free surface: The diffusive flux is equal to the convective flux of the moisture content on the surface of the oblate spheroid.

$$\left(\vec{J}_\ell + \vec{J}_v\right)\bigg|_{\xi = \frac{L_1}{L}} = h_m(M - M_e) \text{ with } T = T_a \text{ and } H = H_a. \qquad (21)$$

(b) Heat
- Initial
$$T(\xi, \eta, t = 0) = T_o = \text{cte}. \qquad (22)$$

- Symmetry plane: Heat angular and radial gradients are equal to zero.

$$\frac{\partial T(\xi, 1, t)}{\partial \eta} = 0 \qquad (23)$$

$$\frac{\partial T(\xi, 0, t)}{\partial \eta} = 0 \qquad (24)$$

$$\frac{\partial T(0, \eta, t)}{\partial \xi} = 0 \qquad (25)$$

- Free surface: The diffusive flux is equal to the heat convective flux on the surface of the solid, more the energy to evaporate the water and the energy to heat the water vapor produced in the evaporation process. Then we can write:

$$\vec{J}_q\bigg|_{\xi = \frac{L_1}{L}} = h_c(T_s - T_a) + h_{fg} \vec{J}_\ell + \left(\vec{J}_\ell + \vec{J}_v\right)c_v(T_s - T_a). \qquad (26)$$

The average moisture content of the porous body during the drying process is given by [32]:

$$\overline{M} = \frac{1}{V}\int_V M(\xi, \eta, t)\, dV. \tag{27}$$

The average temperature of the porous body during the drying process could be obtained as follows:

$$\overline{T} = \frac{1}{V}\int_V T(\xi, \eta, t)\, dV, \tag{28}$$

where V is the volume of the solid.

2.2. Numerical Solution of the Governing Equations

In this work, the finite-volume method was used to numerically solve the diffusion equation in oblate spheroidal coordinates. In this method, the nodal points are centered on the control volume and the mesh adopted has entire volumes throughout the domain [31,33].

The numerical formulation adopted begins with the identification of the domain of interest and, from there, its division into a finite number of subdomains. Figure 2 shows the constant lines ξ and η delimiting the control volume associated with point nodal P. Points N, S, E, and W are the nodal points neighboring P, which represent the nodal points north, south, east, and west, respectively. The distance between the nodal point P and its neighbors ($\delta\xi$ and $\delta\eta$) is also observed, as well as the control volume dimensions ($\Delta\xi$ and $\Delta\eta$).

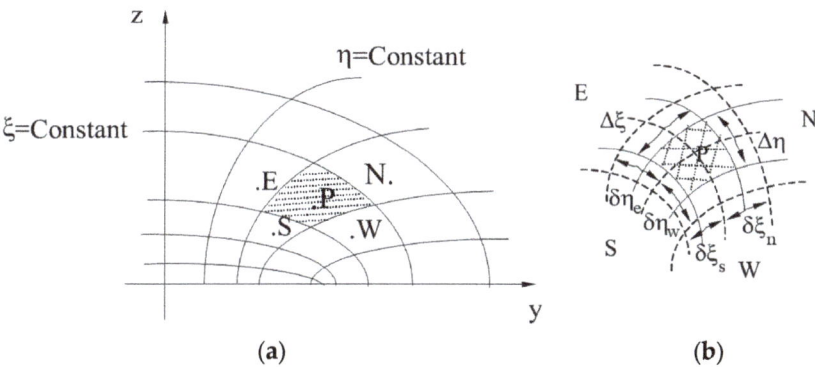

Figure 2. Continuous surface subdivided into nodal points. (**a**) Identification of the control volume and (**b**) geometric parameters of the control volume in the oblate spheroidal coordinates system.

Applying the finite-volume method to Equations (15) and (16), considering a fully implicit formulation, i.e., all diffusive terms of the equation are evaluated at the instant t + Δt, integrating on the control volume (see Figure 2), which corresponds to the internal points of the domain for a time t, and rearranging the terms of the resulting equation, the following linear equation, for mass transfer, is obtained in the following discretized form:

$$A_{1P}\Phi_{1P} = A_{1N}\Phi_{1N} + A_{1S}\Phi_{1S} + A_{1E}\Phi_{1E} + A_{1W}\Phi_{1W} + A_{1P}^o \Phi_{1P}^o + \hat{S}_1^\Phi, \tag{29}$$

where:

$$\hat{S}_1^\Phi = B_{1N}\Phi_{2N} + B_{1S}\Phi_{2S} + B_{1E}\Phi_{2E} + B_{1W}\Phi_{2W} - B_{1P}\Phi_{2P}, \tag{30}$$

where the coefficients A_i are given in Appendix A. In Equations (29) and (30) the parameters $\Phi_1 = M$ and $\Phi_2 = T$.

For energy balance ($\Phi_2 = T$), it results in:

$$A_{2P}\Phi_{2P} = A_{2N}\Phi_{2N} + A_{2S}\Phi_{2S} + A_{2E}\Phi_{2E} + A_{2EW}\Phi_{2W} + A_{2P}^0\Phi_{2P}^0 + S_2^\Phi, \tag{31}$$

with:

$$\hat{S}_2^\Phi = S_{C2}^\Phi + S_{P2}^\Phi, \tag{32}$$

and

$$S_{P2}^\Phi = -\left[\Gamma_{7n}^\Phi\left(\xi^2 + 1\right)\Delta\eta\Delta\xi\left(\frac{\Phi_{2N}^* - \Phi_{2P}^*}{\delta\xi_n}\right)\right] - \left[\Gamma_{7n}^\Phi(1-\eta^2)\Delta\eta\Delta\xi\left(\frac{\Phi_{2E}^* - \Phi_{2P}^*}{\delta\eta_e}\right)\right] + \\ \left[\Gamma_{8n}^\Phi\left(\xi^2+1\right)\Delta\eta\Delta\xi\left(\frac{\Phi_{1N} - \Phi_{1P}}{\delta\xi_n}\right)\right] + \Gamma_{8e}^\Phi(1-\eta^2)\Delta\eta\Delta\xi\left(\frac{\Phi_{1E} - \Phi_{1P}}{\delta\eta_e}\right) \tag{33}$$

$$S_{C2}^\Phi = B_{star} + B_{2N}\Phi_{1N} + B_{2S}\Phi_{1S} + B_{2E}\Phi_{1E} + B_{2W}\Phi_{1W} - B_{2P}\Phi_{1P} - B_{P2}^0\Phi_{P2}^0. \tag{34}$$

The coefficients A_K and $A^0{}_P$, with $K \neq P$, describe the contribution of the different nodes due to the diffuse transport of (from the neighboring points in the direction of the node P) and the influence of the variable Φ in the previous time on its value at the present time, respectively.

The main advantage of the use of the implicit procedure is that it is unconditionally stable [33]; however, the use of this formulation does not mean working with any time interval because the coupling problem can limit with great intensity the value of Δt.

Equations (29) and (31) are applied to all internal points in the discrete domain, except at the border points (see Figure 3), which are the control volumes adjacent to the body surface (boundary volumes). For these volumes, the procedure adopted is the integration of conservation equations, considering the boundary conditions existing on the surface of the porous solid.

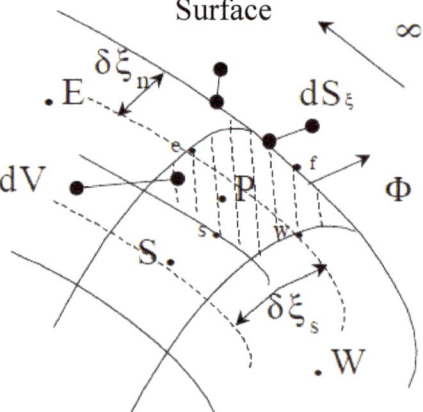

Figure 3. Sketch of the control volume on the surface of the solid and its characteristic dimensions.

In this case, the following discretized equations are valid:

- Mass:

$$\frac{\Delta V}{J_P}\left[\frac{\lambda_P\Phi_{1P} - \lambda_P^0\Phi_{1P}^0}{\Delta t}\right] = \left[\Phi''ds_\xi - \left(D_{11}\frac{\partial\Phi_1}{\partial\xi}\right)\Big|_s\right] + \\ \left[\left(D_{22}\frac{\partial\Phi_1}{\partial\eta}\right)\Big|_e - \left(D_{22}\frac{\partial\Phi_1}{\partial\eta}\right)\Big|_w\right] + \hat{S}_1^\Phi\Delta V, \tag{35}$$

- Heat:

$$\frac{\Delta V}{J_P}[\lambda_{2P}\Phi_{2P} - \lambda_{2P}^0\Phi_{2P}^0] = \left[\Phi'' ds_\xi - \left(D_{aa}\frac{\partial \Phi_2}{\partial \xi}\right)s\right]\Delta t + \left(D_{bb}\frac{\partial \Phi_2}{\partial \eta}\right)|e - \left[\left(D_{bb}\frac{\partial \Phi_2}{\partial \eta}\right)|w\right]\Delta t + \left[\left(D_{cc}\frac{\partial \Phi_2}{\partial \xi}\right)|n - \left(D_{cc}\frac{\partial \Phi_2}{\partial \xi}\right)s\right]\Delta t + \left(D_{dd}\frac{\partial \Phi_2}{\partial \eta}\right)e - \left[\left(D_{dd}\frac{\partial \Phi_2}{\partial \eta}\right)|w\right]\Delta t + \hat{S}_2^\Phi \quad (36)$$

where $\hat{S}_1^\Phi$ and $\hat{S}_2^\Phi$ are given in Appendix A. In this equation, Φ'' is the flux of Φ per unit of the area obtained from the boundary condition imposed on the physical problem.

According to the boundary conditions (convection), it is necessary to specify the diffusive fluxes in the face f of the control volume. For this situation, the diffusive flux crossing the border is equal to the convective flux in its vicinity and outside the domain under study:

$$\Phi'' dS_\xi = D_{ij}\frac{\partial \Phi}{\partial \xi}|_{face} = \text{convective flux (function of } \Phi_n). \quad (37)$$

- Mass

$$M'' = \left(\vec{J}_\ell + \vec{J}_v\right)\bigg|_{\xi=\frac{L_1}{L}} \quad (38)$$

$$\left(\vec{J}_\ell + \vec{J}_v\right)\bigg|_{\xi=\frac{L_1}{L}} = h_m(M_f - M_e) \quad (39)$$

where M_f is the boundary moisture content.

Replacing $\vec{J}_\ell$ and $\vec{J}_v$ in Equation (39), isolating M_f, and replacing them into Equation (38), the following expression is obtained:

$$M'' = \frac{M_p - M_e}{\frac{1}{\left(\frac{\Gamma_1^\Phi}{\delta \xi_n L}\right)\sqrt{\frac{\xi_n^2+1}{\xi_n^2+\eta_p^2}}} + \frac{1}{h_m \rho_s}} + \frac{\left(\Gamma_2^\Phi \nabla T\right)}{\frac{1}{\left(\frac{\Gamma_1^\Phi}{\delta \xi_n L}\right)\sqrt{\frac{\xi_n^2+1}{\xi_n^2+\eta_p^2}}} + \frac{1}{h_m \rho_s}} \quad (40)$$

- Heat

For energy transfer, the following equation is valid:

$$q'' = \vec{J}_q\bigg|_{\xi=\frac{L_1}{L}} \quad (41)$$

$$\vec{J}_q\bigg|_{\xi=\frac{L_1}{L}} = h_c(T_a - T_f) + h_{fg}\vec{J}_\ell + \left(\vec{J}_\ell + \vec{J}_v\right)c_v(T_a - T_f) \quad (42)$$

Replacing $\vec{J}_\ell$, $\vec{J}_v$ and $\vec{J}_q$ in Equation (42), isolating T_f, and replacing them into Equation (41), the following expression is obtained:

$$q'' = \frac{1}{\left(\frac{(L\delta \xi_n)}{\Gamma_{3P}^\Phi \sqrt{\frac{\xi_n^2+1}{\xi_n^2+\eta_p^2}}}\right) + \frac{1}{h_c}} \left\{ \begin{array}{l} T_a - T_p - \frac{\Gamma_{4P}^\Phi}{\frac{\Gamma_{3P}^\Phi}{\delta \xi_n}}\left(\frac{M_f-M_p}{\delta \xi_n}\right) - \frac{h_{fg}\rho_l k_l R_v \ln H}{h_c}\frac{1}{L}\sqrt{\frac{\xi_n^2+1}{\xi_n^2+\eta_p^2}}\left(\frac{T_f-T_p}{\delta \xi_n}\right) - \\ \frac{h_{fg}\rho_l k_l \frac{R_v T}{H}\frac{\partial H}{\partial M}}{h_c}\frac{1}{L}\sqrt{\frac{\xi_n^2+1}{\xi_n^2+\eta_p^2}}\left(\frac{M_f-M_p}{\delta \xi_n}\right) - \\ \frac{\sqrt{\frac{\xi_n^2+1}{\xi_n^2+\eta_p^2}}}{h_c} \times \left(\left[\frac{\Gamma_L^\Phi}{L}\left(\frac{T_f-T_p}{\delta \xi_n}\right)\right] + \frac{\Gamma_{1P}^\Phi}{L}\left(\frac{M_f-M_p}{\delta \xi_n}\right)\right)c_v(T_a - T_f) \end{array} \right\} \quad (43)$$

The numerical solution of the proposed mathematical model is obtained by the solution of Equations (29), (31), (35) and (36) applied to internal and boundary volumes. Thus, the nodal points of symmetry are not considered in the equations to be solved. After that, the system of equations has been solved, the estimation of the moisture content and temperature value at these points of symmetry is made. For this, it is assumed that the moisture or heat flux that coming out of the point adjacent to the point of symmetry is equal to the moisture or heat flux that reaches this point.

In this work, the variable Γ^Φ is dependent on the temperature, relative humidity, and moisture content. In this case, the procedure to obtain its value in the control volume interfaces is to assume a variation of Γ^Φ between the points P and its neighbor, in any direction (N, S, E or W), given by [33]:

$$\Gamma_i^\Phi = \frac{2\Gamma_P^\Phi \Gamma_E^\Phi}{\Gamma_P^\Phi + \Gamma_E^\Phi}. \tag{44}$$

This formulation is the most effective since, if $\Gamma^\Phi{}_P$ or $\Gamma^\Phi{}_E$ are zero, there will be no flux of Φ and therefore $\Gamma^\Phi{}_i$ will be null, which is physically realistic.

2.3. Application to Lentil Grain Drying

The lentil grain has approximately an oblate ellipsoidal form. The following thermophysical and geometric parameters were used in numerical simulations:

- Density of the saturated vapor [34]:

$$\rho_{v0} = \left(2.54 \times 10^8/T\right) \mathrm{Exp}(-5200/T) \; \left(\mathrm{kg/m^3}\right). \tag{45}$$

- Density of water [34]: $\rho = 1000$ kg/m3
- Dry solid density [5]: $\rho s = 1375$ kg/m3
- Latent heat of vaporization [34]:

$$h_{fg} = h_o + h_w \, (\mathrm{J/kg}), \tag{46}$$

$$h_o = 3.11 \times 10^6 - 2.38 \times 10^3 T \; (\mathrm{J/kg}), \tag{47}$$

$$h_w = \frac{R_v T^2}{H} \times \frac{\partial H}{\partial T} \; (\mathrm{J/kg}). \tag{48}$$

- Sorption isotherm (modified Henderson equation) [35]:

$$H = 1 - \mathrm{Exp}\left[-0.000207 \times (T + 21.63811)M^{1.73806}\right] \; (\mathrm{decimal}), \tag{49}$$

valid at the following intervals: $5 \leq T \leq 60$ °C and $4 \leq M \leq 26\%$ (dry base).
- Thermal conductivity of lentil [5]: kT = 0.15 W/m.K
- Conductivity of the liquid and vapor [34]:

$$k_\ell = a_1 \times 4.366 \times 10^{-18} \times H^3 \times \mathrm{Exp}\,(-1331/T) \; (\mathrm{s}), \tag{50}$$

$$k_v = a_2 \times 6.982 \times 10^{-9}(T-273.16)^{0.41} \times \left(H^{0.1715} - H^{1.1715}\right) \; \left(\mathrm{m^2/s}\right) \tag{51}$$

The values of the parameters a_1, a_2 e h_m (mass transfer coefficients) of Equations (39), (50) and (51) were obtained by adjustment between numerical and experimental data of moisture content [36] using the least square error technique, as follows:

$$\mathrm{ERMQ} = \sum_{n-1}^{n} \left(M_{i,\mathrm{Num}} - M_{i,\mathrm{Exp}}\right)^2 \tag{52}$$

$$\overline{S}^2 = \frac{ERMQ}{(n-\overline{n})} \qquad (53)$$

where ERMQ is the minimum square error, $\overline{S}^2$ is the variance, n is the number of experimental points and $\overline{n}$ is the number of parameters [37]. The procedure is started establishing h_m infinity and $a_1 = a_2 = 1.0$. From this condition, h_m is varied until the best value of the ERMQ is reached. Following, the parameters h_m and a_2 are fixed, and the parameter a_1 varied until the best value of the ERMQ is reached. Now, the new parameters h_m and a_1 are fixed, and the parameter a_2 is varied, according to the same procedure. This procedure is used until that all optimized parameters are obtained.

For the calculation of the correlation coefficient (R^2), it was used the software LAB Fit Curve Fitting Software V 7.2.46 [38].

- Specific heat of lentil, liquid water, and water vapor [39,40]:

$$c_b = 0.5773 + 0.00709T + (6.22 - 9.14M) \times M; \qquad (54)$$

$$c_\ell = 4185 \text{ J/kg; } c_v = 1916 \text{ J/kg}.$$

- Lentil grain dimension [5]: $L_1 = 1.4 \times 10^{-3}$ m and $L_2 = 3.4 \times 10^{-3}$ m

The convective heat transfer coefficient was obtained by considering lentil grain as a sphere with a volume of an ellipsoid, as follows:

$$h_c = \frac{k_a}{d_p}\left(2 + 0.6 R_e^{1/2} P_r^{1/3}\right). \qquad (55)$$

where R_e is the Reynolds number and P_r is the Prandtl number.

Table 1 shows the values of drying air parameters and the initial (M_o), equilibrium (M_e), and final (M_f) moisture contents of lentil grain used in this work, based on data reported by Tang and Sokhansanj [36].

Table 1. Conditions of the drying air and lentil grain [36].

Air					Lentil		
T_a (°C)	H_a (%)	v_a (m/s)	M_o (% b.s)	M_e (% b.s)	T_o (°C)	T_e (°C)	t (s)
40	50	0.3	24.5	12.1	25	40	86,400
60	50	0.3	24.5	10.1	25	60	86,400

For the numerical simulation, a computational code was developed in the Software Mathematica®. The results regarding heat and mass transfer were generated through a numerical mesh 20 × 20 nodais points with a time step of $\Delta t = 1.0$ s obtained after rigorous refinement studies. These refinement procedures were proposed based on studies carried out by Carmo et al. [27] and Oliveira et al. [26], who observed that under these conditions, the control volume dimensions and the time interval were sufficient for a suitable refinement of the mesh and time step. The system of equations generated from the discretized equations applied to each control volume was solved using the iterative Gauss–Siedel method, with a convergence criterion of 10^{-8} kg/kg for moisture content and 10^{-8} °C for temperature.

3. Results and Discussion
3.1. Drying and Heating Kinetics

Figure 4 shows a comparison between the drying kinetics of the lentil grain obtained numerically and experimentally [36] throughout the drying process. It is possible to observe a suitable agreement between the values of the average moisture content, indicating that the proposed model is adequate to describe the drying phenomenon of this type of porous material. This is due to the fact that the convective boundary condition on the surface of the material is more physically realistic than a hygroscopic balance boundary condition as described by Oliveira and Lima [24] and Melo et al. [29]. The drying rate behavior of the lentil grain indicates a decrease in velocity of the moisture removal (liquid and water vapor) with the drying process time. At the beginning of the drying process, the drying rate was 0.103 kg water/kg dry solid/h for T = 40 °C, and 0.137 kg water/kg dry solid/h for T = 60 °C. Upon analyzing the drying kinetics in both air temperatures (40 and 60 °C) and the same values of the initial moisture content (24.5 d.b) and relative humidity (50%), it can be seen that the higher temperature, the higher the moisture removal and the lower the total drying time.

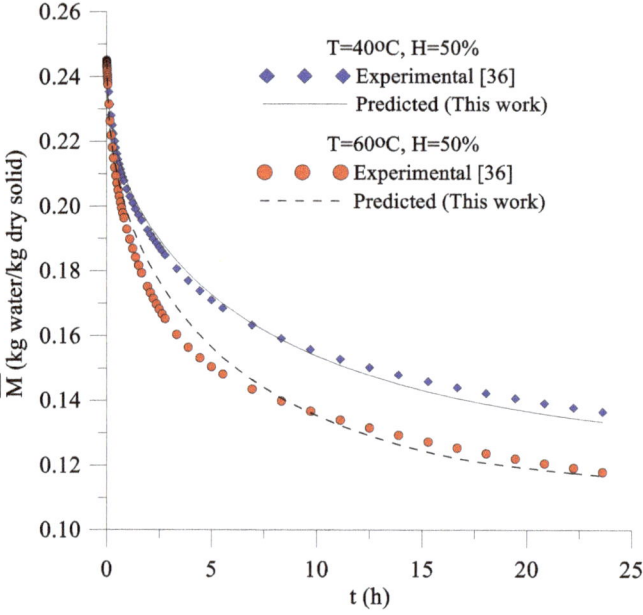

Figure 4. Predicted and experimental average moisture contents of lentils during drying at 40 °C and 60 °C and a relative humidity of 50%.

Figure 5 shows the average heating of lentil grain over the drying time. It is possible to observe that the average temperature of lentil grain rises rapidly in the first 400 s of the drying process and reaches the equilibrium temperature in a time interval of approximately 1 h, with a smooth growth (moderate heating rate) for long times, due to the convective boundary condition used in this model. Furthermore, the higher the air temperature, the higher the heating rate. A comparison of Figures 5 and 6 shows that the moisture removal rate is much lower than the heating rate of lentil grain. However, for low drying temperature, it is expected that the grain does not present thermal and hydric damage, maintaining its post-drying quality.

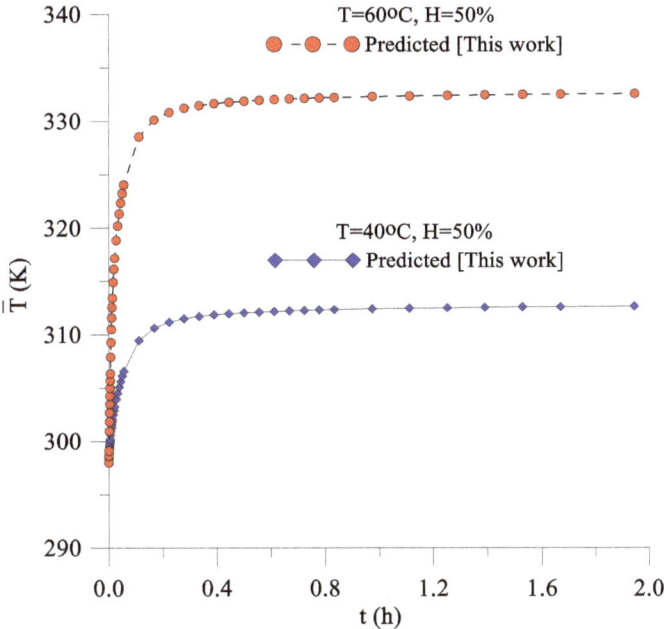

Figure 5. Average temperature of lentil grain over time during drying at 40 °C and 60 °C and a relative humidity of 50%.

3.2. Fluxes of Liquid, Water Vapor, and Heat

Figure 6a–d present the behavior of the liquid, vapor, and liquid plus vapor fluxes and the relationship between the vapor and liquid fluxes on the surface of the lentil grain obtained during the drying process.

It is possible to observe, in Figure 6, that the liquid and vapor fluxes decrease rapidly in the first two hours of the drying process, tending to zero (hygroscopic equilibrium condition), for long process times (t > 20 h.) It is also noted that in the initial drying times, there are differences between the values of liquid and vapor fluxes, with predominance for the vapor flux, and during the majority of the drying process, the behavior was practically the same, at both the drying air temperature. In the beginning, the value of the vapor flux on the surface of the lentil grain is 14.7 times higher than the liquid flux for T = 40 °C, and 23.9 times higher than the liquid flux for T = 60 °C, growing with the drying time, especially at T = 60 °C, due to the decrease in moisture content in the lentil grain, which evidences the predominance of moisture removal in the form of water vapor. The results also prove that for materials with low initial moisture content, the predominance is for the vapor flux. Probably, for hygroscopic materials with high initial moisture content, such as fruits and vegetables, the relationship between vapor and liquid fluxes in the initial drying times is much lower, tending to increase with the decrease in moisture content throughout the drying process. It is important to note that this result could not have been detected when using the theory of liquid diffusion to describe the drying process of lentil grains; this theory assumes that moisture migrates only in the liquid phase inside the material, regardless of the drying temperature.

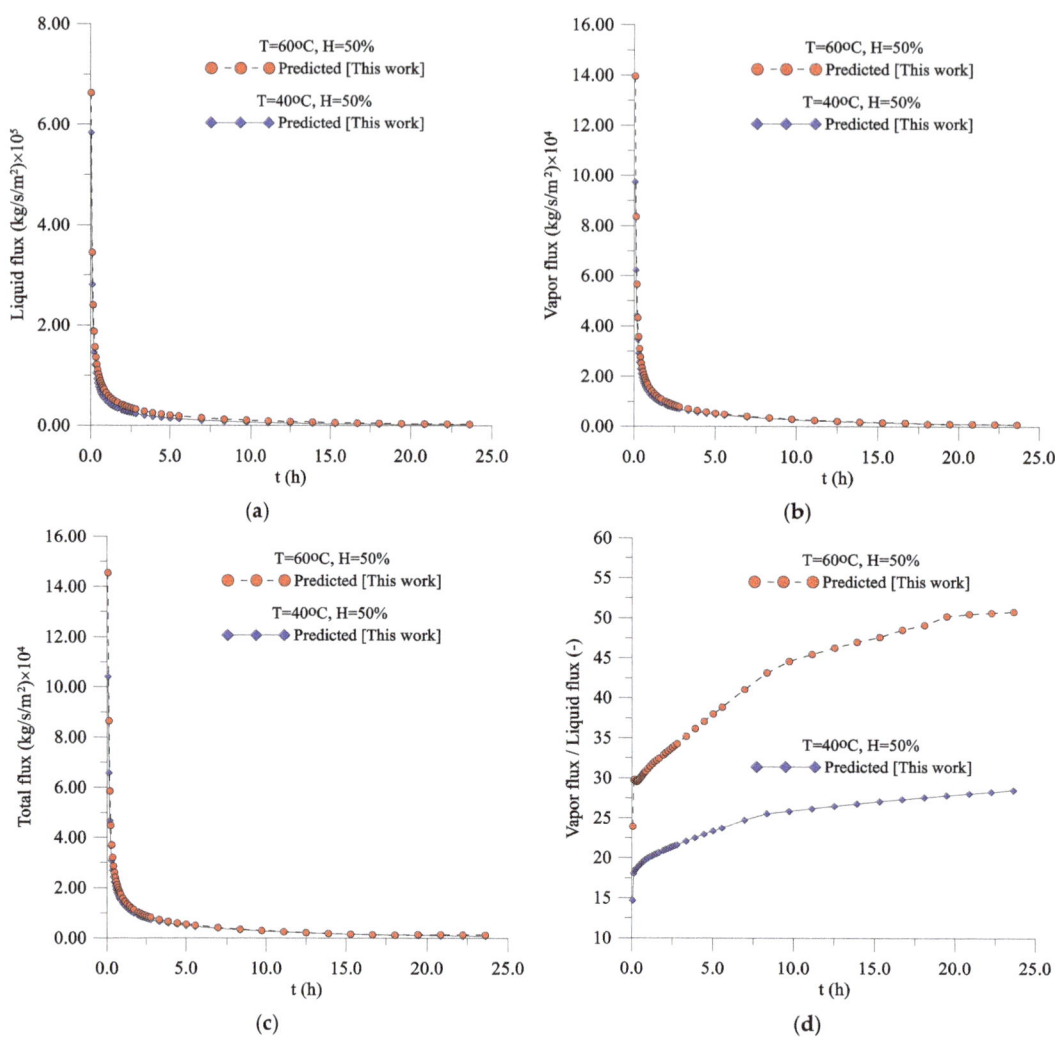

Figure 6. (a) Liquid flux, (b) vapor flux, (c) total flux, and (d) relationship between the vapor flux and the liquid flux, on the surface of the lentil grain for drying at 40 °C and a relative humidity of 50%.

Figure 7 shows the heat flux on the surface of the lentil grain during the drying process. It is possible to observe that, during the drying time, the heat flux decreased significantly up to 2 h of drying; after this time, the heat flux decays more slowly, tending to remain constant until the end of the drying process (thermal equilibrium). This heat is used both to heat the grain and water in the liquid phase up to the saturation temperature (sensitive heat) and to evaporate, at saturation temperature, the moisture of the grain that reaches the surface (latent heat). For both the air temperature, heat flux presented almost the same transient behavior. However, at 200 s, heat flux reached the value 3996.14 W/m² at T = 40 °C, and 5409.48 W/m² at T = 60 °C.

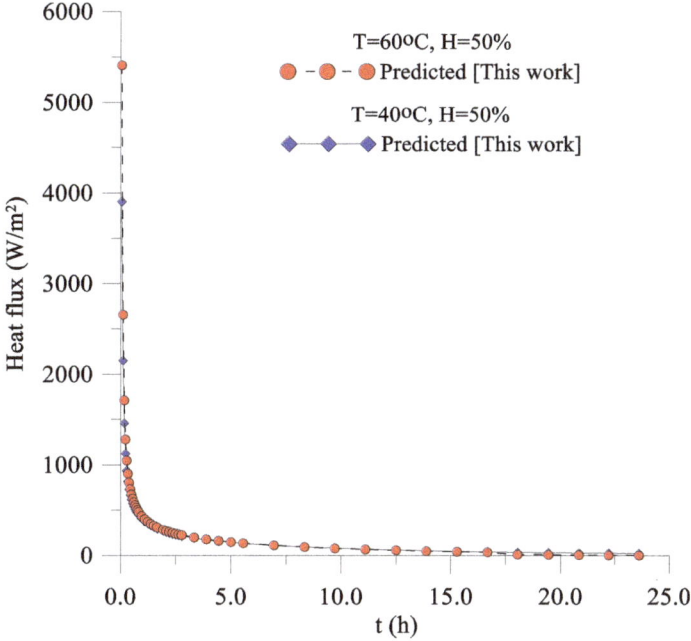

Figure 7. Heat flux on the surface of lentil grain for drying at 40 °C and a relative humidity of 50%.

3.3. Distribution of Moisture Content and Temperature Inside the Lentil Grain

Figure 8 shows the distributions of moisture content inside the lentil grain in six drying process times (1000, 5000, 10,000, 20,000, 40,000, and 85,000 s). In these figures, the moisture contents inside the solid are represented by the iso-moisture lines. Analyzing these figures, it is possible to observe that the moisture (in the forms of liquid and vapor) migrates from the center to the surface of the lentil grain, and the higher gradients are close to the surface of the grain, especially in the region located near the focal point (y = L_2).

The temperature distribution inside the lentil grain is presented in Figure 9, at times 1000, 5000, 10,000, 20,000, 40,000 and 85,000 s. The constant temperature lines (isotherms) showed that the heating is faster on the surface of the solid, and this heating process occurs from the surface to the center of the body, unlike that of the moisture content. However, unlike what has been verified for moisture content, the temperature distribution inside the lentil grain is practically uniform at any drying time, a typical condition of low convective effects.

From the physical point of view, lower hydric and thermal gradients inside the grain are desirable, since it significantly reduces the damage caused by hydric and thermal stresses, which leads to an increase in the quality of the post-drying product.

3.4. Transport Coefficients Evaluation (k_l, k_v, and h_m)

Table 2 presents the values of parameters a_1 e a_2 that appear in the transport coefficients (k_v and k_l), the convective mass and heat transfer coefficients (h_m and h_c), the relative error (ERMQ), variance ($\bar{S}^2$) and the coefficient of correlation (R^2), respectively, determined after the nonlinear regression process. It is possible to observe that the proposed model presented a suitable fit due to the small discrepancy presented between the experimental and numerical data (lower values of ERMQ and $\bar{S}^2$) and a high value of the correlation coefficient.

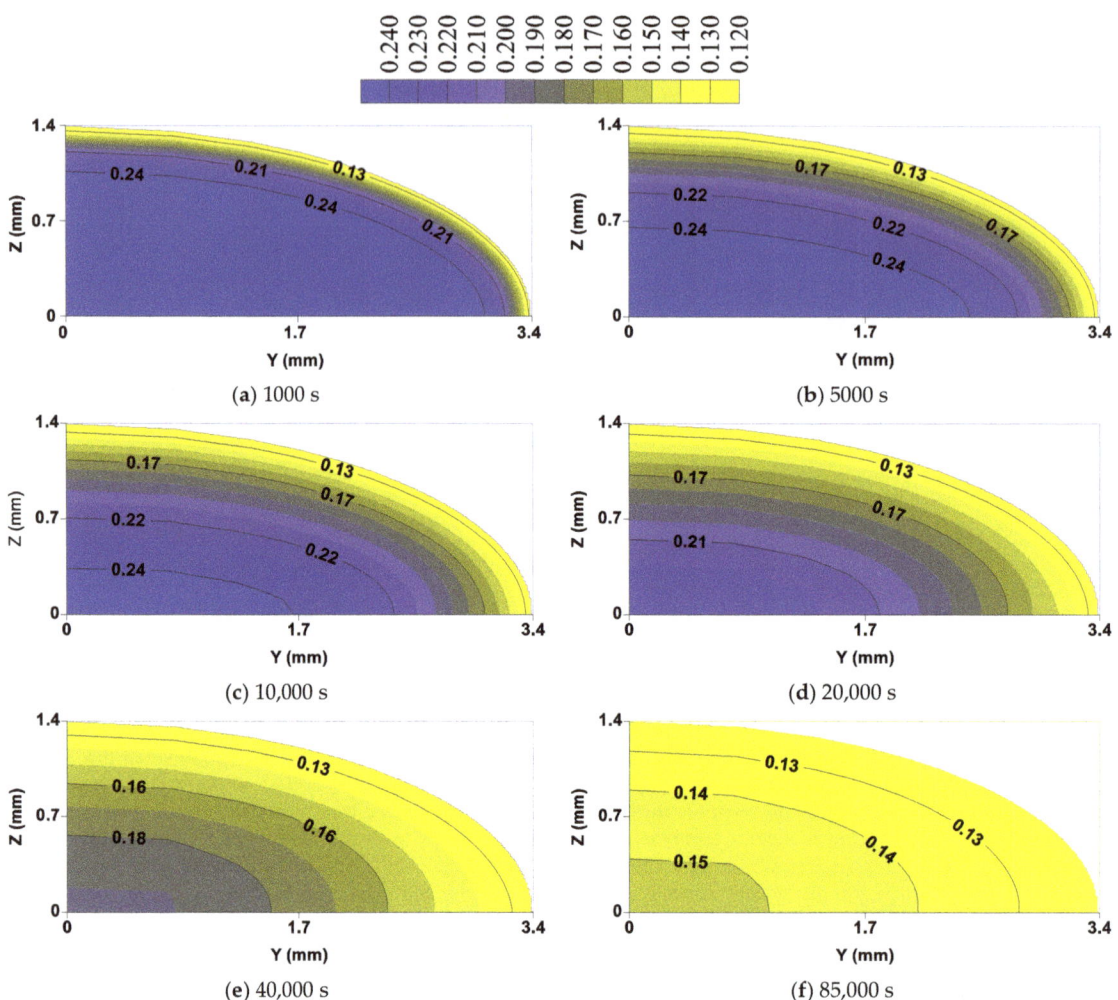

Figure 8. Distribution of moisture content inside lentil grain at different process moments, for drying at 40 °C and a relative humidity of 50%.

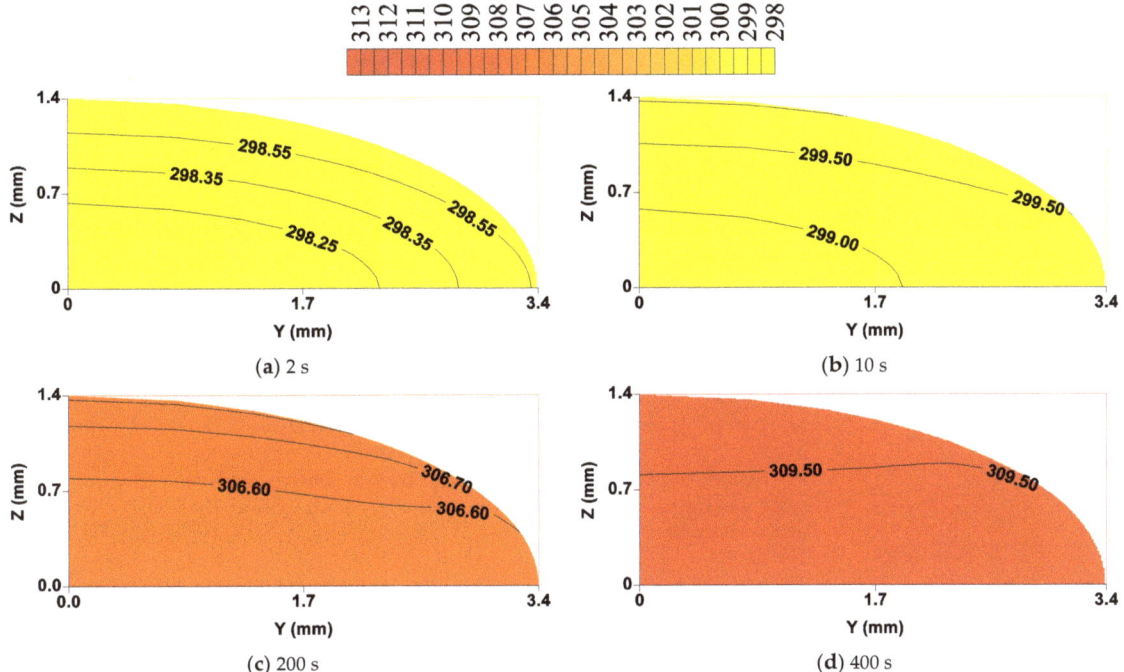

Figure 9. Temperature distribution inside the lentil grain at different process moments, for drying at 40 °C and a relative humidity of 50%.

Table 2. Comparison between the liquid and vapor conductivity and residual error.

Case		a_1 (---)	a_2 (---)	h_m (m/s)	h_c (W/m²K)	ERMQ (kg/kg)²	$\bar{S}$ (kg/kg)²	R^2
T (°C)	H (%)							
40	50	2.33×10^4	25.07×10^2	6.20×10^{-6}	37.31	0.413×10^3	0.61×10^5	0.996
60	50	1.43×10^4	10.74×10^2	12.80×10^{-6}	37.81	1.374×10^3	2.02×10^5	0.991

Applying the coefficients a_1 and a_2 into Equations (44) and (45), respectively, it is possible to observe that the values of liquid and vapor conductivity have inverse behavior during the drying process. While the liquid conductivity decreases during drying, the vapor conductivity increases with the decrease in the moisture content, reaching a maximum vapor conductivity value of 2.35×10^{-5} m²/s at 40 °C and 1.13×10^{-5} m²/s at 60 °C. According to Fortes [22], the liquid conductivity increases with moisture content at a given temperature, and, during this increase, there should occur, simultaneously, molecular diffusion of the liquid and vapor, capillary flow, and filtration. For low moisture content values, the migration mechanisms will consist of vaporization-condensation processes. The same author also observed that the vapor conductivity presents a maximum value at low moisture content due to the liquid discontinuity and, consequently, the emptying of pores and capillaries.

Melo et al. [29] obtained, for the same constants, the following values of $a_1 = 2.00 \times 10^4$ and $a_2 = 22.66 \times 10^2$, considering the drying process of lentils under the drying conditions T = 40 °C and H = 50%. However, the authors considered an equilibrium boundary condition on the solid surface, which evidences the dependence of these parameters with the boundary condition specified at the surface of the physical domain.

About the fitting process, it is important to notice that the values of the estimated parameters a_1, a_2, and h_m are strongly dependent on the considerations and boundary conditions adopted in the model. For example, relationships between the values of the parameter a_1 obtained when considering convective boundary conditions at the surface of the lentil grain and that considering equilibrium condition is around 1.165 for T = 40 °C.

4. Conclusions

In this research, heat and mass transfer in a porous body with a non-conventional shape was studied. For that, a new and advanced mathematical formulation and its numerical solution (finite volumes) were proposed. From the analysis of the results obtained with the numerical simulation of drying in porous bodies with oblate spheroidal geometry (lentil grain), it is possible to conclude:

(a) The mathematical model proposed to predict transient diffusion in oblate spheroidal solids, and its numerical solution using the finite-volumes method with the convective condition on the surface was adequate since the predicted values of the average moisture content of the lentil grain along the process presented low deviation and variance when compared with the data of average moisture content obtained experimentally;

(b) Inside the lentil grain, the dominant mass transfer mechanism is the vapor flux, on the surface of the solid, with a vapor flux/liquid flux ratio greater than 14.7 for T = 40 °C, and greater than 23.9 for T = 60 °C, growing with the drying time, especially at T = 60 °C;

(c) The areas of lentil grain more susceptible to cracks are located on the surface and around the focal point due to the existence of higher thermal and hydric stresses originated by higher moisture and temperature gradients;

(d) The liquid conductivity increases with the increase in the moisture content and decreases with the increase in temperature, while the vapor conductivity increases with the decrease in the moisture content and decreases with the increase in temperature due to the behavior of the saturation pressure in the pore inside the porous material.

Author Contributions: All the authors contributed to the development, analysis, writing, and revision of the paper: conceptualization, J.C.S.M. and R.S.G.; methodology, J.C.S.M., W.P.S., and W.M.P.B.L.; software, W.P.S. and A.G.B.L.; validation, J.C.S.M., I.B.S., and M.C.N.M.; formal analysis, J.C.S.M., J.M.P.Q.D., and A.G.B.L.; investigation, J.C.S.M., J.P.G., R.M.F.F., and A.J.M.Q.; writing—original draft preparation, J.C.S.M., J.M.P.Q.D., and A.G.B.L.; writing—review and editing, J.M.P.Q.D., A.G.B.L., and R.S.G.; visualization, J.C.S.M. and J.E.F.C.; supervision, A.G.B.L., W.P.S., and J.M.P.Q.D. All authors have read and agreed to the published version of the manuscript.

Funding: This work was supported by base funding (UIDB/04708/2020) and programmatic funding (UIDP/04708/2020) of the CONSTRUCT—Instituto de I&D em Estruturas e Construções, which is funded by national funds through the Fundação para a Ciência e a Tecnologia (FCT/MCTES), Central Government Investment and Expenditure Program (PIDDAC) and CNPq, CAPES, and FINEP (Brazilian Research Agencies).

Institutional Review Board Statement: Not applicable.

Informed Consent Statement: Not applicable.

Data Availability Statement: The data that support the findings of this study are available upon request from the authors.

Acknowledgments: The authors are grateful to the Federal University of Campina Grande (Brazil) for the research infrastructure and the references cited in the manuscript.

Conflicts of Interest: The authors declare no conflict of interest.

Appendix A

(a) Parameters of Equations (7) and (8).

$$\Gamma_1^\varphi = \left[\rho_\ell k_\ell \frac{R_v T}{H}\left(\frac{\partial H}{\partial M}\right) + k_v \rho_{v0}\left(\frac{\partial H}{\partial M}\right)\right]; \quad (A1)$$

$$\Gamma_2^\varphi = \left[\rho_\ell k_\ell R_v \ln(H) + k_v\left(\rho_{v0}\frac{\partial H}{\partial T} + H\frac{d\rho_{v0}}{dT}\right)\right]; \quad (A2)$$

$$\Gamma_3^\Phi = k_T \quad (A3)$$

$$\Gamma_4^\Phi = \left[\rho_\ell k_\ell R_v \ell n(H) + k_v\left(\rho_{v0}\frac{\partial H}{\partial T} + H\frac{d\rho_{v0}}{dT}\right)\right]\left(\frac{R_v T^2}{H}\frac{\partial H}{\partial M}\right) \quad (A4)$$

$$\Gamma_5^\Phi = h_{fg} k_v\left(\rho_{v0}\frac{\partial H}{\partial T} + H\frac{d\rho_{v0}}{dT}\right) \quad (A5)$$

$$\Gamma_6^\Phi = h_{fg} k_v \rho_{v0}\left(\frac{\partial H}{\partial M}\right) \quad (A6)$$

$$\Gamma_7^\Phi = c_\ell \rho_\ell k_\ell R_v \ell n(H) + k_v c_v\left[\rho_{v0}\frac{\partial H}{\partial T} + H\left(\frac{d\rho_{v0}}{dT}\right)\right] \quad (A7)$$

$$\Gamma_8^\Phi = c_\ell \rho_\ell k_\ell \frac{R_v}{H}\left(\frac{\partial H}{\partial M}\right) + k_v c_v \rho_{v0}\frac{\partial H}{\partial M} \quad (A8)$$

(b) Source terms of Equations (15) and (16)

$$S_1^\Phi = \frac{\partial}{\partial \xi}\left(\alpha_{11} J \Gamma_2^\Phi \frac{\partial \Phi_2}{\partial \xi}\right) + \frac{\partial}{\partial \eta}\left(\alpha_{22} J \Gamma_2^\Phi \frac{\partial \Phi_2}{\partial \eta}\right); \quad (A9)$$

$$S_2^\Phi = \alpha_{11} J \Gamma_7^\Phi \left(\frac{\partial \Phi_2}{\partial \xi}\right)^2 + \alpha_{22} J \Gamma_7^\Phi \left(\frac{\partial \Phi_2}{\partial \eta}\right)^2 + \alpha_{11} J \Gamma_8^\Phi \left(\frac{\partial \Phi_2}{\partial \xi}\frac{\partial \Phi_1}{\partial \xi}\right) + \alpha_{22} J \Gamma_8^\Phi \left(\frac{\partial \Phi_2}{\partial \eta}\frac{\partial \Phi_1}{\partial \eta}\right) + \frac{\partial}{\partial \xi}\left(\alpha_{11} J \Gamma_4^\Phi \frac{\partial \Phi_1}{\partial \xi}\right) + \frac{\partial}{\partial \eta}\left(\alpha_{22} J \Gamma_4^\Phi \frac{\partial \Phi_1}{\partial \eta}\right) + \frac{\partial}{\partial \xi}\left(\alpha_{11} J \Gamma_6^\Phi \frac{\partial \Phi_1}{\partial \xi}\right) + \frac{\partial}{\partial \eta}\left(\alpha_{22} J \Gamma_6^\Phi \frac{\partial \Phi_1}{\partial \eta}\right) + \frac{\partial}{\partial t}\left(\frac{\lambda_1 \Phi_1}{J}\right) \quad (A10)$$

(c) Coefficients of Equations (29) and (30)

$$A_N = \frac{\Gamma_{1n}^\varphi\left(\xi_n^2 + 1\right)\Delta \eta}{\delta \xi_n} \quad (A11)$$

$$A_S = \frac{\Gamma_{1S}^\varphi\left(\xi_s^2 + 1\right)\Delta \eta}{\delta \xi_S} \quad (A12)$$

$$A_E = \frac{\Gamma_{1e}^\varphi\left(1 - \eta_e^2\right)\Delta \xi}{\delta \eta_e} \quad (A13)$$

$$A_W = \frac{\Gamma_{1w}^\varphi\left(1 - \eta_w^2\right)\Delta \xi}{\delta \eta_w} \quad (A14)$$

$$A_P = A_N + A_S + A_E + A_W + \frac{\lambda_p \Delta \xi \Delta \eta L^2\left(\xi_P^2 + \eta_P^2\right)}{\Delta t} \quad (A15)$$

$$B_{1N} = \frac{\Gamma_{2n}^\varphi\left(\xi_n^2 + 1\right)\Delta \eta}{\delta \xi_n} \quad (A16)$$

$$B_{1S} = \frac{\Gamma_{2s}^\varphi\left(\xi_s^2 + 1\right)\Delta \eta}{\delta \xi_s}; \quad (A17)$$

$$B_{1E} = \frac{\Gamma^{\varphi}_{2e}(1-\eta_e^2)\Delta\xi}{\delta\eta_e} \tag{A18}$$

$$B_{1W} = \frac{\Gamma^{\varphi}_{2w}(1-\eta_w^2)\Delta\xi}{\delta\eta_w} \tag{A19}$$

$$B_{1P} = B_N + B_S + B_E + B_W \tag{A20}$$

$$A_P^0 = \frac{\lambda_P^0 \Delta\xi \Delta\eta L^3 (\xi_P^2 + \eta_P^2)}{\Delta t} \tag{A21}$$

(d) Coefficients of Equation (31)

$$A_{2N} = \frac{\left(\Gamma^{\Phi}_{3n} + \Gamma^{\Phi}_{5n}\right)(\xi_n^2 + 1)\Delta\eta}{\delta\xi_n}; \tag{A22}$$

$$A_{2S} = \frac{\left(\Gamma^{\Phi}_{3s} + \Gamma^{\Phi}_{5s}\right)(\xi_n^2 + 1)\Delta\eta}{\delta\xi_s} \tag{A23}$$

$$A_{2E} = \frac{\left(\Gamma^{\Phi}_{3e} + \Gamma^{\Phi}_{5e}\right)(1 - \eta_e^2)\Delta\xi}{\delta\eta_e} \tag{A24}$$

$$A_{2W} = \frac{\left(\Gamma^{\Phi}_{3w} + \Gamma^{\Phi}_{5w}\right)(1 - \eta_w^2)\Delta\xi}{\delta\eta_w} \tag{A25}$$

$$A_{2P}^0 = \frac{\lambda_{2P}^0 \Delta\xi \Delta\eta L^2 (\xi_P^2 + \eta_P^2)}{\Delta t} \tag{A26}$$

$$A_{2P} = A_N + A_S + A_E + A_W + \frac{\lambda_{2P} \Delta\xi \Delta\eta L^2 (\xi_P^2 + \eta_P^2)}{\Delta t} - S^{\Phi}_{P2} \tag{A27}$$

(e) Coefficients of Equation (34)

$$B_{2N} = \frac{\left(\Gamma^{\Phi}_{42n} + \Gamma^{\Phi}_{62n}\right)(\xi_n^2 + 1)\Delta\eta}{\delta\xi_n} \tag{A28}$$

$$B_{2S} = \frac{\left(\Gamma^{\Phi}_{42s} + \Gamma^{\Phi}_{62s}\right)(\xi_s^2 + 1)\Delta\eta}{\delta\xi_s} \tag{A29}$$

$$B_{2E} = \frac{\left(\Gamma^{\Phi}_{42e} + \Gamma^{\Phi}_{62e}\right)(1 - \eta_e^2)\Delta\xi}{\delta\xi_e} \tag{A30}$$

$$B_{2W} = \frac{\left(\Gamma^{\Phi}_{42w} + \Gamma^{\Phi}_{62w}\right)(1 - \eta_w^2)\Delta\xi}{\delta\xi_w} \tag{A31}$$

$$B_{P2}^0 = \frac{\rho_P^0 h_w^0 \Delta\xi \Delta\eta L^2 (\xi_P^2 + \eta_P^2)}{\Delta t} \tag{A32}$$

$$B_{2P} = B_{2N} + B_{2S} + B_{2E} + B_{2W} - \frac{\rho_P h_w \Delta\xi \Delta\eta L^2 (\xi_P^2 + \eta_P^2)}{\Delta t} \tag{A33}$$

$$\begin{aligned}B_{star} = &\left[\Gamma^{\varphi}_{7n}(\xi^2+1)\Delta\eta\Delta\xi\left(\frac{\Phi^*_{2N}-\Phi^*_{2P}}{\delta\xi_n}\right)\frac{\Phi^*_{2N}}{\delta\xi_n}\right] + \left[\Gamma^{\varphi}_{7n}(1-\eta^2)\Delta\eta\Delta\xi\left(\frac{\Phi^*_{2E}-\Phi^*_{2P}}{\delta\eta_e}\right)\frac{\Phi^*_{2E}}{\delta\eta_e}\right] + \\ &\left[\Gamma^{\varphi}_{8n}(\xi^2+1)\Delta\eta\Delta\xi\left[\left(\frac{\Phi_{1N}-\Phi_{1P}}{\delta\xi_n}\right)\frac{\Phi^*_{2N}}{\delta\xi_n} - 2\left(\frac{\Phi_{1N}-\Phi_{1P}}{\delta\xi_n}\right)\frac{\Phi^*_{2P}}{\delta\xi_n}\right]\right] + \\ &\Gamma^{\varphi}_{8e}(1-\eta^2)\Delta\eta\Delta\xi\left[\left(\frac{\Phi_{1E}-\Phi_{1P}}{\delta\eta_e}\right)\frac{\Phi^*_{2P}}{\delta\xi_e} - 2\left(\frac{\Phi_{1E}-\Phi_{1P}}{\delta\eta_e}\right)\frac{\Phi^*_{2P}}{\delta\xi_e}\right]\end{aligned} \tag{A34}$$

(f) Coefficients of Equation (33)

$$\hat{S}_1^\Phi = \left[\left(D_{33}\frac{\partial \Phi_2}{\partial \xi}\right)\Big|_n - \left(D_{33}\frac{\partial \Phi_2}{\partial \xi}\right)\Big|_s\right] + \left[\left(D_{44}\frac{\partial \Phi_2}{\partial \eta}\right)\Big|_e - \left(D_{44}\frac{\partial \Phi_2}{\partial \eta}\right)\Big|_w\right] \quad \text{(A35)}$$

$$\begin{aligned}
\hat{S}_2^\Phi &= \left[D_{ee}\left(\frac{\Phi_{2N}^* - \Phi_{2P}^*}{\delta\xi_n}\right)\frac{\Phi_{2N}^*}{\delta\xi_n} - D_{ee}\left(\frac{\Phi_{2N}^* - \Phi_{2P}^*}{\delta\xi_n}\right)\frac{\Phi_{2P}^\circ}{\delta\xi_n}\right]\Delta t + \\
&+ \left[D_{ff}\left(\frac{\Phi_{2e}^* - \Phi_{2P}^*}{\delta\eta_e}\right)\frac{\Phi_{2E}^*}{\delta\eta_e} - D_{ff}\left(\frac{\Phi_{2e}^* - \Phi_{2P}^*}{\delta\eta_e}\right)\frac{\Phi_{2P}^\circ}{\delta\eta_e}\right]\Delta t + \\
&+ \left[D_{gg}\left(\frac{\Phi_{1N} - \Phi_{1P}}{\delta\xi_n}\right)\frac{\Phi_{2N}^*}{\delta\xi_n} - D_{gg}\left(\frac{\Phi_{1N} - \Phi_{1P}}{\delta\xi_n}\right)\frac{\Phi_{2P}}{\delta\xi_n}\right]\Delta t + \\
&\left[D_{hh}\left(\frac{\Phi_{1E} - \Phi_{1P}}{\delta\eta_e}\right)\frac{\Phi_{2E}^*}{\delta\eta_e} - D_{hh}\left(\frac{\Phi_{1E} - \Phi_{1P}}{\delta\eta_e}\right)\frac{\Phi_{2P}}{\delta\eta_e}\right]\Delta t + \\
&\left[\left(D_{kk}\frac{\partial \Phi_1}{\partial \xi}\right)|n - \left(D_{kk}\frac{\partial \Phi_1}{\partial \xi}\right)s\right]\Delta t + \left[\left(D_{ll}\frac{\partial \Phi_1}{\partial \eta}\right)|e - \left(D_{ll}\frac{\partial \Phi_1}{\partial \eta}\right)|w\right]\Delta t \\
&+ \frac{\Delta V}{J_p}\left[\lambda_{1P}\Phi_{1P} - \lambda_{1P}^0\Phi_{1P}^0\right]
\end{aligned} \quad \text{(A36)}$$

References

1. Silva, W.P.; Silva, C.D.P.S.; Lima, A.G.B. Uncertainty in the determination of equilibrium moisture content of agricultural products. *Rev. Bras. Prod. Agroind. Gd. Espec.* **2005**, *7*, 159–164. (In Portuguese)
2. Park, K.J.; Antonio, G.C.; Oliveira, R.A.; Park, K.J.B. Process selection and drying equipments. In *Brazilian Congress of Agricultural Engineering*; 2006; Volume 35. SBEA/UFCG, Jaboticabal, Brazil (In Portuguese)
3. Fellows, P.J. *Food Processing Technology: Principles and Practice*; CRC Press: Boca Raton, FL, USA, 2009; ISBN 1845696344.
4. Park, K.J.; Yado, M.K.M.; Brod, F.P.R. Drying studies of sliced pear bartlett (Pyrus sp.) sliced. *Food Sci. Technol.* **2001**, *21*, 288–292. (In Portuguese) [CrossRef]
5. Tang, J.; Sokhansanj, S. Moisture diffusivity in laird lentil seed components. *Trans. ASAE* **1993**, *36*, 1791–1798. [CrossRef]
6. Tang, J.; Sokhansanj, S. Geometric changes in lentil seeds caused by drying. *J. Agric. Eng. Res.* **1993**, *56*, 313–326. [CrossRef]
7. Wang, J.; Mujumdar, A.S.; Mu, W.; Feng, J.; Zhang, X.; Zhang, Q.; Fang, X.-M.; Gao, Z.-J.; Xiao, H.-W. Grape Drying: Current Status and Future Trends. In *Grape and Wine Biotechnology*; Morata, A., Loira, I., Eds.; IntechOpen: London, UK, 2016. [CrossRef]
8. Hatamipour, M.S.; Mowla, D. Shrinkage of carrots during drying in an inert medium fluidized bed. *J. Food Eng.* **2002**, *55*, 247–252. [CrossRef]
9. Karim, M.A.; Hawlader, M.N.A. Drying characteristics of banana: Theoretical modelling and experimental validation. *J. Food Eng.* **2005**, *70*, 35–45. [CrossRef]
10. Li, C.; Li, B.; Huang, J.; Li, C. Energy and exergy analyses of a combined infrared radiation-counterflow circulation (IRCC) corn dryer. *Appl. Sci.* **2020**, *10*, 6289. [CrossRef]
11. Carvalho, E.R.; Francischini, V.M.; Avelar, S.A.G.; Costa, J.C. Temperatures and periods of drying delay and quality of corn seeds harvested on the ears. *J. Seed Sci.* **2019**, *41*, 336–343. [CrossRef]
12. Arora, S.; Bharti, S.; Sehgal, V.K. Convective drying kinetics of red chillies. *Dry. Technol.* **2006**, *24*, 189–193. [CrossRef]
13. Gonelli, A.L.D.; Corrêa, P.C.; Resende, O.; dos Reis Neto, S.A. Study of moisture diffusion in wheat grain drying. *Food Sci. Technol.* **2007**, *27*, 135–140. (In Portuguese) [CrossRef]
14. Resende, O.; Corrêa, P.C.; Goneli, A.L.D.; Ribeiro, D.M. Physical properties of edible bean during drying: Determination and modelling. *Ciência Agrotecnologia* **2008**, *32*, 225–230. (In Portuguese) [CrossRef]
15. Doymaz, İ. Convective drying kinetics of strawberry. *Chem. Eng. Process. Process. Intensif.* **2008**, *47*, 914–919. [CrossRef]
16. Kajiyama, T.; Park, K.J. Influence of feed initial moisture content on spray drying time. *Rev. Bras. Prod. Agroind.* **2008**, *10*, 1–8. (In Portuguese)
17. Almeida, D.P.; Resende, O.; Costa, L.M.; Mendes, U.C.; de Fátima Sales, J. Drying kinetics of adzuki bean (Vigna angularis). *Glob. Sci. Technol.* **2009**, *2*, 72–83. (In Portuguese)
18. Lima, W.M.P.B.; Lima, E.S.; Lima, A.R.C.; Oliveira Neto, G.L.; Oliveira, N.G.N.; Farias Neto, S.R.; Lima, A.G.B. Applying phenomenological lumped models in drying process of hollow ceramic materials. *Defect Diffus. Forum* **2020**, *400*, 135–145. [CrossRef]
19. Fortes, M.; Okos, M.R. A non-equilibrium thermodynamics approach to transport phenomena in capillary porous media. *Trans. ASAE* **1981**, *24*, 756–760. [CrossRef]
20. Luikov, A.V. *Heat and Mass Transfer in Capillary-Porous Bodies*; Pergamon Press: New York, NY, USA, 1966; ISBN 0080108326.
21. Luikov, A.V. Systems of differential equations of heat and mass transfer in capillary-porous bodies. *Int. J. Heat Mass Transf.* **1975**, *18*, 1–14. [CrossRef]
22. Fortes, M. A Non-Equilibrium Thermodynamics Approach to Transport Phenomena in Capillary Porous Media with Special Reference to Drying of Grains and Foods. Ph.D. Thesis, Perdue University, West Lafayette, IN, USA, 1978.

23. Fortes, M.; Okos, M.R. Drying theories: Their bases and limitations as applied to foods and grains. In *Advances in Drying*; Hemisphere Publishing Corporation: Washington, DC, USA, 1980; Volume 1, pp. 119–154.
24. de Oliveira, V.A.B.; Lima, A.G.B. de Drying of wheat based on the non-equilibrium thermodynamics: A numerical study. *Dry. Technol.* **2009**, *27*, 306–313. [CrossRef]
25. Oliveira, V.A.B.; Lima, W.; Farias Neto, S.R.; Lima, A.G.B. Heat and mass diffusion and shrinkage in prolate spheroidal bodies based on non-equilibrium thermodynamics: A numerical investigation. *J. Porous Media* **2011**, *14*, 593–605. [CrossRef]
26. Oliveira, V.A.B.; Lima, A.G.B.; Silva, C.J. Drying of wheat: A numerical study based on the non-equilibrium thermodynamics. *Int. J. Food Eng.* **2012**, *8*, 8. [CrossRef]
27. Carmo, J.E.F.; Lima, A.G.B. Mass transfer inside oblate spheroidal solids: Modelling and simulation. *Braz. J. Chem. Eng.* **2008**, *25*, 19–26. [CrossRef]
28. Melo, J.C.S.; Lima, A.G.B.; Pereira Silva, W.; Lima, W.M.P. Heat and Mass Transfer during Drying of Lentil Based on the Non-Equilibrium Thermodynamics: A Numerical Study. *Defect Diffus. Forum* **2015**, *365*, 285–290. [CrossRef]
29. Melo, J.C.S.; Gomez, R.S.; Silva, J.B., Jr.; Queiroga, A.X.; Dantas, R.L.; Lima, A.G.B.; Silva, W.P. Drying of Oblate Spheroidal Solids via Model Based on the Non-Equilibrium Thermodynamics. *Diffus. Found.* **2020**, *25*, 83–98. [CrossRef]
30. Magnus, W.; Oberhettinger, F.; Soni, R.P. *Formulas and Theorems for the Special Functions of Mathematical Physics*; Springer: Berlim, Germany, 1966; Volume 52, ISBN 3662117614.
31. Maliska, C.R. *Heat Transfer and Computational Fluid Mechanics*; Grupo Gen-LTC: Rio de Janeiro, Brazil, 2017; ISBN 8521633351. (In Portuguese)
32. Whitaker, S. Heat and mass transfer in granular porous media. *Adv. Dry.* **1980**, *1*, 23–61.
33. Patankar, S. *Numerical Heat Transfer and Fluid Flow*; CRC Press: Boca Raton, FL, USA, 1980; ISBN 1482234211.
34. Fortes, M.; Okos, M.R.; Barrett, J.R., Jr. Heat and mass transfer analysis of intra-kernel wheat drying and rewetting. *J. Agric. Eng. Res.* **1981**, *26*, 109–125. [CrossRef]
35. Menkov, N.D. Moisture sorption isotherms of lentil seeds at several temperatures. *J. Food Eng.* **2000**, *44*, 205–211. [CrossRef]
36. Tang, J.; Sokhansanj, S. A model for thin-layer drying of lentils. *Dry. Technol.* **1994**, *12*, 849–867. [CrossRef]
37. Figliola, R.S.; Beasley, D.E. *Theory and Design for Mechanical Measurements*; John Wiley & Sons: New York, NY, USA, 2020; ISBN 1119723450.
38. Silva, W.P.; Silva, C.M.D.P.S. LAB Fit Curve Fitting Software (Nonlinear Regression and Treatment of Data Program) V. 7.2.46 2009. Available online: www.labfit.net (accessed on 17 April 2021). (In Portuguese).
39. Tang, J.; Sokhansanj, S.; Yannacopoulos, S.; Kasap, S.O. Specific heat capacity of lentil seeds by differential scanning calorimetry. *Trans. ASAE* **1991**, *34*, 517–522. [CrossRef]
40. Cengel, Y. *Heat and Mass Transfer: Fundamentals and Applications*; McGraw-Hill: New York, NY, USA, 2014; ISBN 0077654765.

Article

Drying and Heating Processes in Arbitrarily Shaped Clay Materials Using Lumped Phenomenological Modeling

Elisiane S. Lima [1], João M. P. Q. Delgado [2,*], Ana S. Guimarães [2], Wanderson M. P. B. Lima [1], Ivonete B. Santos [3], Josivanda P. Gomes [4], Rosilda S. Santos [5], Anderson F. Vilela [6], Arianne D. Viana [6], Genival S. Almeida [7], Antonio G. B. Lima [1] and João E. F. Franco [1]

1. Department of Mechanical Engineering, Federal University of Campina Grande, Campina Grande, PB 58429-900, Brazil; limaelisianelima@hotmail.com (E.S.L.); wan_magno@hotmail.com (W.M.P.B.L.); antonio.gilson@ufcg.edu.br (A.G.B.L.); jevan.franco@gmail.com (J.E.F.F.)
2. CONSTRUCT-LFC, Department of Civil Engineering, Faculty of Engineering, University of Porto, 4200-465 Porto, Portugal; anasofia@fe.up.pt
3. Department of Physics, State University of Paraiba, Campina Grande, PB 58429-500, Brazil; ivoneetebs@gmail.com
4. Department of Agricultural Engineering, Federal University of Campina Grande, Campina Grande, PB 58429-900, Brazil; josivanda@gmail.com
5. Department of Science and Technology, Federal Rural University of the Semi-Arid Region, Caraúbas, RN 59780-000, Brazil; rosilda.santos@ufersa.edu.br
6. Department of Agro-Industrial Management and Technology, Federal University of Paraíba, Bananeiras, PB 58220-000, Brazil; prof.ufpb.anderson@gmail.com (A.F.V.); arianneviana@hotmail.com (A.D.V.)
7. Federal Institute of Education, Science and Technology of Paraiba, Pedras de Fogo, PB 58328-000, Brazil; genival.almeida@ifpb.edu.br
* Correspondence: jdelgado@fe.up.pt; Tel.: +351-225-081-404

Abstract: This work aims to study the drying of clay ceramic materials with arbitrary shapes theoretically. Advanced phenomenological mathematical models based on lumped analysis and their exact solutions are presented to predict the heat and mass transfers in the porous material and estimate the transport coefficients. Application has been made in hollow ceramic bricks. Different simulations were carried out to evaluate the effect of drying air conditions (relative humidity and speed) under conditions of forced and natural convection. The transient results of the moisture content and temperature of the brick, and the convective heat and mass transfer coefficients are presented, discussed and compared with experimental data, obtaining a good agreement. It was found that the lower the relative humidity is and the higher the speed of the drying air is, the higher the convective heat and mass transfer coefficients are at the surface of the brick and in the holes, and the faster the moisture removal material and heating is. Based on the predicted results, the best conditions for brick drying were given. The idea is to increase the quality of the brick after the process, to reduce the waste of raw material and energy consumption in the process.

Keywords: drying; ceramic materials; industrial brick; lumped model; simulation

1. Introduction

The need for investment in improving quality and productivity is a growing concern for the ceramics sector worldwide. Currently, the industrial ceramics sector is well diversified, consisting of several segments, such as red ceramics, coating materials, refractory materials, electrical porcelain insulators, tableware, artistic ceramics, ceramic filters for domestic use, technical ceramics and thermal insulation.

Red ceramics are a class of material with a reddish color. This raw material has been used in civil construction to manufacture products of different geometries, such as bricks, blocks, tiles, floors, hollow elements, slabs, tubes, decorative vases, household and adornment artifacts, among others.

The red ceramic sector only uses common clay as a raw material for the manufacture of its products. The clays contain a wide variety of mineral substances of a clay nature, such as quaternary alluvial clays, mudstones, siltstones, shales and rhythms, which burn in reddish colors, at temperatures between 800 and 1250 °C. These clays generally have a very fine granulometry, a characteristic that gives them, with the incorporated organic matter, different degrees of plasticity when added with certain percentages of water, and the workability and resistance to green, dry and after the firing process, which are important aspects for the manufacture of a wide variety of ceramic products [1].

As it is a sector that produces solid and pierced bricks, structural blocks, tiles and floors, which has great application in civil construction, the production of red ceramic products, also known as structural ceramics, is considered a basic activity, due to the fact that it can be applied to construct homes, buildings, etc. Thus, due to its importance, in terms of energy and environmental impact, it is necessary to know the manufacture of products originated from clay.

The manufacturing process of clayey ceramic products comprises several stages, such as the exploration of deposits, the pre-treatment of raw materials, homogenization, drying, firing and dispatch [2]. After being extracted from the soil, the clay is sieved and then moistened with water. These procedures give the material greater plasticity [3], which facilitates the molding of wet clay in the shape of the desired piece and increases its mechanical resistance, still in the green state, to be subjected to drying.

Drying is a thermodynamic process whereby the moisture in the solid is reduced by providing a large amount of thermal energy to evaporate the water and heats the humid solid. This drying occurs due to the transport of moisture from the center to the surface of the material, and heat conduction from the surface to the center of the material, when it comes to convective drying, for example. Moisture transport can occur in the form of liquid and/or vapor, depending on the type of the solid, their characteristics in terms of material (shape, composition, porosity, granulometry, etc.) and the percentage of moisture present in them.

Drying is the stage of the process that precedes firing. In the industry, convective drying is the technique usually employed. It is an artificial drying that is carried out in drying chambers or oven. In this type of drying, temperature, relative humidity and air speed (the thermodynamic state of the drying air) and the characteristics of the raw material, the shape of the pieces and the type of dryer are essential conditions for the duration of drying.

In general, the water is evaporated in drying chambers at temperatures ranging from approximately 38 to 204 °C. The total drying time is usually between 24 and 48 h. Although heat can be generated specifically for the drying chambers, it is usually supplied from the exhaust heat of the kiln used to fire ceramic products, with the aim of maximizing the thermal efficiency of the plant.

In any physical situation, heat and moisture must be carefully regulated to avoid or minimize drying problems, such as cracks in the brick. During the drying stage, shrinkage affects the mechanical strength and shape of clay bricks. Cracks can be seen when very rapid drying has been allowed, causing high mechanical stresses within the brick that can produce cracks or local breakage.

The mechanisms of moisture migration and heating in the material during drying can be affected by internal and external conditions and are not yet well known. In this sense, some authors [4–9] state that during the drying of clayey material, the main moisture migration mechanism is liquid diffusion. However, other researchers such as the authors of [10,11] consider that a diffusive transport of liquid and vapor occurs, predominantly in certain moments of drying.

With the drying performed incorrectly, the removal of water from the piece is left uncontrolled, which can cause structural damages such as cracks, deformations, warping and, consequently, a great loss of products. That is why drying operations are important industrial processes and knowing the mechanism of moisture and heat transfer is of fundamental importance for the ceramic industry. Additionally, this can be conducted through refined and precise experimental and theoretical studies, in order to determine the process parameters and their effects on the quality of the final product.

Therefore, control of the drying process and knowledge of the main mechanism of moisture migration inside the material play important roles. Further, since with simulation and/or experimental data, the best conditions of the process (minimum product losses and energy consumption) can be verified, an improvement in the product quality and a less expensive product can be obtained.

Drying experiments, although not less important, require a lot of time to perform and require infrastructure (well-equipped laboratories), which is not always used by the industrial sector. On the other hand, numerical simulation requires little infrastructure for its realization, which makes it less expensive and makes it possible to obtain quick results, which is desirable for the industrial sector.

Mathematical modeling and numerical simulation present a series of useful aspects from a scientific point of view. Depending on the material studied, its shape and arrangements, mathematical models applied to drying can be divided into lumped or distributed models [12–17]. The lumped models describe the rates of heat and mass transfer of the material, ignoring the internal resistance of heat and mass transfer, that is, the effects of temperature and moisture variation inside the material are neglected. In contrast, the distributed models describe the rates of heat and mass transfer as a function of the position within the material and the drying time. Thus, it is clear that these models consider the external and internal resistances to the transport of heat and mass inside the product.

The distributed models based on the diffusion of liquid and/or vapor within the solid describe the rates of heat and mass transfer as a function of the position within the solid and the drying time, thus considering the resistances (external and internal) to the heat and mass fluxes. When applied to the drying of ceramic materials, the following works related to the diffusive model are reported in the literature [9,18,19].

Lumped models can be classified as empirical, semi-empirical and theoretical. In the empirical model, the moisture content is expressed only as a function of the drying time, while the semi-empirical model considers that the drying rate is proportional to the difference between the moisture content of the product and its respective equilibrium moisture content. On the other hand, theoretical models are derived from the distributed models when severe considerations are previously established [17].

In this context, some research related to the drying models applied to ceramic materials, based on a lumped analysis, is reported in the literature. In some of these works, drying experiments are carried out, empirical models are proposed and a linear or non-linear regression is performed to predict the removal of moisture and heating of the product and, subsequently, to estimate the convective mass and heat transfer coefficients on the surface of the product using distributed models [20]. In others, the authors develop a lumped phenomenological mathematical modeling and use previously established heat and mass transfer coefficients to perform simulations and evaluate the effects of these parameters on the drying process [21,22].

Thus, the need to develop more complete (phenomenological) lumped models that involve external and internal parameters to the porous and humid solid was perceived, aiming to make a greater contribution in the prediction of the phenomena of heat and mass transport that occur during the drying of a ceramic material. Then, some researchers have published studies related to predictions of moisture adsorption and desorption in building material under isothermal condition using lumped model, such as the effective moisture penetration depth model [23] and the moisture buffering model [24]. The effective moisture penetration depth model is not suitable when the hygroscopic material is thin and limited, while the moisture buffering model considers that the mass of moisture buffering of the materials is in balance with the room humidity. These assumptions reduce the accuracy of the models provoking deviation between predicted and experimental data. Moreover, these models are commonly applied for physical situations at room temperature.

In view of the above, this work aims to study the heat and mass transfer in porous and arbitrarily shaped solids using the method of lumped analysis with particular reference to the drying of industrial hollow ceramic bricks. The idea is to show the development of a new and advanced phenomenological mathematical model and its analytical solution to predict the drying process and estimate the convective heat and mass transfer coefficients at different process operational conditions, without the need to perform linear or non-linear mathematical regression from experimental data and apply for an arbitrarily shaped body. It is, therefore, an innovative formulation that will certainly help engineers, industrialists, academics and people interested in the subject in making decisions aimed at improving and optimizing the drying process, with regard to energy savings, product quality and an increase in industrial competitiveness. Furthermore, the model has great potential to be applied in predicting moisture adsorption and desorption at room temperature without restrictions or modifications.

2. Methodology

2.1. The Physical Problem and Geometry

To assist in understanding the lumped approach, consider a porous body of arbitrary and hollow shape, as shown in Figure 1. The material can receive (or provide) heat and/or moisture fluxes per unit of area on its surface and has an evenly distributed internal generation of mass and/or energy per unit of volume. The lumped analysis method [25] admits that the moisture and/or temperature of the solid have a uniform distribution at any time, in such a way that the moisture and/or temperature gradients of the solid are negligible. Therefore, the moisture content and temperature of the material vary only with time. Therefore, it is a transient physical problem.

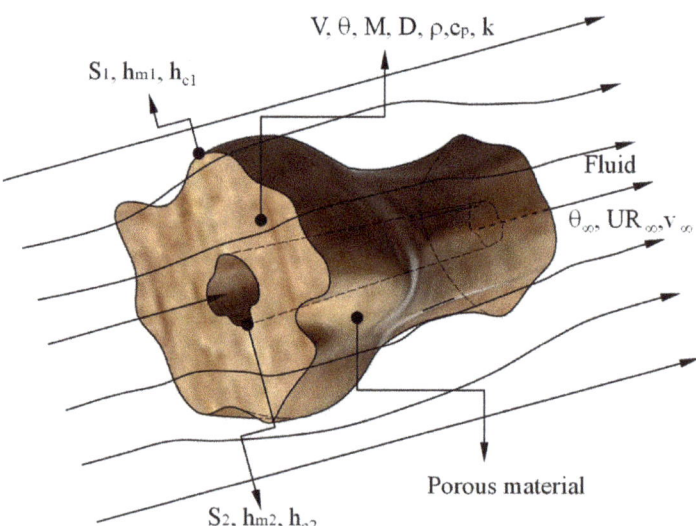

Figure 1. Schematic representation of a solid with arbitrary geometry immersed in a heated fluid.

2.2. Mathematical Modeling

In order to mathematically model the drying process of arbitrary-shaped porous material (Figure 1), the following assumptions were adopted:

(a) The solid is homogeneous and with constant thermophysical properties;
(b) Moisture content and temperature distributions inside the brick are uniform along the drying process;

(c) Drying process occurs by heat and mass diffusion inside the solid and heat and mass convection and evaporation occur on its surface;
(d) Dimensions of the brick are constant along the drying process.

2.2.1. Mass Conservation Equation

To predict the transfer of mass in the holed porous product and with arbitrary shape, based on the adopted considerations, the following mass conservation equation is proposed for all physical domain presented in Figure 1:

$$V \frac{d\overline{M}}{dt} = -h_{m1} S_1 (\overline{M} - \overline{M}_e) - h_{m2} S_2 (\overline{M} - \overline{M}_e) + \dot{M} V \tag{1}$$

Using the initial condition $M (t = 0) = M_0$, separating the variables from Equation (1) and integrating it from the initial condition, we have the following as a result:

$$\frac{(\overline{M} - \overline{M}_e) - \frac{\dot{M} V}{h_{m1} S_1 + h_{m2} S_2}}{(\overline{M}_0 - \overline{M}_e) - \frac{\dot{M} V}{h_{m1} S_1 + h_{m2} S_2}} = \mathrm{Exp}\left[\left(\frac{-h_{m1} S_1 - h_{m2} S_2}{V}\right) t\right] \tag{2}$$

where, in Equations (1) and (2), $\overline{M}$ is the moisture content on a dry basis (kg/kg); h_{m1} and h_{m2} are the external and internal convective mass transfer coefficients (m/s), respectively; V is the volume of the homogeneous solid (m^3); S_1 and S_2 are the external and internal surface areas of the homogeneous solid (m^2), respectively; $\dot{M}$ is the generation of moisture (kg/kg/s); M_e is the equilibrium moisture content on a dry basis (kg/kg) and t is the time (s).

2.2.2. Thermal Energy Conservation Equation

For the analysis of heat transfer, mass transfer analogy can be made and we can assume that on the surface of the solid thermal convection, the evaporation and heating of the produced vapor occur simultaneously. Therefore, the following energy conservation equation is valid for all the physical domains presented in Figure 1:

$$\rho_u V c_p \frac{d\overline{\theta}}{dt} = h_{c1} S_1 (\overline{\theta}_\infty - \overline{\theta}) + h_{c2} S_2 (\overline{\theta}_\infty - \overline{\theta}) + \rho_s V \frac{d\overline{M}}{dt} \left[h_{fg} + c_v (\overline{\theta}_\infty - \overline{\theta})\right] + \dot{q} V \tag{3}$$

where ρ_u is the specific density of the wet solid (kg/m^3); c_p is the specific heat (J/kgK); $\overline{\theta}$ is the instantaneous temperature of the solid (K or °C); $\overline{\theta}_\infty$ is the temperature of the external medium (K or °C); $\overline{\theta}_0$ is the initial temperature of the solid (K or °C); h_{c1} and h_{c2} are the external and internal convective heat transfer coefficients (W/m^{2}K), respectively; ρ_s is the specific density of the dry solid (kg/m^3); c_v is the specific heat of the vapor (J/kgK); h_{fg} is the latent heat of water vaporization (J/kg) and $\dot{q}$ is the heat generation per unit volume (W/m^3).

Equation (3) is an ordinary differential equation of the first order, it is non-linear and inhomogeneous; therefore, it cannot be solved analytically. Thus, for the simplification of Equation (3), the energy needed to heat water vapor from the temperature on the surface of the solid to the temperature of the fluid is disregarded. Thus, after simplifying and replacing Equations (1) and (2) in Equation (3), the result is as follows:

$$\rho V c_p \frac{d\bar{\theta}}{dt} = (h_{c1}S_1 + h_{c2}S_2)(\bar{\theta}_\infty - \bar{\theta})$$
$$+ \rho_s h_{fg}\left\{(-h_{m1}S_1 - h_{m2}S_2)\left[\left[(\overline{M}_0 - \overline{M}_e) - \frac{\dot{M}V}{h_{m1}S_1 + h_{m2}S_2}\right]\text{Exp}\left[\left(\frac{-h_{m1}S_1 - h_{m2}S_2}{V}\right)t\right]\right.\right. \quad (4)$$
$$\left.\left.+ \frac{\dot{M}V}{h_{m1}S_1 + h_{m2}S_2}\right] + \dot{M}V\right\} + \dot{q}V$$

or yet

$$\frac{d\bar{\theta}}{dt} = \left(\frac{(h_{c1}S_1 + h_{c2}S_2)}{\rho V c_p}\right)(\bar{\theta}_\infty - \bar{\theta}) + \quad (5)$$

$$\frac{\rho_s h_{fg}}{\rho V c_p}\left\{\left[(-h_{m1}S_1 - h_{m2}S_2)(\overline{M}_0 - \overline{M}_e) + \dot{M}V\right]\text{Exp}\left[\left(\frac{-h_{m1}S_1 - h_{m2}S_2}{V}\right)t\right]\right\} + \frac{\dot{q}}{\rho c_p} \quad (6)$$

Assuming $y = \bar{\theta}_\infty - \bar{\theta}$, then $\frac{dy}{dt} = -\frac{d\bar{\theta}}{dt}$. Thus, Equation (5) can be written as follows:

$$y' + a = -b e^{-ct} - d \quad (7)$$

where

$$a = \frac{(h_{c1}S_1 + h_{c2}S_2)}{\rho V c_p} \quad (8)$$

$$b = \frac{\rho_s h_{fg}}{\rho c_p}\left[(-h_{m1}S_1 - h_{m2}S_2)(\overline{M}_0 - \overline{M}_e) + \dot{M}V\right] \quad (9)$$

$$c = \frac{h_{m1}S_1 - h_{m2}S_2}{V} \quad (10)$$

$$d = \frac{\dot{q}}{\rho c_p} \quad (11)$$

Using the initial condition, $\bar{\theta}(t=0) = \bar{\theta}_0$, and solving Equation (6), we obtain as a result the following:

$$\bar{\theta} = \bar{\theta}_\infty - \left[(\bar{\theta}_\infty - \bar{\theta}_0) + \left(\frac{b}{a-c} + \frac{d}{a}\right)\right]e^{-at} + \left(\frac{b}{a-c}e^{-ct} + \frac{d}{a}\right) \quad (12)$$

Equation (11) describes the transient behavior of the temperature of the solid throughout the drying process. It is important to note that Equation (1) and (3) incorporate several thermophysical parameters of the heating fluid and the material, in such a way, that the model proposed to describe the drying of the material can be considered as phenomenological, differently from the models commonly used in the literature, for such purpose.

2.3. Applications to Drying of Industrial Ceramic Bricks
2.3.1. Volume and Surface Area of the Brick

The research was applied to describe the drying of industrial ceramic bricks (Figure 2). For this solid, the equations for determining the surface area of the brick are defined by the following:

$$A_C = A_L + A_I \quad (13)$$

where

$$A_L = 2[(R_x R_z) + (R_x R_y) + (R_y R_z) - 8(a_h a_v)] \quad (14)$$

$$A_I = 16[(a_h R_z) + (a_v R_z)] \quad (15)$$

$$a_h = \frac{R_x - 2a_2 - a_4}{2} \quad (16)$$

$$a_v = \frac{R_y - 2a_1 - 3a_3}{4} \quad (17)$$

where A_C is the total surface area of the brick, A_L corresponds to the lateral area (faces) of the bricks, A_I is the internal surface area (internal faces determined by the holes) and a_h and a_v are the height and width of a hole, respectively.

The calculation of the volume of the brick was made based on the value of its dimensions, width (Rx), height (Ry), length (Rz) and the dimensions that characterize the brick holes, a_1, a_2, a_3 and a_4, using the following equations:

$$V = V_T - V_f \tag{18}$$

where

$$V_T = R_x R_y R_z \tag{19}$$

$$V_F = 8 a_v a_h R_z \tag{20}$$

where V is the total volume of the brick, V_T is the volume of the solid brick (with the holes) and V_F is the volume of the holes.

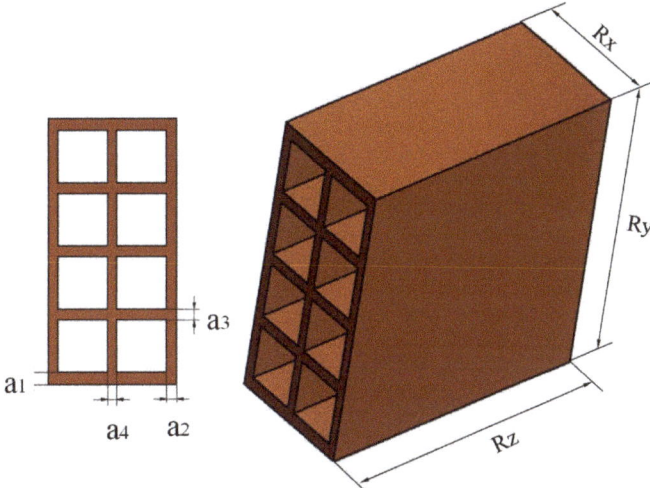

Figure 2. Scheme of the brick used in the research.

2.3.2. Thermophysical Properties of Air and Water

Specific heat of the air, latent heat of water vaporization, air density, absolute temperature, universal air constant, relative humidity, saturation vapor pressure and local atmospheric pressure [26,27], are determined by the following:

$$c_{p_a} = 1.00926 \times 10^3 - 4.04033 \times 10^{-2} T + 6.17596 \times 10^{-4} T^2 - 4.0972 \times 10^{-7} T^3 \tag{21}$$

$$k_a = 2.425 \times 10^{-2} + 7.889 \times 10^{-5} T - 1.790 \times 10^{-8} T^2 - 8.570 \times 10^{-12} T^3 \tag{22}$$

$$\mu_a = 1.691 \times 10^{-5} + 4.984 \times 10^{-8} T - 3.187 \times 10^{-11} T^2 + 1.3196 \times 10^{-14} T^3 \tag{23}$$

$$\rho_a = \frac{P_{atm} \overline{MM}_a}{R_a T_{abs}} \; (kg/m^3) \tag{24}$$

$$T_{abs} = T_a + 273.15 \; K \tag{25}$$

$$R_a = 8314.34 \; (J/kmol \; K) \tag{26}$$

$$RH = \frac{P_{atm} \, x_a}{(x_a + 0.622) \, P_{vs}} \tag{27}$$

$$h_{fg} = 352.8 \times (374.14 - T)^{0.33052} \text{ (kJ/kg)} \tag{28}$$

$$P_{vs} = 22105649.25 \, \text{Exp} \left\{ \begin{array}{c} [-27405.53 + 97.5413 T_{abs} - 0.146244 T_{abs}^2 + 0.12558 \times 10^{-3} T_{abs}^3 - 0.48502 \times 10^{-7} T_{abs}^4] \, / \\ [4.34903 T_{abs} - 0.39381 \times 10^{-2} T_{abs}^2] \end{array} \right\} \text{ (Pa)} \tag{29}$$

$$P_{atm} = 101325 \text{ Pa} \tag{30}$$

where, in these equations, x_a is the absolute humidity of the air, RH is the relative humidity of the air, $\overline{MM}_a = 28.966$ kg/kmol is the molecular weight of the gas, R_a is the universal constant of the gases, T_{abs} is the absolute temperature in Kelvin and P is the pressure on Pascal.

The specific heats of water in the liquid and vapor phases are determined by the following [26]:

$$c_w = 2.82232 + 1.18277 \times 10^{-2} T_{abs} - 3.5047 \times 10^{-5} T_{abs}^2 + 3.6010 \times 10^{-8} T_{abs}^3 \tag{31}$$

$$c_v = 1.8830 - 0.16737 \times 10^{-3} T_{abs} + 0.84386 \times 10^{-6} T_{abs}^2 - 0.2696610^{-9} T_{abs}^3 \tag{32}$$

2.3.3. Estimation of Convective Heat and Mass Transfer Coefficients

In this research, the brick was considered to be in a lateral position to the fluid that flows on the material surface, as illustrated in Figure 3. For this situation, depending on the speed of the drying air, the drying process can occur by natural, forced convection or a combination of them on all external faces of the brick and by natural convection in the internal walls of the hole.

In forced convection, the tendency of a particular system is based mathematically on the Reynolds number (Re) of the fluid, which is the ratio of the inertial forces to the viscous forces. This dimensionless parameter is defined as follows:

$$Re_{L_c} = \frac{\rho_a v_a L_c}{\mu_a} \tag{33}$$

where L_c represents a characteristic length of the porous solid.

In natural convection, the tendency of a particular system is based mathematically on the Grashof number (Gr) of the fluid, which is the ratio between the buoyancy and viscous forces. This dimensionless parameter is given as follows:

$$Gr_{L_c} = \frac{g\beta(T_s - T_\infty)L_c^2}{\nu^2} \tag{34}$$

where g is the gravitational acceleration, $\beta = \frac{1}{T_f}$ is the thermal expansion coefficient, $T_f = \frac{T_s + T_\infty}{2}$ is the film temperature on the absolute scale, T_s is the plate temperature, T_∞ is the fluid temperature and $\nu = \frac{\mu}{\rho}$ is the kinematic viscosity.

The product of the Grashof number by the Prandtl number (Pr) is called the Rayleigh number (Ra) of the fluid. Thus, the Rayleigh number is given by the following:

$$Ra_{L_c} = Gr_{L_c} Pr = \frac{\delta\beta(T_s - T_\infty)L_c^3}{\nu^2} Pr \tag{35}$$

where $Pr = \frac{c_a \mu_a}{k_a}$.

The relative magnitudes of the Grashof and Reynolds numbers determine which form of convection dominates the phenomenon. For example, if $\frac{Gr}{Re^2} \gg 1$, forced convection can be neglected, whereas if $\frac{Gr}{Re^2} \ll 1$, natural convection can be neglected. For other situations, both forced convection and natural convection must be considered in a combined manner.

The average Nusselt number of the fluid on the surface of the porous solid can be determined as follows:

$$\overline{Nu} = \frac{\overline{h_c} L_c}{k_a} \quad (36)$$

Looking to Figure 3, it can be seen that the brick is composed of vertical and horizontal plates, upper and lower, both on the external surface and in the holes, all at a lower temperature than the drying fluid. Thus, based on this geometry, one can determine the Nusselt numbers for heat transfer and, thus, determine the convective heat and mass transfer coefficients for each physical situation, under air flow regime [25,28].

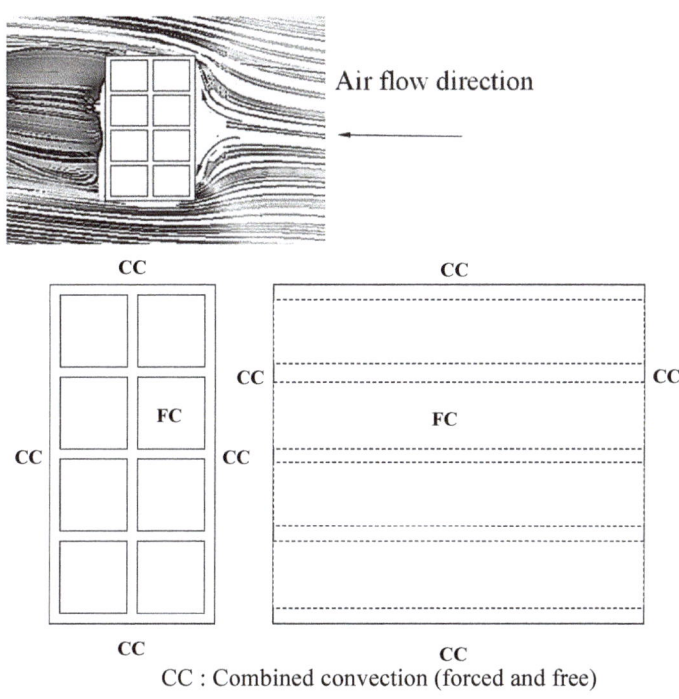

CC : Combined convection (forced and free)

FC : Free convection

Figure 3. Brick with side facing the air flow.

Natural Convection

For natural convection on flat surfaces, the following empirical correlations for the average Nusselt number are given:

(a) Vertical plate

$$\overline{Nu}_{pvt} = \left\{ 0.825 + \frac{0.387 R_{aL}^{1/6}}{\left[1 + (0.492/Pr)^{9/16}\right]^{8/27}} \right\}^2 \quad (37)$$

where, in the calculation of the number of Rayleigh (R_{aL}), Lc = Ry, for the front and rear plates.

$$\overline{Nu}_{pvl} = \left\{ 0.825 + \frac{0.387 R_{aL}^{1/6}}{\left[1 + (0.492/Pr)^{9/16}\right]^{8/27}} \right\}^2 \quad (38)$$

where, in the calculation of the Rayleigh number (R_{aL}), Lc = Ry − 4 ((Ry − 2a1 − 3a3)/4), for the side plates.

(b) Horizontal plate (Top surface)

$$\overline{Nu}_{phs} = 0.27(Gr_L Pr)^{1/4}, \quad 10^5 < Gr_L\,Pr < 10^{11} \tag{39}$$

where, in calculating the number of Rayleigh (R_{aL}), Lc = RxRz/(2Rx + 2Rz).

(c) Horizontal plate (bottom surface)

$$\overline{Nu}_{phi} = 0.54\,(Gr_L\,Pr)^{1/4}, \quad 10^4 < Gr_L Pr < 10^7 \tag{40}$$

$$\overline{Nu}_{phi} = 0.15\,(Gr_L\,Pr)^{1/3}, \quad 10^7 < Gr_L Pr < 10^{11} \tag{41}$$

where, in calculating the number of Rayleigh (R_{aL}), Lc = RxRz/(2Rx + 2Rz).

(d) Vertical Plate (Hole)

$$\overline{Nu}_{pvf} = \left\{ \frac{576}{[R_{aL}(L_c/Rz)]^2} + \frac{2.873}{[R_{aL}(L_c/Rz)]^{0.5}} \right\}^{-0.5} \tag{42}$$

where Lc = (Rx − 2a2 − 2a4)/2.

(e) Horizontal plate (Hole)

$$\overline{Nu}_{phf} = \left\{ \frac{576}{[R_{aL}(L_c/Rz)]^2} + \frac{2.873}{[R_{aL}(L_c/Rz)]^{0.5}} \right\}^{-0.5} \tag{43}$$

where Lc = (Ry − 2a1 − 3a3)/4.

From Equations (34)–(40), the convective heat transfer coefficients were calculated on the flat external surfaces and on the brick hole, using the following equation:

$$\overline{h}_c = \frac{\overline{Nu}\,k_a}{L_c} \tag{44}$$

With Equation (41) applied to each of the outer flat surfaces and the brick hole, the average external and internal convective heat transfer coefficients were obtained, as follows:

$$h_{c1} = \overline{h}_c\,_{ext}\Big|_{nc} = \left[\frac{\left(2\overline{h}_{cpvt} + 2\overline{h}_{cpvl} + \overline{h}_{cphs} + \overline{h}_{cphi}\right)}{6} \right]_{nc} \tag{45}$$

$$h_{c2} = \overline{h}_c\,_{int}\Big|_{nc} = \left[\frac{\left(\overline{h}_{cpvf} + \overline{h}_{cphf}\right)}{2} \right]_{nc} \tag{46}$$

where the nc sign indicates free convection.

Once the external and internal convective heat transfer coefficients were determined, the convective mass transfer coefficients were determined on the flat outer and hole surfaces of the brick, using the Chilton–Colburn analogy for heat and mass transfer, as follows:

$$\overline{h}_m = \frac{\overline{h}_c D_{va} Le^{1/3}}{k_a} \tag{47}$$

where $Le = \frac{\alpha}{D_{va}} = \frac{k_a}{D_{va}\rho_a c_{pa}}$ is the Lewis number, $\alpha = \frac{k_a}{\rho_a c_{pa}}$ is the thermal diffusivity of the air and D_{va} is the diffusion coefficient of water vapor in the air, given by the following:

$$D_{va} = 1.87 \times 10^{-10} \frac{T(K)^{2.072}}{P(atm)} \tag{48}$$

With Equation (44) applied to each of the external flat surfaces and the brick hole, the average external and internal convective mass transfer coefficients were obtained, as follows:

$$\overline{h}_{m\ ext}\Big|_{nc} = \left[\frac{\left(2\overline{h}_{mpvt} + 2\overline{h}_{mpvl} + \overline{h}_{mphs} + \overline{h}_{mphi}\right)}{6}\right]_{nc} \tag{49}$$

$$\overline{h}_{m\ int}\Big|_{nc} = \left[\frac{\left(\overline{h}_{mpvf} + \overline{h}_{mphf}\right)}{2}\right]_{nc} \tag{50}$$

Since the convective heat and mass transfer coefficients, determined by Equations (45) and (46), are calculated with the drying air parameters, and these same parameters that are present in Equations (2) and (11) are dependent on the brick parameters, it is necessary to recalculate these parameters for this new physical situation. This is performed using the following equations:

$$h_{m1} = \frac{\rho_a}{\rho_s} \frac{(x_{bu} - x_0)}{(M_0 - M_e)} \overline{h}_{m\ ext}\Big|_{nc} \tag{51}$$

$$h_{m2} = \frac{\rho_a}{\rho_s} \frac{(x_{bu} - x_0)}{(M_0 - M_e)} \overline{h}_{m\ int}\Big|_{nc} \tag{52}$$

where ρ_s is the density of the brick, x_o is the absolute humidity of the air at the drying air temperature and x_{bu} is the absolute humidity of the air at the wet bulb temperature of the drying air, given by the following:

$$x_{bu} = \left(\frac{P_{vwb}}{P_{atm} - P_{vwb}}\right)\left(\frac{\overline{MM}_v}{\overline{MM}_a}\right) \tag{53}$$

where P_{vwb} is the vapor pressure in the air at the wet bulb temperature of the air (which is equal to the saturation pressure of the water vapor at the wet bulb temperature) and $\overline{MM}_v$ is the molecular weight of the water vapor.

Forced Convection

For forced convection on the external walls of the brick, depending on the air flow regime, the Nusselt number can be determined as follows:

(a) Laminar flow (Re < 5 × 10^5)

$$\overline{Nu}_i = 0.664 Re_{Lc}^{0.5} Pr^{1/3} \tag{54}$$

(b) Turbulent flow (5 × 10^5 < Re < 1 × 10^7)

$$\overline{Nu}_i = 0.037 Re_{Lc}^{0.8} Pr^{1/3} \tag{55}$$

where subscript i refers to pvt, pvl, phs and phi. In calculating the Reynolds number, Lc = Ry for the front and rear plates; Lc = RxRz/(2Rx + 2Rz) for the upper and lower plates and Lc = Ry $-$ 4 [(Ry $-$ 2a$_1$ $-$ 3a$_3$)/4] for the side plates.

In the hole, the condition remains natural convection, the Nusselt number being calculated as defined in Equations (34) and (40). From Equations (34), (40), (50) and (52), the convective heat transfer coefficients were calculated on the flat external surfaces and on the brick hole, using Equation (41). From these coefficients, the average external and internal convective heat transfer coefficients were obtained, as follows:

$$h_{c1} = \overline{h}_{c\ ext}\Big|_{fc} = \left[\frac{\left(2\overline{h}_{cpvt} + 2\overline{h}_{cpvl} + \overline{h}_{cphs} + \overline{h}_{cphi}\right)}{6}\right]_{fc} \tag{56}$$

$$h_{c2} = \overline{h}_{c\ int}\Big|_{nc} = \left[\frac{\left(\overline{h}_{cpvf} + \overline{h}_{cphf}\right)}{2}\right]_{nc} \tag{57}$$

where the subscript fc indicates forced convection.

Once the external and internal convective heat transfer coefficients were determined, the convective mass transfer coefficients were determined on the flat outer and surfaces of the brick hole, using the Chilton–Colburn analogy for heat and mass transfer, as defined by Equation (43). From these coefficients, the average external and internal convective mass transfer coefficients were obtained, as follows:

$$\overline{h}_{m\ ext}\Big|_{fc} = \left[\frac{\left(2\overline{h}_{mpvt} + 2\overline{h}_{mpvl} + \overline{h}_{mphs} + \overline{h}_{mphi}\right)}{6}\right]_{fc} \tag{58}$$

$$\overline{h}_{m\ int}\Big|_{nc} = \left[\frac{\left(\overline{h}_{mpvf} + \overline{h}_{mphf}\right)}{2}\right]_{nc} \tag{59}$$

Since the convective heat and mass transfer coefficients, determined by Equations (54) and (55), are calculated with the drying air parameters, and these same parameters that are present in Equations (2) and (11) are dependent on the brick parameters, it is necessary to recalculate these parameters for this new situation. This is performed using the following equations:

$$h_{m1} = \frac{\rho_a}{\rho_s} \frac{(x_{bu} - x_0)}{(M_0 - M_e)} \overline{h}_{m\ ext}\Big|_{fc} \tag{60}$$

$$h_{m2} = \frac{\rho_a}{\rho_s} \frac{(x_{bu} - x_0)}{(M_0 - M_e)} \overline{h}_{m\ int}\Big|_{nc} \tag{61}$$

Combined Natural and Forced Convection

In case both natural and forced convection are important, the Nusselt number is given by a combination of effects as follows:

$$\overline{Nu_i} = \left[\overline{Nu}_{i,\ nc}^3 + \overline{Nu}_{i,\ fc}^3\right]^{\frac{1}{3}} \tag{62}$$

where subscript i refers to pvt, pvl, phs and phi for external surface plates and pvf and phf, for hole plates.

From the Nusselt numbers calculated using Equation (58), the convective heat transfer coefficients on the flat external surfaces and the brick hole were calculated, using Equation (41). In turn, from these coefficients, the average external and internal convective heat transfer coefficients were obtained, as follows:

$$h_{c1} = \overline{h}_{c\ ext}\Big|_{cc} = \left[\frac{\left(2\overline{h}_{cpvt} + 2\overline{h}_{cpvl} + \overline{h}_{cphs} + \overline{h}_{cphi}\right)}{6}\right]_{cc} \tag{63}$$

where the cc subscript indicates combined convection.

Once the external and internal convective heat transfer coefficients were determined, the convective mass transfer coefficients were determined on the flat outer and hole surfaces of the brick, using the Chilton–Colburn analogy for heat and mass transfer, as defined by Equation (43). From these coefficients, the average external and internal convective mass transfer coefficients were obtained, as follows:

$$\overline{h}_{m\ ext}\Big|_{cc} = \left[\frac{\left(2\overline{h}_{mpvt} + 2\overline{h}_{mpvl} + \overline{h}_{mphs} + \overline{h}_{mphi}\right)}{6}\right]_{cc} \quad (64)$$

$$\overline{h}_{m\ int}\Big|_{cc} = \left[\frac{\left(\overline{h}_{mpvf} + \overline{h}_{mphf}\right)}{2}\right]_{cc} \quad (65)$$

Since the convective heat and mass transfer coefficients, determined by Equations (61) and (62), are calculated with the drying air parameters, and these same parameters that are present in Equations (2) and (11) are dependent on the brick parameters, it is necessary to recalculate these parameters for this new situation. This is performed using the following equations:

$$h_{m1} = \frac{\rho_a}{\rho_s}\frac{(x_{bu} - x_0)}{(M_0 - M_e)}\overline{h}_{m\ ext}\Big|_{cc} \quad (66)$$

$$h_{m2} = \frac{\rho_a}{\rho_s}\frac{(x_{bu} - x_0)}{(M_0 - M_e)}\overline{h}_{m\ int}\Big|_{cc} \quad (67)$$

2.3.4. Cases Studied

Validation

Silva [8] developed an experimental study on the drying of industrial clay bricks. To carry out the drying experiment, the following procedures were adopted:

- Initially, dimensions, mass, brick temperature, ambient temperature and relative humidity were measured.
- Following that, the bricks were placed inside the oven where drying was carried out. In this process, air temperature inside the oven was established at 100°C through the temperature controller.
- At previously established time intervals, the brick was withdrawn from the oven, making it possible to measure its temperature, mass and dimensions. In principle, measurements were made every 10 min for up to 30 min. Then, the measurements were made every 30 min, with the next measurements being made every 60 min until the mass was approximately constant.
- Soon after, the sample was subjected to drying for 24 h to obtain the equilibrium mass, and then for another 24 h, to obtain the dry mass.

Figure 4 illustrates the specimen model used by the author of [8], the positions where the measurements of temperature, width (Rx), height (Ry), length (Rz) and the dimensions that characterize the holes in the bricks (a_1, a_2, a_3 and a_4) were obtained.

From the data obtained of the average moisture content throughout the drying process, it was possible to verify that in the initial period, the moisture loss rates are higher, requiring the researcher to read these data in shorter intervals, which can be expanded, as the process developed. From the physical point of view, this methodology is extremely satisfactory, as it allows the description of the phenomenon with great precision, especially at the beginning of the process. However, it is statistically more appropriate to perform parameter comparisons from a more uniform distribution of points throughout the process. Thus, the author proposed to fit an exponential equation with 2 terms and 4 parameters to these experimental data, as follows:

$$\left(\frac{\overline{M} - Me}{Mo - Me}\right) = A_1 \mathrm{Exp}(k_1 t) + A_2 \mathrm{Exp}(k_2 t) \tag{68}$$

where t is given in minutes. The estimation of parameters A_1, A_2, k_1 and k_2 in Equation (65) was performed using the numerical method of Rosembrock and quasi-Newton using the Statistica® Software, with a convergence criterion of 0.001. Table 1 summarizes the values of these estimated parameters.

Similar to the procedure adopted by the author of [8], Lima et al. [15] proposed fitting an exponential equation with 2 terms and 4 parameters to the experimental data of the surface temperature of the brick, as follows:

$$\theta = B_1 + B_2 \mathrm{Log}_{10}\left(t^{k_3} + B_3\right) \tag{69}$$

where t is given in minutes. The estimation of parameters B_1, B_2, k_3 and B_3 was performed using the Quasi-Newton numerical method using the Statistica® Software, with a convergence criterion of 0.0001. Table 2 summarizes the values of these estimated parameters.

Table 1. Parameters of Equation (65) obtained after fitting to experimental data of average moisture content [8].

T (°C)	Parameters				R	Proportion of Variance
	A_1 (-)	k_1 (min^{-1})	A_2 (-)	k_2 (min^{-1})		
100	4.875507	−0.008383	−3.827964	−0.007881	0.998297496	0.996597890

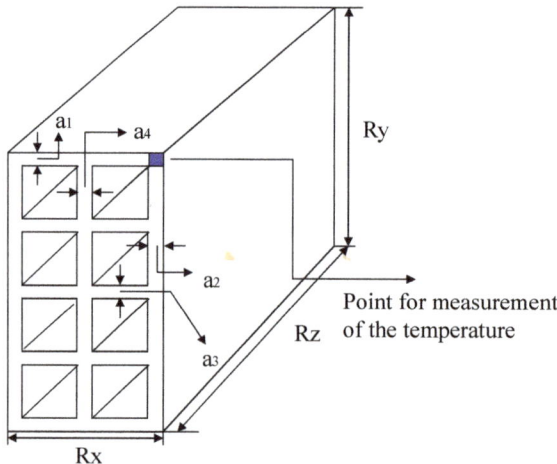

Figure 4. Brick model with the dimensions that characterize the brick and holes, and the location of the brick temperature measurement.

With the equations adjusted, instants of "data taking" were fixed along the process in which the average moisture content and temperature could be determined. This procedure allowed that uniform distribution of these points was reached. Following that, these equations were used in the computer program, to ensure that the predicted and experimental results of these process parameters could be compared.

Table 2. Equation (66) parameters obtained after adjustment to experimental moisture content data [15].

T (°C)	Parameters				R	Proportion of Variance
	B_1 (°C)	B_2 (°C/Log$_{10}$(min))	B_3 (min)	k_3 (-)		
100	−2.86969	15.41788	118.38213	2.234665	0.984632771	0.969501694

Tables 3 and 4 show, for the experiment, the geometric information, temperature and moisture content of the brick and the drying air conditions used in the oven. Table 5 summarizes the thermophysical properties, external and internal surface areas and brick volume, at the beginning of the process. To obtain the predicted results of the average moisture content and temperature at the vertex of the brick, a computational code was developed in the Mathematica®software. The quadratic deviations between the experimental and calculated values and the variance for the average moisture content and surface temperature were obtained as follows:

$$\text{ERMQ}_M = \sum_{i=1}^{n} \left(\overline{M}_{i,\text{Num}} - \overline{M}_{i,\text{Exp}}\right)^2 \tag{70}$$

$$\overline{S}_M^2 = \frac{\text{ERMQ}_M}{(n - \hat{n})} \tag{71}$$

$$\text{ERMQ}_\theta = \sum_{i=1}^{n} \left[\frac{(\theta_{i,\text{Num}} - \theta_{i,\text{Exp}})}{(\theta_e - \theta_0)}\right]^2 \tag{72}$$

$$\overline{S}_\theta^2 = \frac{\text{ERMQ}_\theta}{(n - \hat{n})} \tag{73}$$

where n = 56 is the number of experimental points and $\hat{n} = 0$ is the number of fitted parameters (number of degrees of freedom) [29].

Table 3. Experimental parameters of air and dimensions of hollow ceramic bricks used in the experiments [8].

Air			Brick						
T (°C)	RH (%)	v (m/s)	R_x (mm)	R_y (mm)	R_z (mm)	a_1 (mm)	a_2 (mm)	a_3 (mm)	a_4 (mm)
100	1.8	0.10	92.8	198.0	202.0	11.7	9.41	8.74	8.0

Table 4. Experimental parameters of air and brick for drying test [8].

Air	Brick					t (h)
T (°C)	M_o (d.b.)	M_f (d.b.)	M_e (d.b.)	θ_0 (°C)	θ_f (°C)	
100	0.16903	0.00038	0.00038	26.1	93.2	12.3

Table 5. Thermophysical and geometric parameters of the brick used in the simulation.

Brick						
k (W/mK)	ρ_u (kg/m^3)	ρ_s (kg/m^3)	c_p (J/kgK)	S_1 (mm^2)	S_2 (mm^2)	V (mm^3)
0.833	1754.88	1889.95	545.00	134,651.775	226,514.720	1,734,026.095

Studied Cases

To verify the application of the mathematical model to the drying of industrial bricks, 11 cases were defined for simulation (Table 6). The aim is evaluating the effect of the following different process parameters: temperature and relative humidity of the drying air, in the removal of moisture and temperature of the brick during drying. In all cases studied, the dimensions, initial and equilibrium moisture contents, initial temperature and thermophysical properties of the brick are reported in Tables 3–5.

Table 6. Air and brick conditions used in drying simulations.

Case	Air				x_0 (kg/kg)
	T_∞ (°C)	RH (%)	v (m/s)	T_{wb} (°C)	
1	100	20	0.1	62.46	0.15550
2	100	30	0.1	70.48	0.26660
3	100	40	0.1	76.76	0.41470
4	100	50	0.1	81.98	0.62210
5	100	60	0.1	86.45	0.93330
6	100	70	0.1	90.38	1.45200
7	100	70	0.5	90.38	1.45200
8	100	70	1.0	90.38	1.45200
9	100	70	3.0	90.38	1.45200
10	100	70	5.0	90.38	1.45200
11	100	70	8.0	90.38	1.45200

3. Results and Discussion

3.1. Validation

Figure 5 illustrates a comparison between the predicted and experimental results [8] of the average moisture content of the brick during the drying process. An analysis of this figure shows a good agreement between the results. From the comparison with the experimental data obtained from Equation (50), $ERMQ_M$ = 0.00509741 (kg/kg)^2 and $\bar{S}_M^2$ = 9.10252 × 10^{-5} (kg/kg)^2 were obtained. For this physical situation, the following convective mass transfer coefficients were obtained hm_1 = 6.46958 × 10^{-7} m/s and hm_2 = 6.13473 × 10^{-7} m/s.

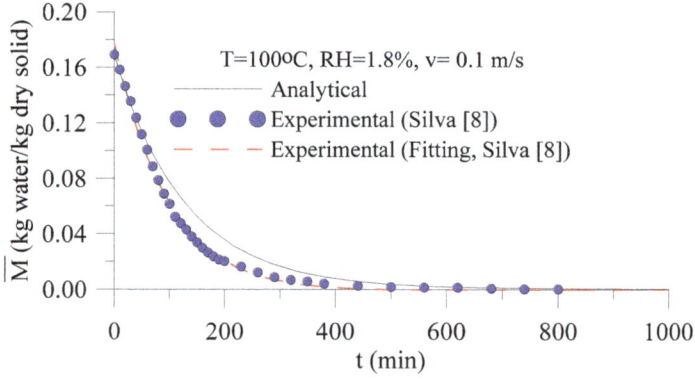

Figure 5. Comparison between the predicted and experimental average moisture content during the drying process of industrial hollow ceramic bricks.

Figure 6 illustrates a comparison between the predicted and experimental results [8,15] of the temperature at the vertex of the brick during the drying process. An analysis of this figure shows a good agreement between the results. From the comparison with the experimental data obtained from Equation (51), ERMQ$_\theta$ = 0.346251426 (°C/°C)^2 and $\overline{S}_\theta^2$ = 0.006183061 (°C/°C)^2 were obtained. For this physical situation, the following convective heat transfer coefficients were obtained: hc$_1$ = 6.89281 W/m^{2}K and hc$_2$ = 6.53606 W/m^{2}K. These low values of the convective heat transfer coefficient are typical of a physical situation of natural convection. In fact, for this case, a ratio Gr/Re^2 = 10705.4 was obtained, which is much greater than 1.0, justifying that the effect of natural convection is much higher than forced convection.

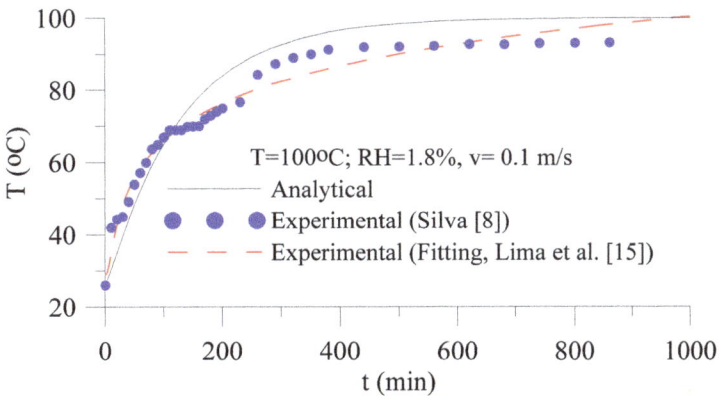

Figure 6. Comparison between predicted and experimental surface temperature (vertex) during the drying process of industrial hollow ceramic bricks.

3.2. Analysis of the Drying Process

3.2.1. Effect of Relative Humidity of Drying Air

Figure 7 shows the drying kinetics of the industrial ceramic brick, in different relative humidities of the drying air, with the air temperature and speed kept constant (Cases 1 to 7, from Table 6). An analysis of this figure shows that the lower the relative humidity of the drying air, the faster the solid dries and, with that, there is a reduction in the drying time, to reach a previously specified moisture content. To give an idea, in t = 500 min (8.33 h) of the process, the average moisture content of the brick is 0.01433 and 0.05341 kg/kg (dry basis), for the relative humidity of 20% and 70%, respectively.

Figure 8 shows the heating kinetics of the industrial ceramic brick, in different relative humidities of the drying air, keeping the temperature and air speed constant. An analysis of this figure shows that the lower the relative humidity of the drying air is, the faster the solid heats up. To give an idea, in t = 500 min (8.33 h) of the process, the brick temperature is 95.49 and 89.09 °C for the relative humidity of 20 and 70%, respectively.

It is worth mentioning that higher drying and heating rates can cause problems in the brick structure and reduce the quality of the product at the end of the drying process. Therefore, it is preferable to have a process with a higher relative humidity, thus preventing defects from the drying process, such as cracks, deformations and fractures in the brick. However, the process time is longer, which reduces the productivity of this type of ceramic product.

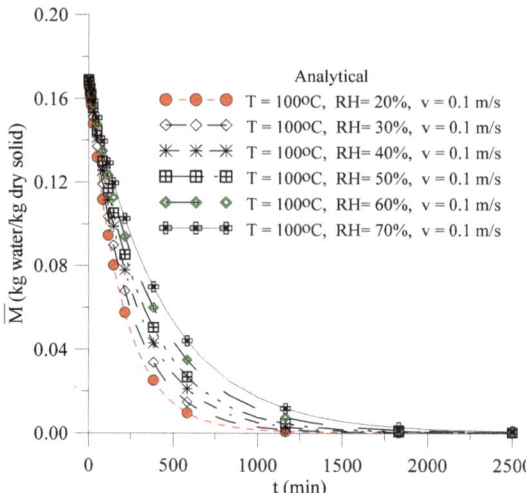

Figure 7. Average moisture content of industrial ceramic brick as a function of time for different relative humidity of the drying air.

Table 7 summarizes the values of the convective heat and mass transfer coefficients obtained under different drying conditions (Cases 1 to 6). An analysis of this table shows that both the convective mass transfer coefficient and the convective heat transfer coefficient decrease with an increase in the relative humidity of the drying air. Its values are small, typical of a natural mass and thermal convection problem. This can be confirmed by the high values of the relationship between the Grashof and Reynolds numbers ($\frac{Gr}{Re^2} \gg 1$), in such a way that the effects of forced convection are negligible. Similar to the effect of the drying air temperature, the convective heat and mass transfer coefficients in the hole of the brick are slightly lower than those on the outer surface of the brick.

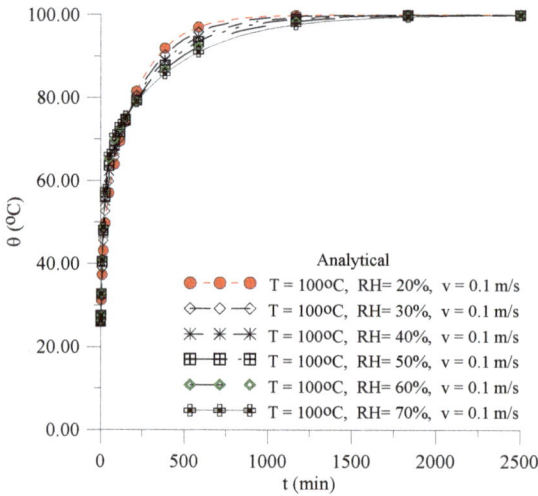

Figure 8. Temperature at the vertex of the industrial ceramic brick as a function of time for different relative humidity of the drying air.

From the above, it can be said that for the drying situation with a lower relative humidity, there is slightly longer drying time, which leads to slightly lower productivity, but with products with good quality. This is because the removal of moisture is controlled and the temperature of the brick, throughout the process, is lower, reducing the thermo-mechanical effects of moisture and thermal gradients inside the brick. Moreover, we can mention the fact that the energy cost is much lower when drying with a low relative humidity (20%) compared to drying with relative humidity 70%, which reduces the final cost of the product.

Table 7. Heat transfer coefficients and convective mass for different relative humidity of the drying air.

Drying Air Condition			Mass Transfer Coefficient		Heat Transfer Coefficient		Gr/Re^2 (-)
T (°C)	RH (%)	v (m/s)	hm_1 (m/s)	hm_2 (m/s)	hc_1 (W/m²K)	hc_2 (W/m²K)	
100	20	0.1	4.10×10^{-7}	3.92×10^{-7}	5.91	5.66	6037.75
100	30	0.1	3.46×10^{-7}	3.32×10^{-7}	5.53	5.31	4694.73
100	40	0.1	2.94×10^{-7}	2.83×10^{-7}	5.19	4.99	3663.88
100	50	0.1	2.59×10^{-7}	2.50×10^{-7}	4.85	4.67	2820.57
100	60	0.1	2.22×10^{-7}	2.14×10^{-7}	4.50	4.34	2107.97
100	70	0.1	1.89×10^{-7}	1.83×10^{-7}	4.13	3.98	1488.59

3.2.2. Effect of Drying Air Speed

Figure 9 shows the drying kinetics of the industrial ceramic brick, at different speeds of the drying air, keeping the temperature and relative humidity of the air constant (Cases 6 to 11, from Table 6). Analyzing this figure, it can be seen that, the higher the speed of the drying air is, the faster the solid dries and, with that, there is a reduction in the drying time, to reach the previously desired moisture content. In addition, it can be seen that for low speeds (up to 0.5 m/s), the drying kinetics is independent of the value of this parameter, substantially modifying the behavior of moisture loss for higher speeds. To give an idea, in t = 500 min (8.33 h) of the process, the average moisture content of the brick is 0.05341 and 0.00091 kg/kg (dry basis), for air speeds of 0.1 and 8.0 m/s, respectively.

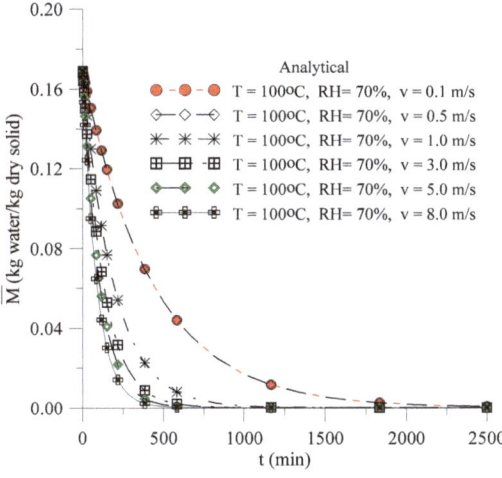

Figure 9. Average moisture content of industrial ceramic brick as a function of time for different drying air speeds.

In Figure 10, it is possible to observe that the heating kinetics of the industrial ceramic brick, at different speeds of the drying air, keeping the temperature and relative humidity of the air constant. From the analysis of this figure, it can be seen that the higher the speed of the drying air, the faster the solid heats up (higher heating rates). The explanation for this fact is related to the dependence of the convective heat transfer coefficient on the speed. To give an idea, in t = 500 min (8.33 h) of the process, the brick temperature is 89.09 and 99.89 °C for the air speeds of 0.1 and 8.0 m/s, respectively.

As previously mentioned if the drying and heating rates are higher, it can cause problems in the brick structure and reduce the quality of the product at the end of the drying process. Therefore, it is preferable to have a process with a lower air speed, thus preventing defects from the drying process, such as cracks, deformations and fractures in the brick. However, the process time is longer, which reduces the productivity of this type of ceramic product.

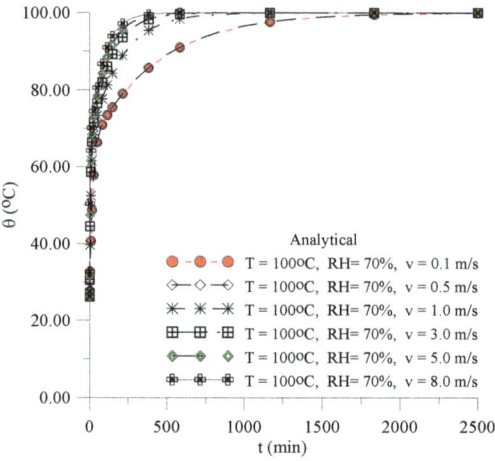

Figure 10. Temperature at the vertex of the industrial ceramic brick as a function of time for different drying air speeds.

Table 8 shows the values of the convective heat and mass transfer coefficients obtained under different drying conditions (Cases 6 to 11). Analyzing the results of this table, it can be seen that both the convective mass transfer coefficient and the convective heat transfer coefficient increase with an increasing drying air speed. For speeds up to 0.5 m/s, the values of hm_1, hm_2, hc_1 and hc_2 are small, typical of a natural mass and a thermal convection problem. This can be confirmed by the high values of the relationship between the Grashof and Reynolds numbers ($\frac{Gr}{Re^2} \gg 1$), in such a way that the effects of forced convection are negligible. However, for a speed between 1.0 and 8.0 m/s, the process occurs by combined natural and forced convection, confirmed by the values of the relationship between the Grashof and Reynolds numbers that are in the range of $0.23 \leq \frac{Gr}{Re^2} \leq 14.89$.

Since the convective effects are more intense on the external surface of the brick, the values of the mass transfer coefficients (hm_1) and heat (hc_1) are much higher than the values of these parameters in the holes, where natural convection predominates.

Comparing the values of these parameters with those obtained when analyzing the effects of the relative humidity of the drying air, it can be observed that drying at a higher speed (Case 11) has a much higher convective mass and heat transfer coefficients. For example, for drying at T = 100 °C, RH = 70% and v=8.0 m/s, at t = 1000 min (16.67 h), the average moisture content of the brick was 0.00038 kg/kg (dry basis) and its vertex temperature reached 100 °C. For drying at T = 100 °C, RH = 20% and v = 0.1 m/s, at that same time, the average moisture content of the brick was 0.0015 kg/kg (dry basis) and its temperature at the apex reached 99.62 °C.

Table 8. Convective heat and mass transfer coefficients for different drying air speeds.

Drying Air Condition			Mass Transfer Coefficient		Heat Transfer Coefficient		Gr/Re^2 (-)
T (°C)	RH (%)	v (m/s)	hm_1 (m/s)	hm_2 (m/s)	hc_1 (W/m²K)	hc_2 (W/m²K)	
100	70	0.1	1.89×10^{-7}	1.83×10^{-7}	4.13	3.98	1488.59
100	70	0.5	1.89×10^{-7}	1.83×10^{-7}	4.13	3.98	59.54
100	70	1.0	7.44×10^{-7}	2.30×10^{-7}	16.22	5.02	14.89
100	70	3.0	12.81×10^{-7}	2.30×10^{-7}	27.93	5.02	1.65
100	70	5.0	16.53×10^{-7}	2.30×10^{-7}	36.02	5.02	0.59
100	70	8.0	20.90×10^{-7}	2.30×10^{-7}	45.55	5.02	0.23

It is worth noting that in the industry, the average moisture content found in the ceramic bricks, at the end of the drying stage, reaches values between 3 and 4% (wet basis) and the process time is around 24 h, for drying being carried out in a crossflow tunnel dryer. In this type of dryer, the length of the equipment can reach 100 m, the temperature varies from 25 to 120 °C and the relative humidity ranges from 60 to 90%, but in some equipment of shorter length (35 m, for example), the value of this parameter can reach 10%. In general, after drying, due to the low moisture content, the brick absorbs water (either in the form of vapor, contained in the air, or in liquid form, from rain or soil, during the storage phase, before firing), which causes slight degradation due to volumetric expansion.

Table 9 summarizes the results that meet the conditions of the final moisture content established by the industrial sector for each drying condition studied. Thus, evaluating the results obtained in the analyzed cases, considering the maximum final moisture content of 3 to 4% (dry basis) and the temperature of the brick, the position of the brick with a hole perpendicular to the air flow, the drying time in these moisture content conditions, as well as the experimental results (in terms of brick quality during drying) presented by the author of [8], T = 100 °C, RH = 50% and v = 0.1 m/s are suggested as the optimal drying conditions.

Table 9. Brick conditions at the end of the drying process.

Case	Air			Brick		
	T_∞ (°C)	RH (%)	v (m/s)	M (kg/kg)	θ (°C)	t (min)
1	100	20	0.1	0.03239	89.65	333.33
2	100	30	0.1	0.03385	90.34	383.33
3	100	40	0.1	0.03398	91.31	450.00
4	100	50	0.1	0.03502	91.60	500.00
5	100	60	0.1	0.03502	92.30	583.33
6	100	70	0.1	0.03644	92.58	666.67
7	100	70	0.5	0.03644	92.58	666.67
8	100	70	1.0	0.03825	92.21	283.33
9	100	70	3.0	0.03167	93.56	216.67
10	100	70	5.0	0.03782	92.30	158.33
11	100	70	8.0	0.03024	93.86	150.00

3.3. Limitations of the Proposed Model

Despite the advantages presented by the proposed mathematical model, it fails in some points, such as:

(a) In physical situations involving the effect of moisture removal on solid heating (simultaneous heat and mass transport);
(b) In physical situations involving heterogeneous solids;
(c) It does not allow for the determination of temperature gradients and the moisture content inside the material; therefore, analysis of internal hydric and thermal stresses during the drying process is not possible.

4. Conclusions

In this work, the physical problem of industrial clay brick drying using a lumped approach has been analyzed. From the results obtained, it can be concluded that:

(a) The phenomenological mathematical modelling based on a lumped analysis to predict the mass and heat transfers in hollow ceramic bricks, was adequate, with small deviations from the experimental data on the average moisture content and temperature of the product over the process;
(b) The lower the relative humidity is and the higher the speed of the drying air is, the faster the brick loses moisture and raises its temperature, which can cause defects in the post-drying product, reducing its quality for the firing step;
(c) The heat and mass transfer coefficients on the external surface and on the brick hole are different, being higher on the external surface, especially under a forced convection condition;
(d) The convective mass transfer coefficient on the external surface of the brick varied from 1.89×10^{-7} m/s in the drying condition at 100 °C, 70% and 0.1 m/s (natural convection) up to 20.90×10^{-7} m/s at 100 °C, 70% and 8.0 m/s (combined and natural convection). The convective mass transfer coefficient in the brick hole varied from 1.83×10^{-7} m/s in the drying condition at 100 °C, 20% and 0.1 m/s (natural convection) up to 2.30×10^{-7} m/s at 100 °C, 70% and 8.0 m/s (natural convection).
(e) The convective heat transfer coefficient on the external surface of the brick varied from 4.13 W/m^{2}K in the drying condition 100 °C, 70% and 0.1 m/s (natural convection) to 45.55 W/m^{2}K at 100 °C, 70% and 8.0 m/s (combined and natural convection). The convective heat transfer coefficient in the brick hole varied from 3.98 W/m^{2}K in the drying condition 100 °C, 70% and 0.1 m/s (natural convection) to 5.02 W/m^{2}K at 100 °C, 70% and 8.0 m/s (natural convection).
(f) Based on the maximum final moisture content of 3 to 4% (dry basis) after drying, commonly used in the industry, the temperature of the brick, the position of the brick with a hole perpendicular to the air flow, the drying time in these moisture content conditions and the experimental results (in terms of brick quality during drying) presented by the author of [8], T = 100 °C, RH = 50% and v = 0.1 m/s are proposed as the optimal drying conditions.

Author Contributions: All the authors contributed to the development, analysis, writing and revision of the paper: conceptualization, E.S.L. and W.M.P.B.L.; methodology, E.S.L., I.B.S. and W.M.P.B.L.; software, J.E.F.F., G.S.A. and A.G.B.L.; validation, E.S.L., I.B.S. and W.M.P.B.L.; formal analysis, E.S.L., J.M.P.Q.D., A.S.G. and A.G.B.L.; investigation, E.S.L., W.M.P.B.L., R.S.S. and J.M.P.Q.D.; writing—original draft preparation, E.S.L., J.M.P.Q.D. and A.G.B.L.; writing—review and editing, J.M.P.Q.D., A.G.B.L. and J.P.G.; visualization, E.S.L., A.F.V. and A.D.V.; supervision, A.G.B.L. and J.M.P.Q.D. All authors have read and agreed to the published version of the manuscript.

Funding: This work was supported by base funding (UIDB/04708/2020) and programmatic funding (UIDP/04708/2020) of the CONSTRUCT—Instituto de I&D em Estruturas e Construções, which is funded by national funds through the Fundação para a Ciência e a Tecnologia (FCT/MCTES). This work was funded by Paraiba State Research Foundation (FAPESQ) (grant number: 324/19), CNPq (grant number: 307414/2018−3), CAPES, and FINEP (Brazilian Research Agencies).

Institutional Review Board Statement: Not applicable.

Informed Consent Statement: Not applicable.

Data Availability Statement: The data that support the findings of this study are available upon request from the authors.

Acknowledgments: The authors would like to thank the Computational Laboratory of Thermal and Fluids, Mechanical Engineering Department, Federal University of Campina Grande (Brazil), for the research infrastructure and the references cited in the manuscript.

Conflicts of Interest: The authors declare no conflict of interest.

Nomenclature

Latin Letters

a_1, a_2, a_3, a_4	Dimensions of the brick holes	(m)
A_C	Total surface area	(m²)
A_L	Lateral area	(m²)
A_I	Internal surface area	(m²)
a_h	Height of a hole	(m)
a_v	Width of a hole	(m)
c_p	Specific heat of the brick	(J/kgK)
c_v	specific heat of the vapor phase	(J/kgK)
c_{p_a}	Specific heat of the air	(J/kgK)
c_w	specific heat of water in the liquid phase	(J/kgK)
D_{va}	diffusion coefficient of water vapor in the air	(m²/s)
$ERMQ_M$	Quadratic deviations for moisture content	(kg²/kg²)
$ERMQ_\theta$	Quadratic deviations for temperature	(°C²/°C²)
Gr	Grashof number	(—)
g	Gravitational acceleration	(m/s²)
h_{fg}	Latent heat of water vaporization	(J/kg)
h_{c1}	External convective heat transfer coefficients	(W/m²K)
h_{c2}	Internal convective heat transfer coefficients	(W/m²K)
h_{m1}	External convective mass transfer coefficients	(m/s)
h_{m2}	Internal convective mass transfer coefficients	(m/s)
Le	Lewis number	(—)
Lc	Characteristic length	(m)
$\overline{MM}_v$	molecular weight of the water vapor	(kg/kmol)
$\dot{M}$	Generation of moisture	(kg/kg/s)
M_e	Equilibrium moisture content	(kg/kg, d.b)
M_0	Initial moisture content	(kg/kg, d.b)
$\overline{M}$	Average moisture content	(kg/kg, d.b)
$\overline{MM}_a$	Molecular weight of the gas	(kg/kmol)
n	Number of experimental points	(—)
$\hat{n}$	Number of fitted parameters	(—)
Nu	Nusselt number	(—)
Pv_{wb}	Vapor pressure in the air at the wet bulb temperature	(Pa)
Pr	Prandtl number	(—)
Pvs	Saturation vapor pressure	(Pa)
Patm	Atmospheric pressure	(Pa)
P	pressure	(Pa)
$\dot{q}$	Heat generation per unit volume	(W/m³)
R_a	Universal constant of the gases	(J/molK)
Ra	Rayleigh number	(—)
Re	Reynolds number	(—)
Rx	Width	(m]

Ry	Height	(m)
Rz	Length	(m)
RH	Relative humidity	(%)
S_1	Internal surface area	(m^2)
S_2	Enternal surface area	(m^2)
$\bar{S}_M^2$	Variance for moisture content	(kg^2/kg^2)
$\bar{S}_\theta^2$	Variance for temperature	($°C^2/°C^2$)
t	Time	(s or min)
T_{abs}	Absolute temperature	(K)
T_f	Film temperature	(K or °C)
Ts	Plate temperature	(K or °C)
T∞	Fluid temperature	(K or °C)
V_T	Volume of the solid brick (with the holes)	(m^3)
V_F	Volume of the holes	(m^3)
V	Volume	(m^3)
x_a	Absolute humidity of the air	(kg/kg)
x_o	Absolute humidity of the air at the drying air temperature	(kg/kg)
x_{bu}	Absolute humidity of the air at the wet bulb temperature	(kg/kg)

Greek Letters

α	Thermal diffusivity of the air	(m^2/s)
β	Thermal expansion coefficient	(K^{-1})
ν	Kinematic viscosity	(m^2/s)
$ρ_s$	Specific density of the dry solid	(kg/m^3)
$ρ_a$	Air density	(kg/m^3)
$ρ_u$	Specific density of the wet solid	(kg/m^3)
$\bar{θ}$	Instantaneous temperature	(K or °C)
$\bar{θ}_∞$	Temperature of the external medium	(K or °C)
$\bar{θ}_0$	Initial temperature of the solid	(K or °C)

References

1. Cabral Junior, M.; Motta, J.F.M.; Almeida, A.S.; Tanno, L.C. RMIs: Clay for red ceramic. Rocks and industrial minerals in Brazil: Uses and specifications. *CETEM* **2008**, *2*, 747–770. (In Portuguese)
2. Sahu, K.M.; Singh, L.; Choudhary, S.N. Critical Review on Bricks. *Int. J. Eng. Manag. Res.* **2016**, *6*, 80–88.
3. Andrade, F.A.; Al-Qureshi, H.A.; Hotza, D. Measuring the plasticity of clays: A review. *Appl. Clay Sci.* **2011**, *51*, 1–7. [CrossRef]
4. Brooker, D.B.; Bakker-Arkema, F.W.; Hall, C.W. *Drying and Storage of Grains and Oilseeds*; AVI Book: New York, NY, USA, 1992.
5. Yang, J.; Lu, H.; Zhang, X.; Li, J.; Wang, W. An experimental study on solidifying municipal sewage sludge through skeleton building using cement and coal gangue. *Adv. Mater. Sci. Eng.* **2017**, *2017*, 5069581. [CrossRef]
6. Musielak, G.; Śliwa, T. Modeling and Numerical Simulation of Clays Cracking During Drying. *Dry Technol.* **2015**, *33*, 1758–1767. [CrossRef]
7. Itaya, Y.; Taniguchi, S.; Hasatani, M. A numerical study of transient deformation and stress behavior of a clay slab during Drying. *Dry Technol.* **1997**, *15*, 1–21. [CrossRef]
8. Silva, J.B. Simulation and Experimentation of the Holed Ceramic Bricks Drying. Ph.D. Thesis, Federal University of Campina Grande, Campina Grande, Brazil, 2009.
9. Silva, J.B.; Almeida, G.S.; Lima, W.M.P.B.; Neves, G.A.; de Barbosa Lima, A.G. Heat and mass diffusion including shrinkage and hygrothermal stress during drying of holed ceramics bricks. *Defect Diffus. Forum* **2011**, *312–315*, 971–976. [CrossRef]
10. Van der Zanden, A.J.J. *Modelling and Simulating Simultaneous Liquid and Vapour Transport in Partially Sutured Porous Materials, Mathematical Modeling and Numerical Techniques in Drying Technology*; Mareei Dekker Inc.: New York, NY, USA, 1997.
11. Van der Zanden, A.J.J.; Schoenmakers, A.M.E.; Kerkof, P.J.A.M. Isothermal Vapour and Liquid Transport Inside Clay During Drying. *Dry Technol.* **1996**, *14*, 647–676. [CrossRef]
12. Zaccaron, A.; Souza Nandi, V.; Bernardin, A.M. Fast drying for the manufacturing of clay ceramics using natural clays. *J. Build. Eng.* **2021**, *33*, 101877. [CrossRef]
13. El Ouahabi, M.; El Boudour El Idrissi, H.; Daoudi, L.; El Halim, M.; Fagel, N. Moroccan clay deposits: Physico-chemical properties in view of provenance studies on ancient ceramics. *Appl. Clay Sci.* **2019**, *172*, 65–74. [CrossRef]

14. Kolesnikov, G.; Gavrilov, T. Modeling the Drying of Capillary-Porous Materials in a Thin Layer: Application to the Estimation of Moisture Content in Thin-Walled Building Blocks. *Appl. Sci.* **2020**, *10*, 6953. [CrossRef]
15. De Barbosa Lima, A.G.; Delgado, J.M.P.Q.; Nascimento, L.P.C.; Lima, E.S.; Oliveira, V.A.B.; Silva, A.V.; Silva, J.V. Clay ceramic materials: From fundamentals and manufacturing to drying process predictions. In *Transport Processes and Separation Technologies*, 1st ed.; Delgado, J.M.P.Q., Lima, A.G.B., Eds.; Springer: Cham, Switzerland, 2021; Volume 133, pp. 1–29.
16. Yan, M.; Song, X.; Tian, J.; Lv, X.; Zhang, Z.; Yu, X.; Zhang, S. Construction of a New Type of Coal Moisture Control Device Based on the Characteristic of Indirect Drying Process of Coking Coal. *Energies* **2020**, *13*, 4162. [CrossRef]
17. Lima, E.S.; Lima, W.M.P.B.; Lima, A.G.B.; Farias Neto, S.R.; Silva, E.G.; Oliveira, V.A.B. Advanced study to heat and mass transfer in arbitrary shape porous materials: Foundations, phenomenological lumped modeling and applications. In *Transport Phenomena in Multiphase Systems*, 1st ed.; Delgado, J.M.P.Q., Lima, A.G.B., Eds.; Springer: Cham, Switzerland, 2018; Volume 93, pp. 181–217.
18. Araújo, M.V.; Correia, B.R.B.; Brandão, V.A.A.; Oliveira, I.R.; Santos, R.S.; Oliveira Neto, G.L.; Silva, L.P.L.; de Barbosa Lima, A.G. Convective Drying of Ceramic Bricks by CFD: Transport Phenomena and Process Parameters Analysis. *Energies* **2020**, *13*, 2073. [CrossRef]
19. Silva, W.P.; Silva, C.M.D.P.S.; Silva, L.D.; Farias, V.S.O. Drying of clay slabs: Experimental determination and prediction by two-dimensional diffusion models. *Ceram. Int.* **2013**, *39*, 7911–7919. [CrossRef]
20. Silva, A.M.V.; Delgado, J.M.P.Q.; Guimarães, A.S.; Lima, W.M.P.B.; Gomez, R.S.; Farias, R.P.; Lima, E.S.; de Barbosa Lima, A.G. Industrial Ceramic Blocks for Buildings: Clay Characterization and Drying Experimental Study. *Energies* **2020**, *13*, 2834. [CrossRef]
21. Lima, W.M.P.B.; Lima, E.S.; Lima, A.R.C.; Oliveira Neto, G.L.; Oliveira, N.G.N.; Farias Neto, S.R.; de Barbosa Lima, A.G. Applying Phenomenological Lumped Models in Drying Process of Hollow Ceramic Materials. *Defect Diffus. Forum* **2020**, *400*, 135–145. [CrossRef]
22. De Lemos, M.J.S. *Turbulence in Porous Media: Modeling and Applications*; Elsevier: Amsterdam, The Netherlands, 2006.
23. Wana, H.; Xua, X.; Gaoa, J.J.; Lia, A. A moisture penetration depth model of building hygroscopic material. *Procedia Eng.* **2017**, *205*, 3235–3242. [CrossRef]
24. Zhang, M.; Qin, M.; Rode, C.; Chen, Z. Moisture buffering phenomenon and its impact on building energy consumption. *Appl. Therm. Eng.* **2017**, *124*, 337–345. [CrossRef]
25. Incropera, F.P.; de Witt, D.P. *Fundamentals of Heat and Mass Transfer*; John Wiley & Sons: New York, NY, USA, 2002.
26. Pakowski, Z.; Bartczak, Z.; Strumillo, C.; Stenstrom, S. Evaluation of equations approximating thermodynamic and transport properties of water, steam and air for use in CAD of drying processes. *Dry Technol.* **1991**, *9*, 753–773. [CrossRef]
27. Jumah, R.Y.; Mujumdar, A.S.; Raghavan, G.S.V. A mathematical model for constantand intermittent batch drying of grains in a novel rotating jet spouted bed. *Dry Technol.* **1996**, *14*, 765–802.
28. Çengel, Y.A. *Heat and Mass Transfer: A Practical Approach*; McGraw Hill: Boston, MA, USA, 2007.
29. Figliola, R.S.; Beasley, D.E. *Theory and Design for Mechanical Measurements*; Wiley: New York, NY, USA, 1995.

Article

Equivalent Parallel Strands Modeling of Highly-Porous Media for Two-Dimensional Heat Transfer: Application to Metal Foam

Nihad Dukhan

Department of Mechanical Engineering, University of Detroit Mercy, Detroit, MI 48221, USA; dukhanni@udmercy.edu

Abstract: A new geometric modeling of isotropic highly-porous cellular media, e.g., open-cell metal, ceramic, and graphite foams, is developed. The modelling is valid strictly for macroscopically two-dimensional heat transfer due to the fluid flow in highly-porous media. Unlike the current geometrical modelling of such media, the current model employs simple geometry, and is derived from equivalency conditions that are imposed on the model's geometry a priori, in order to ensure that the model produces the same pressure drop and heat transfer as the porous medium it represents. The model embodies the internal structure of the highly-porous media, e.g., metal foam, using equivalent parallel strands (EPS), which are rods arranged in a spatially periodic two-dimensional pattern. The dimensions of these strands and their arrangement are derived from equivalency conditions, ensuring that the porosity and the surface area density of the model and of the foam are indeed equal. In order to obtain the pressure drop and heat transfer results, the governing equations are solved on the geometrically-simple EPS model, instead of the complex structure of the foam. By virtue of the simple geometry of parallel strands, huge savings on computational time and cost are realized. The application of the modeling approach to metal foam is provided. It shows how an EPS model is obtained from an actual metal foam with known morphology. Predictions of the model are compared to experimental data on metal foam from the literature. The predicted local temperatures of the model are found to be in very good agreement with their experimental counterparts, with a maximum error of less than 11%. The pressure drop in the model follows the Forchheimer equation.

Keywords: metal foam; graphite foam; geometric modelling; heat transfer; porous media

Citation: Dukhan, N. Equivalent Parallel Strands Modeling of Highly-Porous Media for Two-Dimensional Heat Transfer: Application to Metal Foam. *Energies* **2021**, *14*, 6308. https://doi.org/10.3390/en14196308

Academic Editors: Moghtada Mobedi and Kamel Hooman

Received: 6 July 2021
Accepted: 30 September 2021
Published: 2 October 2021

Publisher's Note: MDPI stays neutral with regard to jurisdictional claims in published maps and institutional affiliations.

Copyright: © 2021 by the author. Licensee MDPI, Basel, Switzerland. This article is an open access article distributed under the terms and conditions of the Creative Commons Attribution (CC BY) license (https://creativecommons.org/licenses/by/4.0/).

1. Introduction

Highly-porous open-cell media include open-cell metal foam (e.g., aluminum, copper, nickel) and graphite foam. These foams can have porosities exceeding 90%. They are composed of cells and ligaments that form pores or windows, which is shown in Figure 1. Many of these highly-porous media, such as well-produced aluminum foam made by casting over polymeric pre-forms, are practically isotropic and have uniform average geometrical properties [1,2]. There are a few thermal advantages of metal foams. The advantages stem from their high solid-phase conductivities, very large surface area (up to 10,000 m^2/m^3) [3–5], and their good permeability, which is in the order of 10^{-8} m^2 [6–9]. Additionally, the foams' internal structure causes vigorous mixing, which augments convection heat transfer. Thermal applications of metal foams have been sought in compact heat exchangers [10–14], the thermal management of fuel cells [15,16], and high-power batteries [17]. Combining aluminum foam with phase change materials caused a 50% temperature drop and provided a uniform temperature of Li-Ion batteries [18]. The combination has been considered for the cooling of portable [19,20] and other electronics [21,22]. The use of metal foams is also envisioned in metal hydride reactors [23], catalytic reactors [24], turbojet engines [25], and geothermal power plants [26].

Studying the foam by direct numerical simulation is computationally demanding; it can require millions of computational elements. For example, 1.5 million elements were used to analyze ten foam cells [27]. The common analytical approach for studying transport in the foam is to average the microscopic governing equations over an elementary volume: the volume-average theory. The smallest representative elementary volume (REV) for metal foam is eight cells (two in each direction) [28]. The resulting macroscopic momentum and energy equations require four closure terms: permeability, inertial coefficient, thermal dispersion conductivity, and interstitial heat transfer coefficient. These closure terms need to be determined experimentally [29] or from correlations [30]. To solve the resulting averaged equations for a full REV, the number of numerical elements will be too large. For example, to analyze an REV of 10-ppi foam with a porosity of 93.2%, the number of elements would be in the order of 200 million [31]. The volume-average theory, conceived for tight porous media, has produced some unacceptable results when applied to highly-porous metal foam. For example, this theory predicts that the hydrodynamic entrance length in any porous medium is of order KV/ν, where K is the permeability, V is the Darcy velocity, and ν is the kinematic viscosity [32,33]. For packed 1-mm spheres, the porosity is 35% and the permeability is 10×10^{-10} m^2 [34], whereas for a metal foam with a porosity of 87.6%, the permeability is 5.3×10^{-8} m^2 [35]; thus, at the same velocity, the entry region in the foam is about two orders of magnitude longer compared to the spheres. Microscopic simulations [36] have shown that the hydrodynamic entry length for air flow in metal foam having a cell diameter of 5.1 mm is 30.5 mm, compared to 1.7 mm predicted by the volume-average theory (at the same Reynolds number). This clearly shows the disparity between the two approaches when it comes to hydrodynamic development. Sachan et al. [37] indicated that the volume-average theory may not be valid for metal foams.

7.6 cm

Figure 1. Photograph of a block of open-cell aluminum foam with 20 pores per inch. (Photo taken by author).

Another approach for capturing the complex structure of metal foam and similar highly-porous media is by using micro-computed tomography (μCT) [38–41]. Governing transport equations are then solved over the reconstructed foam structure. Tomography machines are very expensive. Moreover, μCT data require powerful computers in order to capture and reconstruct the 3D structure of the foam from 2D images, and then to solve the governing equations over this complex structure. Typically, researchers scanned small foam samples and investigated transport on these samples, e.g., 8 mm^3 [38], 9 mm^3 [13], 10 × 11.7 × 11.7 mm [28], which is inadequate for capturing entry/boundary effects, and thus the obtained results are only valid for fully-developed or periodic conditions. Even though larger samples can be scanned, the maximum sample size depends on the resolution of the µCT machines. Also, when using µCT, morphological and transport properties of a specific foam structure are obtained individually on a case-by-case basis. This prevents performing sizable systematic studies on multiple samples in order to yield general relations among various parameters.

In order to save costs and time, researchers have modeled the complex geometry of metal foam using 'idealized' cells that are intended to capture the relevant characteristics of the foam's structure [24,27,42]. The widely used cells [43] are the Kelvin tetradecahedron cell [44–47] and Kelvin-like cells [48]. The foam's structure has also been generated by Laguerre-Voronoi tessellations [49]. Kumar et al. [50] investigated the effect of strut's cross-sectional shape on the thermo-hydraulic properties of metal foams when they are represented by circular, square, diamond, hexagon, and star strut shapes. It was shown that the cross-section shape had an influence on the thermal and hydraulic phenomena in the foam.

In order to calculate the effective thermal conductivity, Bhattacharya et al. [51] represented the foam structure with two-dimensional hexagonal units with circular intersections. The same representation was used by [52], but the intersections were square. Hu et al. [53] represented the foam structure by tetrakaidecahedra to investigate the dehumidifying process of moist air in metal foam. The same structural unit was used by [54] to determine the surface area density of foams. Miwa et al. [55] utilized a geometrical cell that has exterior cubical skeleton with interior struts.

Issues with current geometrical modeling include the complexity of cell structures, the multi-steps needed to create the models, and the effort needed to duplicate cells for large numerical-solution domains. Because of the complex cell structures, solutions usually require considerable computational power and time. Certain models have limitations in terms of the foam than can be represented by them. For example, the body-centered- cube (BCC) model is limited to a foam porosity higher than 93% [55]. Other issues with the BCC model is the difficulty of adjusting the ligament thickness and the pore diameter independently [55]. Existing geometrical-modeling studies have typically assumed fully-developed conditions by using a single (periodic) cell or few cells in their simulations, e.g., in [40,41]; as such, critical boundary and entrance effects are automatically not investigated. Sachan et al. [37] has shown that these effects are considerable in foams.

Modeling efforts seem to be primarily focused on trying to match the geometrical shape of the internal structure of the foam [27,56–58], not on matching pertinent morphological properties (e.g., porosity and surface area). By doing so, they inadvertently compromise the critical matching of key transport properties, e.g., effective thermal conductivity and convection heat transfer coefficient. Thus, not surprisingly, some of the idealized structures poorly predict heat transfer and pressure drop as compared to actual structures of foam [59,60]. Actually, several foam's modeling studies have recorded significant differences between predictions of their models and experimental data, e.g., in [27,56,58].

It is evident that there is a need for robust modeling of highly-porous media such as metal foams, so that cost and time of investigating flow and heat transfer therein can be reduced. To that end, the modeling needs to be manageable in terms of geometric complexity. More importantly, any geometric model must possess the equivalent properties as the foam, such that pressure drop and heat transfer in the model match their counterparts in the actual foam. This kind of matching is far more important than matching geometrical shapes. The current paper is a first attempt at addressing some of these critical points. The novelty and advantages of the current model described in this paper—over existing models and other approaches for studying heat transfer in highly-porous media—can be summarized as follows. The current modeling effort is focused on matching pertinent heat transfer properties, e.g., effective thermal conductivity and convection heat transfer area, in order to better predict the combined conduction/convection heat transfer in metal foam. Unlike previous modeling, the current modeling employs a very simple geometry that is easy to duplicate, and saves computational time. The simple geometry is conducive to constructing a large computational domain quickly. The model presented here mitigates the need for expensive micro-computed tomography. In addition, the present model avoids the use of the volume-average theory, which seems to produce unacceptable results when applied to highly-porous media such as metal foam.

2. New Model Rationale and Development

When a highly-porous cellular material, e.g., open-cell metal foam, is attached to a heated solid surface and subjected to fluid flow, the thin ligaments of the foam conduct heat from the heated surface, which is subsequently convected to the through-moving fluid. The ligaments have small diameters compared to the cell diameter. For example, for a 97%-porous aluminum foam with a 6.90 mm cell diameter, the ligament diameter is 0.41 mm [27]. This foam can be represented by equivalent parallel strands (EPS) made from the same material as the foam, and attached perpendicularly to the same solid surface. This modeling pertains strictly to two-dimensional heat transfer, e.g., a metal-foam heat sink and metal foam sandwiched between two parallel plates with one or both plates heated. To ensure that the EPS and the foam produce the same heat transfer and pressure drop, the geometrical parameters for the EPS will be strategically derived from the actual morphology of foam, as explained below.

In one manifestation of the EPS, the strands can be thin circular cylinders arranged in a staggered fashion. For this case, the foam is actually represented by thin cylinders of equal diameter, as in tube banks or pin fins. The geometry of the EPS is not arbitrary, but is governed by the morphology of the foam. The geometry of an EPS model is fully described by the cylinder's diameter as well as the longitudinal, transverse, and diagonal pitches. Figure 2 shows a top view of an EPS, identifying the width W, length L, and height H (not shown). Assume that the EPS is attached at the bottom to a solid wall to apply heat, as in the case of a heat sink. The geometric parameters of the EPS can be determined using modeling relations (conceived for the first time). To ensure that the EPS and the foam are equivalent in producing the same transport effects, conditions are imposed on the modelling *a priori*. The equivalency conditions will yield equations that will be eventually solved to yield the geometrical parameters of the EPS model, as expounded next.

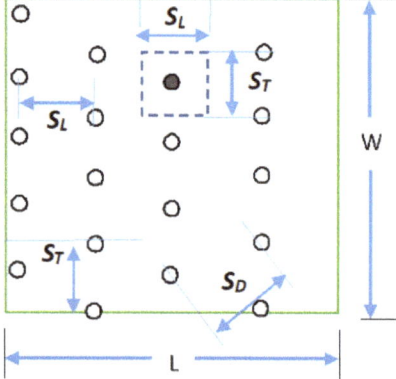

Figure 2. Top view of the proposed EPS model showing its length L and width W. The longitudinal, transverse, and diagonal pitches for the equilateral-triangle arrangement is also shown. The unit for all of these parameters is mm.

- **Conduction Equivalency Condition and First Modelling Relation:**

The conduction in the solid phase of the foam is governed by the effective thermal conductivity, which is primarily a function of porosity [48,60,61]. The gas flow in the metal foam dispersion is negligible [62]. The porosity of the foam ε_f is the ratio of the void fraction to the total volume of the foam. The first equivalency condition is that the porosities of EPS and foam must be equal:

$$\epsilon_f = \varepsilon_m \tag{1}$$

For isotropic foam, the contact area between the foam and the wall is governed by the solid fraction of foam, which is $1 - \varepsilon_f$. Unlike packed spheres, the porosity of foams is constant even close to solid boundaries [57]. The percentage of the contact area %A_{con} (the area where the cylinders contact the wall) to the total wall's area is given by %$A_{con} = 1 - \varepsilon_f$. To take this further and illustrate the development of the EPS model, the cylinders' arrangement of Figure 2 is considered. This is a staggered arrangement with a longitudinal pitch S_L, transverse pitch S_T, and diagonal pitch S_D. One can define a repeating unit: an area having a length S_L and width S_T surrounding a typical EPS cylinder (the dotted small rectangle in Figure 2). If the diameter of a typical cylinder is D, then the contact area is given by %$A_{con} = (\pi D^2/4)/S_T S_L$. Substituting for %$A_{con}$ in the above equation, the following is obtained:

$$1 - \varepsilon_f = \pi D^2/4/S_T S_L \tag{2}$$

This is the first EPS modelling relation. It relates the longitudinal pitch S_L and transverse pitch S_T of the EPS model to the porosity of the foam.

- **Convection Equivalency Condition and Second Modelling Relation:**

The internal surface area of the foam controls the convection in such materials to a large extent. This is, of course, in addition to the heat transfer coefficient, geometry, and temperature difference between the solid and fluid inside the foam. In an attempt to have the EPS model produce the same heat transfer as the foam, the second equivalency condition imposed is that the surface area of the foam and that of its EPS model must be equal:

$$\sigma_f = \sigma_m \tag{3}$$

The surface area of metal foams is often given in the literature [57,58], or can be calculated [48,63,64]. Sometimes it is available by manufacturers, e.g., in [65]. Based on Figure 2, the surface area (per unit volume) of the EPS is given by $\sigma_m = n(\pi DH)/LWH = n(\pi D)/LW$, where n is the number of cylinders in a volume LWH. Based on a single cylinder, the EPS is given by $\sigma_m = (\pi DH)/S_T S_L H = (\pi D)/S_T S_L$. Substituting in the surface area equivalency condition in the second EPS modelling relation is obtained as:

$$\sigma_f = (\pi D)/S_T S_L \tag{4}$$

This is the second modeling relation. It relates the longitudinal pitch S_L and transverse pitch S_T of the EPS model to the surface area density of the foam. Equations (2) and (4) have three unknowns: D, S_T, and S_L.

- **Staggering Option and Closing Set of Modelling Relations:**

The remaining set of modelling relations depends on the choice of strands' staggering arrangement. For example, for the equilateral-triangle arrangement shown in Figure 3, there is a relationship between the three pitches; namely $S_D = S_T = \left(2/\sqrt{3}\right) S_L$. Once this is substituted in Equations (2) and (4), only two unknowns D and S_D remain, and the resulting two equations are solved simultaneously to yield Equations (5)–(8):

$$D = 4(1-\varepsilon_f)/\sigma_f \tag{5}$$

$$S_D = \frac{2}{\sigma_f}\sqrt{\frac{2\pi(1-\varepsilon_f)}{\sqrt{3}}} \tag{6}$$

$$S_L = \frac{1}{\sigma_f}\sqrt{2\sqrt{3}\pi(1-\varepsilon_f)} \tag{7}$$

$$S_T = \frac{2}{\sigma_f}\sqrt{\frac{2\pi(1-\varepsilon_f)}{\sqrt{3}}} \tag{8}$$

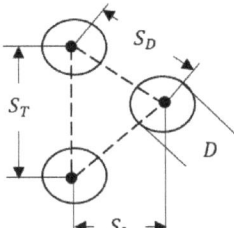

Figure 3. The equilateral-triangle arrangement. The longitudinal, transverse, and diagonal pitches, as well as the diameter, are measured in mm.

Note that in these last four equations, D, S_D, S_L, and S_T (the full description of the EPS) are completely expressed in terms of the foam's porosity ε_f and surface area density σ_f. The following procedure illustrates how the results of this section can be applied to study heat transfer and pressure drop in metal foam:

(a) Start with metal foam with known porosity and surface area density, in which heat transfer and pressure drop are of interest;
(b) use these two known morphological properties of the foam, and find the geometrical properties of the tube bank (EPS) that would produce the same heat transfer and pressure drop for this foam. The geometrical properties of the tube bank are obtained using Equations (5)–(9);
(c) construct this tube bank geometry (EPS) in a numerical package, e.g., ANSYS;
(d) investigate (solve the momentum and energy governing equations) over this tube bank in the numerical package. The resulting heat transfer and pressure drop of this step will be the same as those of the foam.

These steps, and the way the current modeling is employed, are illustrated by the following example.

3. Example and Numerical Solution on EPS

Consider a block of commercial open-cell aluminum foam having 20 pores per inch (ppi). Let the dimensions of the cross-section of the foam be W = 10.16 cm (4 in) and H = 10.16 cm (4 in). The length of the block in the flow direction is L = 5.08 cm (2 in). The block is brazed to a 3-mm thick solid aluminum base. The aluminum block acts like a heat sink: it is heated from the bottom with a constant heat flux of 29,900 W/m^2 and is cooled by the air. The cross-sectional area is open to the flow, while all other sides are insulated. The porosity is 78.2%, which can be determined by comparing the weight of the foam to the weight of an equivalent volume of solid aluminum alloy from which the foam is made. The surface area density for this 20 ppi foam is available directly from the manufacturer [65]:

$$\sigma_f = 442.2 \ln(1-\varepsilon_f) + 2378.6 \; (\text{m}^2/\text{m}^3) \tag{9}$$

It should be noted that the surface area depends on the ppi, so the surface area and the porosity are independent. For this foam block, knowledge of the porosity and surface area density, along with a choice of staggering arrangement, are sufficient to determine an EPS. Employing an equilateral-triangle arrangement, the EPS representing this foam block is obtained using Equations (5)–(8), and is shown in Figure 4. The diameter of each cylindrical strand in the EPS is 0.41 mm, and diagonal S_D, longitudinal S_L, and transverse S_T pitches are 1.37, 1.19, and 1.37 mm, respectively.

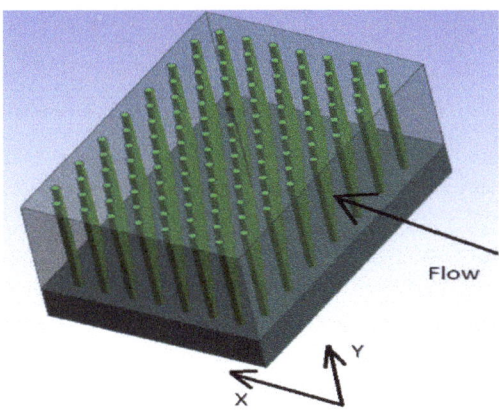

Figure 4. An equivalent-parallel-strands model for a metal foam block heated from the bottom. The x- and y-coordinates are measured in mm.

The EPS was imported into a numerical package, i.e., ANSYS, and the microscopic governing equations for energy and momentum were solved. To save time, and due to symmetry, only a thin representative slice with a width of 1.37 mm, a height of 50 mm, and the full length in the flow direction of 50.8 mm was used as the solution domain. This segment includes 21 EPS cylinders, and is shown after meshing in Figure 5.

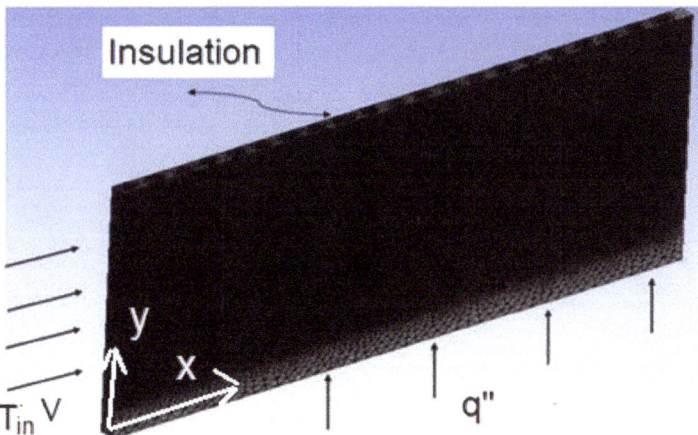

Figure 5. A representative section of the EPS model as a solution domain with meshing. The x- and y-coordinates are measured in mm, while the velocity V is measured in m/s, the inlet temperature T_{in} in K, and the heat flux q'' in W/m².

The CFD solver within ANSYS was used to solved the governing equations for the conservation of mass, momentum, and energy in two dimensions. The velocity components were u and v in the x- and y-directions, respectively. The flow was assumed laminar and incompressible with constant properties. The governing equations [27,39] are presented below in the form of the dependent variables u, v, P, and T:

Continuity:
$$\frac{\partial u}{\partial x} + \frac{\partial v}{\partial y} = 0 \tag{10}$$

Momentum:

$$\text{x} - \text{direction}: \rho\left[u\frac{\partial u}{\partial x} + v\frac{\partial u}{\partial y}\right] = -\frac{\partial p}{\partial x} + \mu\left[\frac{\partial^2 u}{\partial x^2} + \frac{\partial^2 u}{\partial y^2}\right] \tag{11}$$

$$\text{y} - \text{direction}: \rho\left[u\frac{\partial v}{\partial x} + v\frac{\partial v}{\partial y}\right] = -\frac{\partial p}{\partial y} + \mu\left[\frac{\partial^2 v}{\partial x^2} + \frac{\partial^2 v}{\partial y^2}\right] \tag{12}$$

$$\text{Energy equation}: \rho c\left[u\frac{\partial T}{\partial x} + v\frac{\partial T}{\partial y}\right] = k\left[\frac{\partial^2 T}{\partial x^2} + \frac{\partial^2 T}{\partial y^2}\right] \tag{13}$$

where k and c are thermal conductivity and specific heat of the fluid, respectively.

The following boundary conditions were imposed:

$$\text{At y} = 0, \; k\frac{\partial T}{\partial y} = q'', \; u = v = 0 \tag{14}$$

$$\text{At y} = H, \; \frac{\partial T}{\partial y} = 0, \; u = v = 0 \tag{15}$$

$$\text{At x} = 0, \; T = T_\infty, \; u = u_o, \; v = 0 \tag{16}$$

A mesh independence study was conducted in order to establish both the accuracy and independence of the results from the mesh employed. It was also conducted to save time and computational cost. Three types of meshing were used: coarse, medium, and fine. The number of nodes and the simulation time for three meshes are listed in Table 1. There was a minor change in the results (0.17%) between the medium and fine meshes, and the medium mesh was used in all investigations. The number of computation cells was 1,072,143. A laptop with a processor with four cores and eight logical processors at 1.6 GHz and 8 GB of RAM was used. A typical run took 7 h and 15 min.

4. Numerical Predictions of the EPS Model

For an inlet air temperature T_∞ of 300 K and a velocity u_o of 2.5 m/s, Figure 6 shows the air temperature as a function of y (distance from the heated base) at three different axial locations measured form the inlet X = 6.53 mm, 19.05 mm, and 31.75 mm. The same information is given for velocity 2.71 m/s in Figure 7. For these two velocities, the Reynolds numbers based on the permeability of the foam are 37.4 and 40.6, respectively, which indicates that the flow regime is laminar [35,66].

It is clear that the temperature behavior at the qualitative level makes sense and is expected. The temperature decreases as the distance from the heated base increases. It is at its maximum at the heated base, as indicated by the red color in Figures 6 and 7. At a sufficiently far location from the heated base, the air temperature is cooler, and it is equal to the inlet air temperature shown in blue. In addition, the temperature increases in the flow direction due to continuous heating of the cooling air as it travels through the heat sink. As the velocity increases from 2.5 to 2.71 m/s, the cooling of the EPS is more efficient because of higher convection rates at the higher velocity. Also, the colored region (outside the blue region) is shorter, and heat transfer is confined to a smaller region closer to the heated base. While these trends are encouraging, a direct quantitative comparison to experimental heat transfer data in metal foam is needed in order to establish the validity of the modeling approach, which will be shown below.

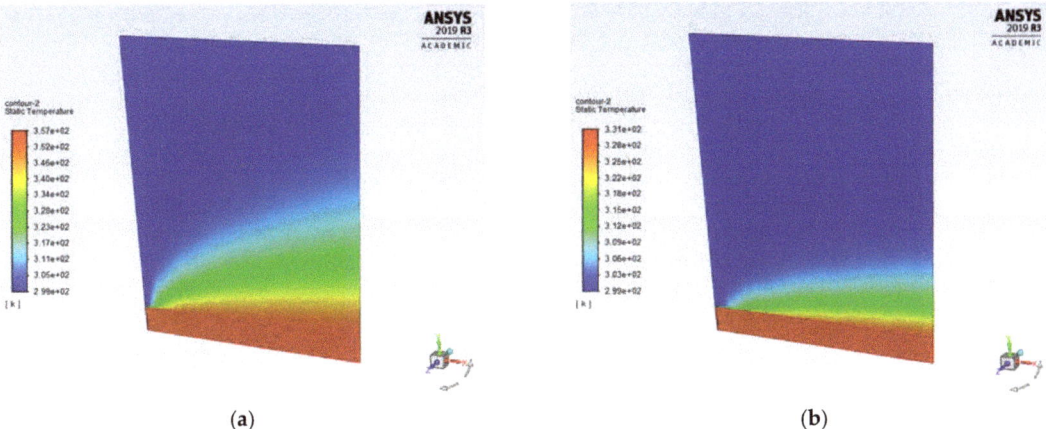

Figure 6. Temperature distribution for: (**a**) u_o = 2.50 m/s and (**b**) u_o = 2.71 m/s. Temperatures are given in K.

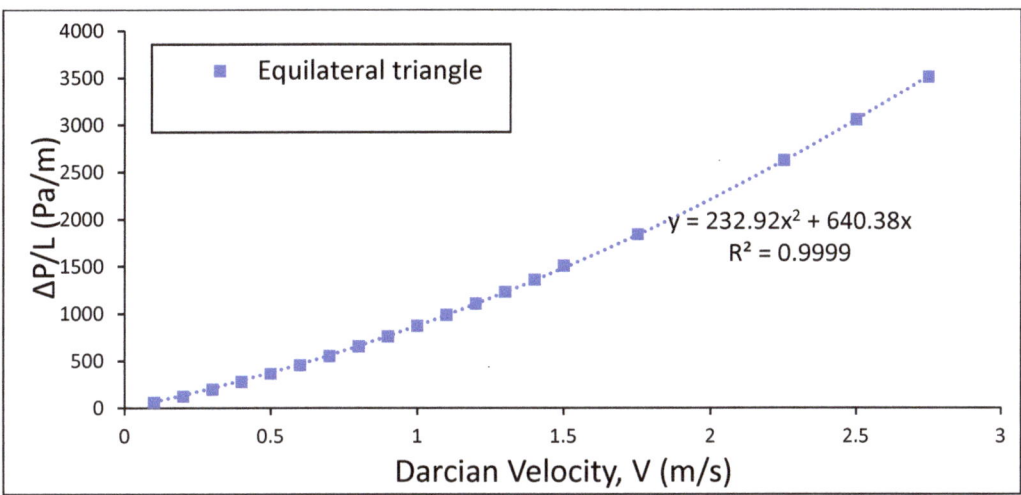

Figure 7. Pressure drop per unit length in flow direction as predicted by the EPS model with an equilateral-triangle arrangement.

One key parameter for investigating engineering designs employing forced convection heat transfer in metal foam is the associated pressure drop. Figure 7 is a plot of the pressure drop predicted by the EPS per unit length (in the flow direction) as a function of average flow velocity. It is clear that the EPS model produces pressure drop data that strongly follow the Forchheimer equation for porous media, including metal foam [8]:

$$\frac{\Delta P}{L} = \frac{\mu}{K} V + \rho C V^2 \qquad (17)$$

In other words, the functional relationship between the pressure drop and the average velocity is quadratic, which is the same behavior found in metal foam. This is encouraging and suggests the possibility that the pressure drop in the EPS model can, albeit with some adjustment, be made to match the pressure drop in metal foam. The adjustment incudes changing the staggering arrangement, while keeping the porosity and surface area density of the EPS model equal to their counterparts in the foam. Nonetheless, experimental pressure drop data is needed in order to directly compare pressure drop and further validate the model.

5. Experimental Validation

Dukhan and Chen [67] measured steady-state local temperatures in a metal foam block subjected to a constant heat flux with air flowing through it to remove the heat. The current EPS modeling is used to model one of the foam blocks of [67], and then the predictions of the model are compared to the experimental temperatures obtained by [67]. The dimensions of the cross section of the foam block of [67] were 10.16 cm (4 in) by 10.16 cm (4 in), and its length in the flow direction was 5.08 cm (2 in). The block was brazed to a 12.7 mm (0.5 in) thick solid aluminum base. The temperature measurements in the foam were performed at the axial locations (flow direction) X = 2.54, 3.81, and 4.44 cm, and at y-locations 1.27, 2.54, 3.81, 6.35, and 8.89 cm (perpendicular to the flow direction), as shown in Figure 8. The problem was assumed two-dimensional by [67], and all temperature measurements were performed in the z = 5.08 cm plane, where z is the coordinate perpendicular to the page (not shown in Figure 8).

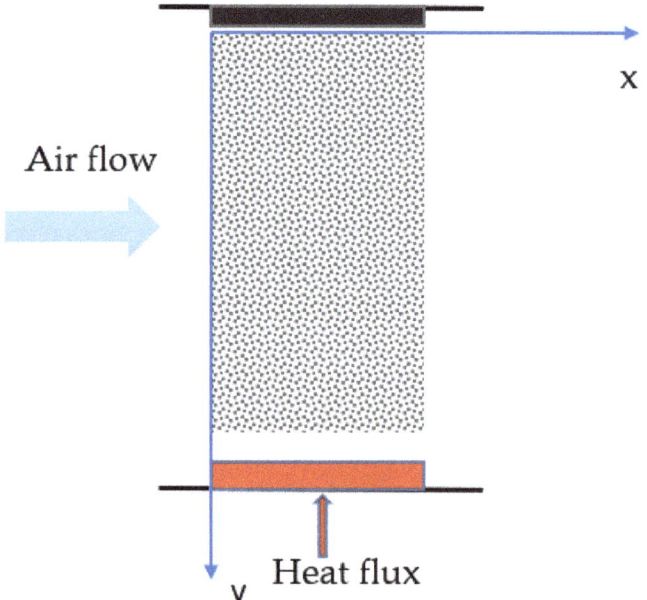

Figure 8. Experimental arrangement showing locations of the local temperature measurement and the coordinate system of [67]. The x- and y-coordinates are measured in mm.

The foam block of [67] had the morphological properties shown in Table 1.

Table 1. Properties of one sample of open-cell aluminum foam used by [67].

Pore Density, ppi	Porosit, ε (%)	Surface Area Density, σ (m^2/m^3)
20	78.2	1705

Since the porosity (78.2%) and the pore density (20 ppi) of this foam are known, an EPS model for this foam can be constructed. The porosity is used to calculate the surface area density according to Equation (9), which is specific to ERG aluminum foam with 20 ppi only. Using the modeling relations Equations (5)–(8), the EPS model representing this particular foam was obtained, for which the geometrical properties are shown in Table 2. The EPS model was constructed in ANSYS, and CFD analysis was performed on it according to Equations (10)–(16). The same average air velocity and inlet temperature was imposed on the EPS as those used by [67] in their experiment. The boundary conditions were also the same for the EPS and the actual foam of [67].

Table 2. Equilateral-triangle equivalent parallel strands. Parameters for the foam sample of Table 1.

D (mm)	S_D (mm)	S_L (mm)	S_T (mm)
0.51	1.04	0.90	1.04

The EPS models' predicted temperatures are compared to their experimental counterparts of [67] in the following figures. Figure 9 represents an average air speed of 2.5 m/s. In this figure, parts (a), (b), and (c) are for axial locations 6.35 cm, 19.05 cm, and 31.75 cm, respectively. At these three axial locations, the experimental and EPS-predicted temperatures are plotted as a function of y (the distance from the heated base). The two temperatures agree very well, and they exhibit the same trend: the temperature increases as the distance from the heated base decreases (note the coordinate system in Figure 8). This is true for all axial locations; however, there are two observations. The first is that the agreement is best at intermediate y locations, and is poorer at the heated base (y = 44.45 cm) and far away from the heated base (y = 6.35 cm). This may be caused by differences in the flow fields of the EPS model and the actual foam close to the solid heated boundary, which may alter convection heat transfer in that region. The other observation is that the agreement gets poorer as the distance from the inlet increases, meaning that the agreement is best close to the inlet (at x = 6.35 cm, Figure 9a), and is worse far away from the inlet (at x = 31.75, Figure 9c). The reason for this may be the existence of an exit region in metal foam. This exit region may be different in the case of the EPS model as compared to the foam due to geometric differences. Nonetheless, it can be stated that the agreement is generally very good, since the average error between the predicted and experimental temperature is rather small, as shown in Table 3. The maximum error is 9.46%.

Table 3. Percentage error between the EPS models' predicted and experimental temperatures at an air speed of 2.50 m/s.

Axial Location, x (cm)	Average Error (%)	Maximum Error (%)
6.35	2.64	4.05
19.05	2.81	5.33
31.75	3.28	9.46

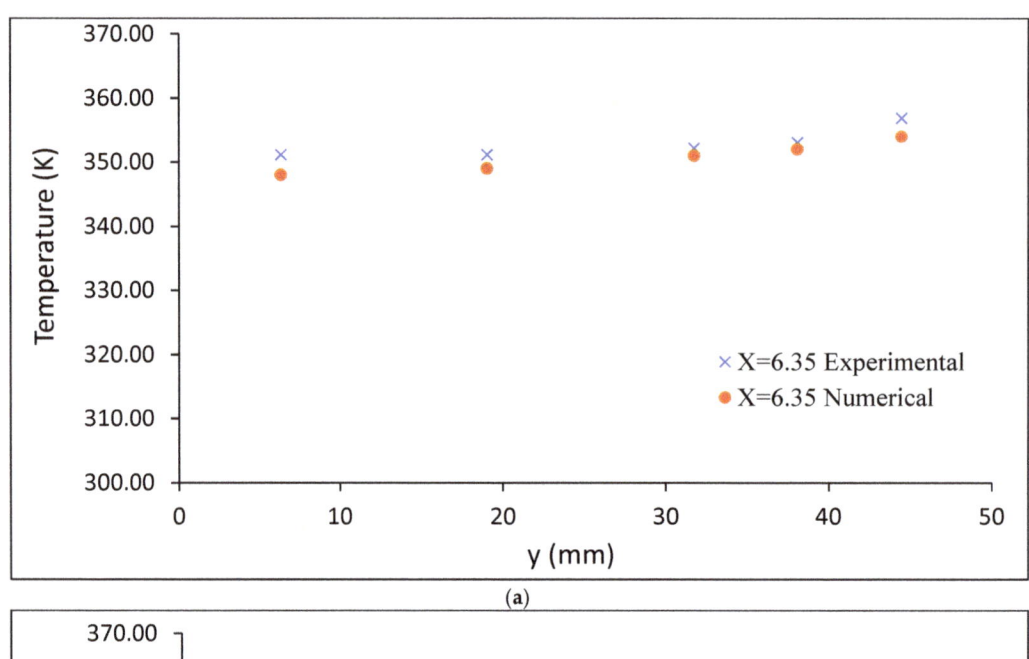

(a)

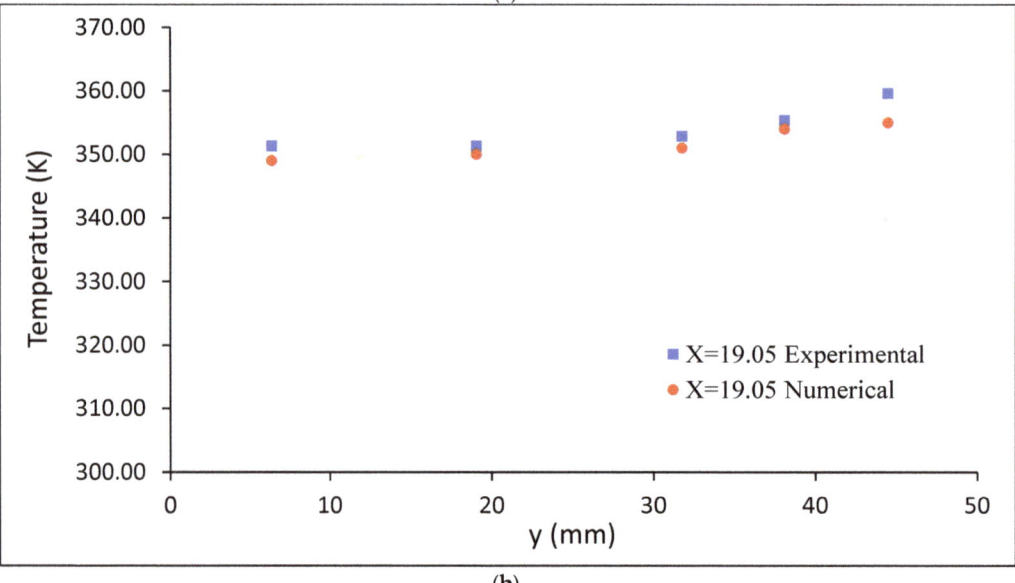

(b)

Figure 9. *Cont.*

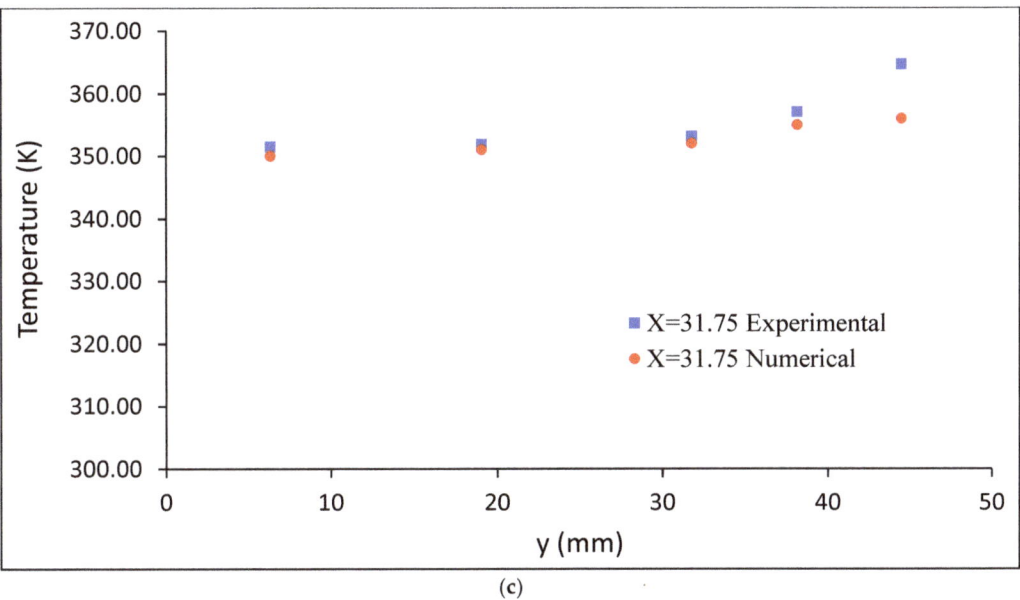

(c)

Figure 9. Comparison between EPS models' prediction and the experimental data of [67] at an average air speed of 2.50 m/s and at the axial location form the inlet: (**a**) 6.35 cm, (**b**) 19.05 cm, (**c**) 31.75 cm.

Figure 10 has a comparison between the EPS model and the actual foam of [67] at an average air speed of 2.71 m/s. Similar observations that were made for the lower velocity of 2.50 m/s can be made for this higher flow velocity. For the two velocities and at all locations, the EPS model is seen to always predict lower temperatures compared to the experimental temperatures of [67]. At this higher velocity, the difference between the models' predictions and the experimental temperatures is higher, as indicated in Table 4. This is due to the fact that geometric differences between the model and the foam are expected to play a stronger role at a higher velocity. The geometry has a stronger influence on the flow field at higher velocities, and thus on convection heat transfer. Another issue is that the parallel strands of the EPS are not cross-connected, while the ligaments of the foam connect randomly in three dimensions. This contributes to a more efficient heat transfer in the case of the foam.

Table 4. Percentage error between the EPS models' predicted and experimental temperatures at an air speed of 2.71 m/s.

Axial Location, x (cm)	Average Error (%)	Maximum Error (%)
6.35	3.05	3.85
19.05	4.05	8.38
31.75	3.55	10.67

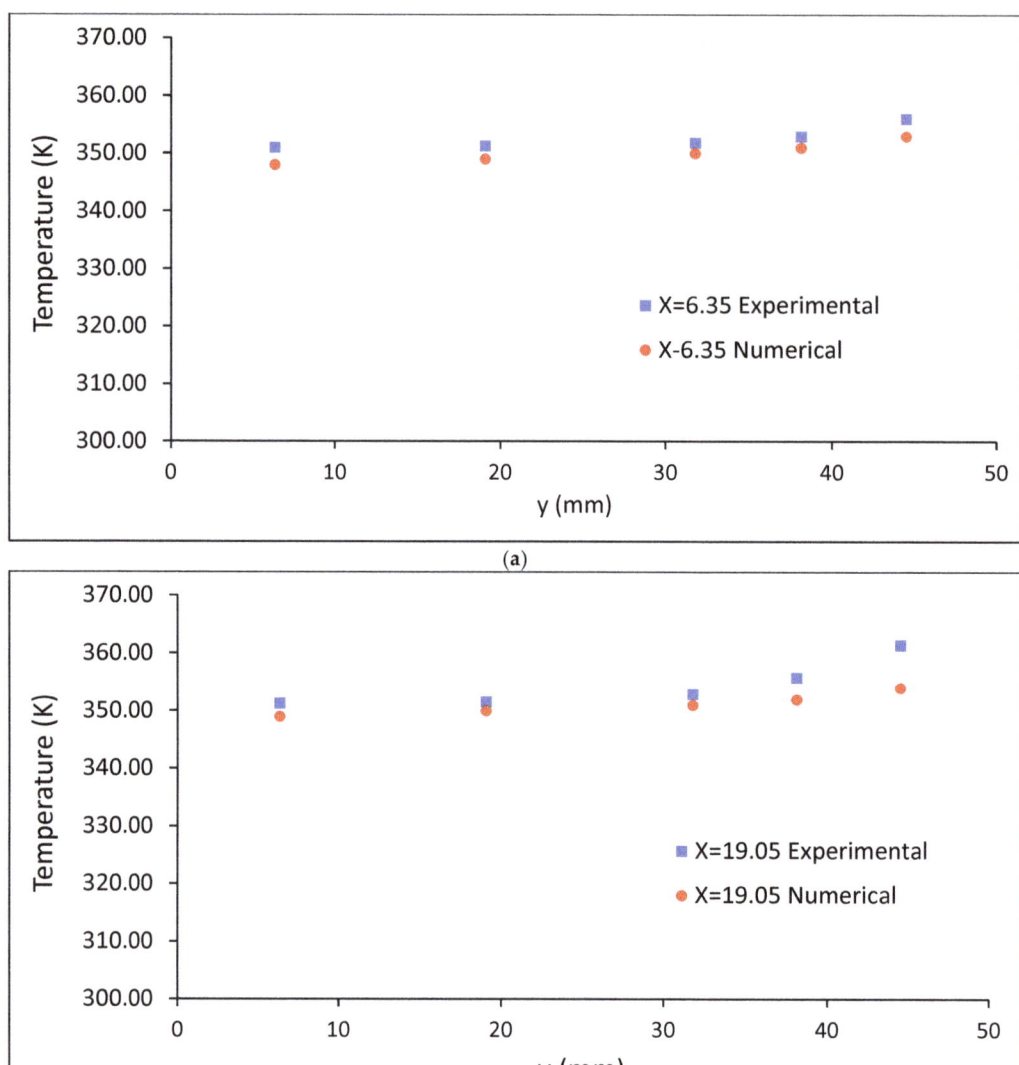

Figure 10. Cont.

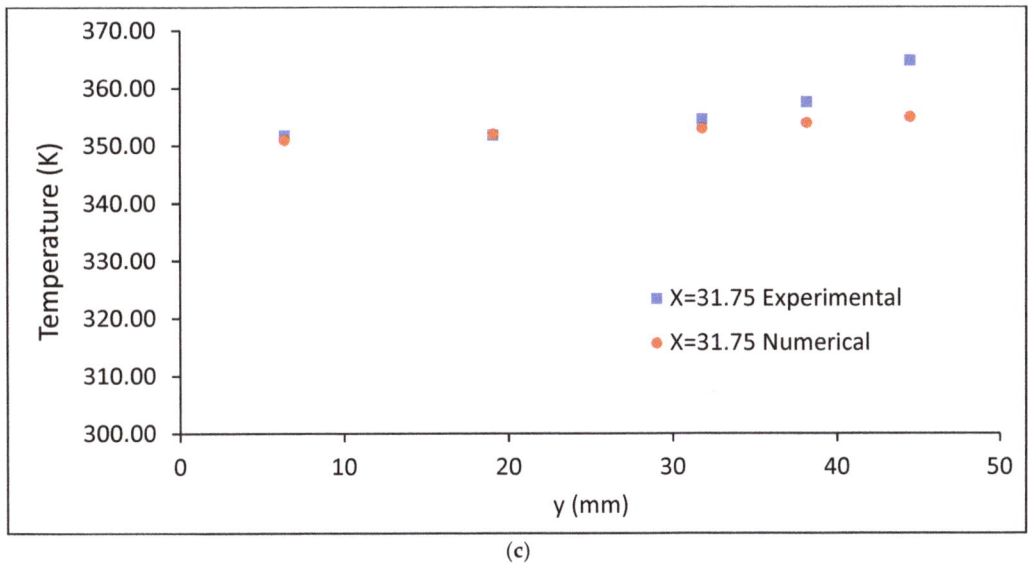

(c)

Figure 10. Comparison between EPS models' prediction and the experimental data of [67] at an average air speed of 2.71 m/s, at the axial location form the inlet: (**a**) 6.35 cm, (**b**) 19.05 cm, (**c**) 31.75 cm.

The solution domain employed in this study was 1.37 mm × 50 mm × 50.8 mm, which represented 1574 foam cells. For this large domain, only 1,072,143 computational elements were needed, and the obtained numerical results were in very good agreement with their experimental counterparts. This represents considerable savings in the computational time and power needed. For example, 188,885 computational cells were needed to investigate one body-centered-cube model of a single foam cell [57], while for 20 body-centered-cube cells representing metal foam, a total 918,016 computational elements we needed in the numerical solution of [36]. As for solving the governing equations over a domain made of 20 Kelvin cells representing metal foam, three million numerical cells were needed [56], and for the Laguerre-Voronoi structure representing 10 foam cells, eight million computational cells were needed [49]. In the case of solving a domain of 2 mm × 2 mm × 10 mm of foam captured by μCT, eight million computational cells were required [51]. These comparisons show the efficiency of the current EPS modeling.

6. Conclusions

A new modelling technique for highly-porous metal foam and similar porous media has been presented. The modelling is valid strictly for macroscopically two-dimensional heat transfer due to the fluid flow in highly-porous media; for example, metal-foam heat sinks and metal foam packed between two parallel plates, with one or both plates heated. The modeling is designed to produce a simpler geometry than the foam's complex structure, and thus save time and computational costs. The new paradigm stresses that, instead of focusing on geometrical shape similarity, the modeling should strive to match key morphological and transport properties of the foam, so that pressure drop and heat transfer in the model are as close as possible to those of the foam. The modeling technique was illustrated by an example, and partially validated by comparison to experimental local temperatures in actual foam from the literature. Good agreement was obtained for all temperatures. The case considered in the comparison was for cylindrical strands in an equilateral-triangle staggered arrangement, and one metal foam with one pore density was considered. Further verification and experimental data are needed to establish the robustness of the modelling and to identify its limitations. The pressure drop in the equivalent parallel strands model obeyed the Forchheimer equation, suggesting that the strands followed the Forchheimer equation for porous media just like open-cell metal foam. However, pressure-drop predictions of the model need to be directly compared to actual pressure-drop data from experiments on actual foam. Such a comparison may lead to ways of improving the model and/or calibrating it.

Funding: This research received no external funding.

Institutional Review Board Statement: Not applicable.

Informed Consent Statement: Not applicable.

Data Availability Statement: Not applicable.

Conflicts of Interest: Authors declare no conflict of interest.

Nomenclature

A_{con}	contact area
c	specific heat of fluid (kJ/kg·K)
C	form drag coefficient (m^{-1})
D	diameter (mm)
EPS	equivalent parallel strands
H	height of model or foam (cm)
k	thermal conductivity (W/m K)
K	permeability (m^2)
L	length of model or foam in flow direction (cm)
n	the number of cylinders in EPS
ppi	pore per inch
q''	heat flux (W/m^2)
S_D	diagonal pitch (mm)
S_L	longitudinal pitch (mm)
S_T	transverse pitch (mm)
T	temperature (K)
u	velocity in x-direction
V	Darcian velocity (m/s)
v	velocity in y-direction
W	width of model or foam (cm)
x	axial coordinate along the flow direction
y	coordinate perpendicular to flow direction
Δp	pressure drop (Pa)
ρ	density of fluid (kg/m^3)
ε	porosity
σ	surface area density (m^2/m^3)
μ	viscosity (kg/m s)
f	foam
m	model

References

1. Ranut, P.; Nobile, E.; Mancini, L. High resolution X-ray microtomography-based CFD simulation for the characterization of flow permeability and effective thermal conductivity of aluminum metal foams. *Exp. Therm. Fluid Sci.* **2015**, *67*, 30–36. [CrossRef]
2. Magnico, P. Analysis of permeability and effective viscosity by CFD on isotropic and anisotropic metallic foams. *Chem. Eng. Sci.* **2009**, *64*, 3564–3575. [CrossRef]
3. Ashby, F. *Metal Foam, A Design Guide*; Butterworth-Heinemann: Oxford, UK, 2000.
4. Leong, K.C.; Jin, L.W. Characteristics of oscillating flow through a channel filled with open-cell metal foam. *Int. J. Heat Fluid Flow* **2006**, *27*, 144–153. [CrossRef]
5. Inayat, A.; Schwerdtfeger, J.; Freund, H.; Körner, C.; Singer, R.F.; Schwieger, W. Periodic open-cell foams: Pressure drop measurements and modeling of an ideal tetrakaidecahedra packing. *Chem. Eng. Sci.* **2011**, *66*, 2758–2763. [CrossRef]
6. Antohe, B.; Lage, J.L.; Price, D.C.; Weber, R.M. Experimental determination of the permeability and inertial coefficients of mechanically compressed aluminum metal layers. *J. Fluids Eng.* **1997**, *11*, 404–412. [CrossRef]

7. Bonnet, J.-P.; Topin, F.; Tadrist, L. Flow laws in metal foams: Compressibility and pore size effect. *Transp. Porous Media* **2008**, *73*, 233–254. [CrossRef]
8. Mancin, S.; Zilio, C.; Cavallini, A.; Rossetto, L. Pressure drop during air flow in aluminum foam. *Int. J. Heat Mass Transf.* **2010**, *53*, 3121–3130. [CrossRef]
9. Depois, J.-F.; Mortensen, A. Permeability of open-pore microcellular materials. *Acta Mater.* **2005**, *53*, 1381–1388. [CrossRef]
10. Boomsma, K.; Poulikakos, D.; Zwick, F. Metal foams as compact high performance heat exchangers. *Mech. Mater.* **2003**, *35*, 1161–1176. [CrossRef]
11. Tadrist, L.; Miscevic, M.; Rahli, O.; Topin, F. About the use of fibrous materials in compact heat exchangers. *Exp. Therm. Fluid Sci.* **2004**, *28*, 193–196. [CrossRef]
12. Chumpia, A.; Hooman, K. Performance evaluation of single tubular aluminum foam heat exchanges. *Appl. Therm. Eng.* **2014**, *66*, 266–273. [CrossRef]
13. Tamayol, A.; Hooman, K. Thermal assessment of forced convection through metal foam heat exchangers. *J. Heat Transf.* **2011**, *133*, 111801-1–111801-7. [CrossRef]
14. Mahjoob, S.; Vafai, K. A Synthesis of fluid and thermal transport models for metal foam heat exchangers. *Int. J. Heat Mass Transf.* **2008**, *51*, 3701–3711. [CrossRef]
15. Odabaee, M.; Mancin, S.; Hooman, K. Metal foam heat exchangers for thermal management of fuel cell systems- an experimental study. *Exp. Therm. Fluid Sci.* **2013**, *51*, 214–219. [CrossRef]
16. Hossain, M.S.; Shabani, B. Metal foams application to enhance cooling for open cathode polymer electrolyte membrane fuel cells. *J. Power Sources* **2015**, *295*, 275–291. [CrossRef]
17. Sullines, D.; Daryabeige, K. Effective thermal conductivity of high porosity open cell nickel foam. In Proceedings of the 35th AIAA Thermophysics Conference, Anaheim, CA, USA, 11–14 June 2001; Volume 12, p. 2819.
18. Khatee, S.A.; Amiuddin, S.; Farid, M.; Selman, J.R.; Al-Hallaj, S. Thermal management of Li-ion battery with phase change material for electric scooters: Experimental validation. *J. Power Sources* **2005**, *142*, 345–353. [CrossRef]
19. Alawadhi, E.M.; Amon, C.H. PCM thermal control unit for portable electronic devices: Experimental and numerical studies. *IEEE Trans. Compon. Packag. Technol.* **2003**, *26*, 116–125. [CrossRef]
20. Vesligaj, M.J.; Amon, C.H. Transient thermal management of temperature fluctuations during time varying workloads on portable electronic. *IEEE Trans. Compon. Packag. Technol.* **1999**, *22*, 541–550. [CrossRef]
21. Bhattacharya, A.; Mahajan, R.L. Finned metal foam heat sinks for electronics cooling in forced convection. *J. Electron. Packag.* **2002**, *124*, 155–163. [CrossRef]
22. Zhang, Y.C.L.W.; Xie, H.; Mahajan, R.L. Cooling of a FCHIP pakage with 100 W, 1 cm^2 chip. In Proceedings of the 1993 ASME International Electronic Packaging Conference, San Diego, CA, USA, 12–15 September 1993; Volume 1, pp. 419–423.
23. Laurencelle, F.; Goyette, J. Simulation of heat transfer in a metal hydride reactor with aluminum foam. *Int. J. Hydrogen Energy* **2007**, *32*, 2957–2964. [CrossRef]
24. Horneber, T.; Rauh, C.; Delgado, A. Fluid dynamic characterization of porous solid in catalytic fixed-bed reactors. *Microporous Mesoporous Mater.* **2012**, *154*, 170–174. [CrossRef]
25. Azzi, W.; Roberts, W.L.; Rabiei, A. A study on pressure drop and heat transfer in open cell metal foam for jet engine applications. *Mater. Des.* **2007**, *28*, 569–574. [CrossRef]
26. Ejlali, A.; Ejlali, A.; Hooman, K.; Gurgenci, H. Application of high porosity metal foams as air-cooled heat exchangers to high heat load removal systems. *Int. Commun. Heat Mass Transf.* **2009**, *36*, 674–679. [CrossRef]
27. Kopanidis, A.; Theodorakakos, E.; Gavaises, D. Bouris, 3D numerical simulation of flow and conjugate heat transfer through a pore scale model of high porosity open cell metal foam. *Int. J. Heat Mass Transf.* **2010**, *53*, 2539–2550. [CrossRef]
28. Brun, E. De l'imagérie 3D Des Structures a L'étude Des Mécanismes De Transport En Milieux Cellulaires. Ph.D. Thesis, Aix Marseille Université, Marseille, France, 2009.
29. Taheri, M. Analytical and Numerical Modeling of Fluid Flow and Heat Transfer Through Open-Cell Metal Foam Heat Exchangers. Ph.D. Thesis, University of Toronto, Toronto, ON, Canada, 2015.
30. Feng, S.; Shi, M.; Li, Y.; Lu, T.J. Pore-scale and volume-averaged numerical simulations of melting phase change heat transfer in finned metal foam. *Int. J. Heat Mass Transf.* **2015**, *90*, 838–847. [CrossRef]
31. De Schampheleire, S.; De Jaeger, P.; De Kerpel, K.; Ameel, B.; Huisseune, H.; De Paepe, M. How to study thermal applications of open-cell metal foam: Experiments and computational fluid dynamics. *Materials* **2016**, *9*, 94. [CrossRef]
32. Vafai, K.; Tien, C.L. Boundary and inertia effects of flow and heat transfer in porous media. *Int. J. Heat Mass Transf.* **1981**, *34*, 195–203. [CrossRef]
33. Nield, D.; Bejan, A. *Convection in Porous Media*, 2nd ed.; Springer: New York, NY, USA, 1999; p. 73.
34. Bağcı, Ö.; Dukhan, N.; Özdemir, M. Flow regimes in packed beds of spheres from pre-Darcy to turbulent. *Transp. Porous Media* **2014**, *104*, 501–520. [CrossRef]
35. Dukhan, N.; Bağcı, Ö.; Özdemir, M. Metal foam hydrodynamics: Flow regimes from pre-Darcy to turbulent. *Int. J. Heat Mass Transf.* **2014**, *77*, 114–123. [CrossRef]
36. Dukhan, N.; Suleiman, A.S. Simulation of entry-region flow in open-cell metal foam and experimental validation. *Transp. Porous Media* **2014**, *101*, 229–246. [CrossRef]

37. Sachan, S.; Kumar, S.; Krishnan, S.; Ramamoorthy, S. Impact of entry–exit loss on the measurement of flow resistivity of porous materials. *AIP Adv.* **2020**, *10*, 105031. [CrossRef]
38. Otaru, A.J.; Morvan, H.P.; Kennedy, A.R. Measurement and simulation of pressure drop across replicated porous aluminium in the Darcy-Forchheimer regime. *Acta Mater.* **2018**, *149*, 265–273. [CrossRef]
39. Tabor, G.; Yeo, O.; Young, P. CFD simulation of flow through an open cell foam. *Int. J. Mod. Phys. C* **2008**, *19*, 703–715. [CrossRef]
40. Ranut, P.; Nobile, E.; Mancini, L. High resolution microtomography-based CFD simulation of flow and heat transfer in aluminum metal foams. *Appl. Therm. Eng.* **2014**, *69*, 230–240. [CrossRef]
41. Bianchi, E.; Heidig, T.; Visconti, C.G.; Groppi, G.; Freund, H.; Tronconi, E. An appraisal of the heat transfer properties of metallic open-cell foams for strongly exo-/endo thermic catalytic processes in tubular reactors. *Chem. Eng. J.* **2012**, *198*, 512–528. [CrossRef]
42. Liu, H.; Yu, Q.N.; Qu, Z.G.; Yang, R.Z. Simulation and analytical validation of forced convection in open-cell metal foams. *Int. J. Therm. Sci.* **2017**, *111*, 234–245. [CrossRef]
43. Xu, W.; Zhang, H.; Yang, Z.; Zhang, J. Numerical investigation on the flow characterizations and permeability of three-dimensional reticulated foam materials. *Chem. Eng. J.* **2008**, *140*, 562–569. [CrossRef]
44. Zhaodong, C.; Kuncan, Z.; Xiao, W.; Fuling, H.; Chong, G.; Juanjuan, B. Derivation and validation of a new porous media resistance formula based on a tube-sphere model. *Ind. Eng. Chem. Res.* **2020**, *59*, 18170–18179. [CrossRef]
45. Yang, H.; Li, Y.; Yang, Y.; Chen, D.; Zhu, Y. Effective thermal conductivity of high porosity open-cell metal foams. *Int. J. Heat Mass Transf.* **2020**, *147*, 118974. [CrossRef]
46. Nie, Y.; Lin, Q.; Tong, Q. Modeling structures of open cell foams. *Comput. Mater. Sci.* **2017**, *131*, 160–169. [CrossRef]
47. Hu, C.; Sun, M.; Xie, Z.; Yang, L.; Song, Y.; Tang, D.; Zhao, J. Numerical simulation on the forced convection heat transfer of porous medium for turbine engine heat exchanger applications. *Appl. Therm. Eng.* **2020**, *80*, 115845. [CrossRef]
48. Jobic, Y.; Kumar, P.; Topin, F. Transport properties of solid foams having circular strut cross section using pore scale numerical simulations. *Heat Mass Transf.* **2018**, *54*, 2351–2370. [CrossRef]
49. Nie, Z.; Lin, Y.; Tong, Q. Numerical investigation of pressure drop and heat transfer through open cell foams with 3D Laguerre-Voronoi model. *Int. J. Heat Mass Transf.* **2017**, *113*, 819–839. [CrossRef]
50. Kumar, P.; Topin, F.; Tadrist, L. Geometrical characterization of Kelvin-like metal foams for different strut shapes and porosity. *J. Porous Media* **2015**, *18*, 637–652. [CrossRef]
51. Jafarizade, A.; Panjepour, M.; Meratain, M.; Emami, M.D. Numerical simulation of gas/solid heat transfer in metallic foams: A general correlation for different porosities and pore sizes. *Trans. Porous Media* **2018**. [CrossRef]
52. Calmidi, V.V.; Mahajan, R.L. The effective thermal conductivity of high porosity fibrous metal foam. *J. Heat Transf.* **1999**, *121*, 466–471. [CrossRef]
53. Hu, H.; Lai, Z.; Weng, X.; Ding, G.; Zhuang, D. Numerical model for the dehumidifying process of wet air flow in open-cell metal foam. *Appl. Therm. Eng.* **2017**, *133*, 309–321. [CrossRef]
54. Inayat, A.; Freund, H.; Zeiser, T.; Schwieger, W. Determining the specific surface area of ceramic foams: The Tetradecahedra model revisited. *Chem. Eng. Sci.* **2011**, *266*, 1179–1188. [CrossRef]
55. Miwa, S.; Kane, C.; Revankar, S. Microscopic fluid dynamics simulation of the metal foam using idealized cell structure. *Transp. Porous Media* **2016**, *115*, 35–51. [CrossRef]
56. Iasiello, M.; Cunsolo, S.; Biano, N.; Chiu, W.K.S.; Naso, V. Developing thermal flow in open-cell foam. *Int. J. Therm. Sci.* **2017**, *111*, 129–137. [CrossRef]
57. Krishnan, S.; Murthy, J.Y.; Garimella, S.V. Direct simulation of transport in open-cell metal foam. *J. Heat Transf.* **2006**, *128*, 793–799. [CrossRef]
58. Boomsma, K.; Poulikakos, D.; Ventikos, Y. Simulation of flow through open cell metal foams using an idealized periodic cell structure. *Int. J. Heat Fluid Flow* **2003**, *24*, 825–834. [CrossRef]
59. Iasiello, M.; Cunsolo, S.; Oliviero, M.; Harris, W.M.; Bianco, N.; Chiu, W.K.S.; Naso, V. Numerical analysis of heat transfer and pressure drop in metal foams for different morphological models. *J. Heat Transf.* **2014**, *136*, 11260. [CrossRef]
60. Bhattacharya, A.; Calmidi, V.V.; Mahajan, R.L. Thermophysical properties of high porosity metal foams. *Int. J. Heat Mass Transf.* **2002**, *45*, 1017–1031. [CrossRef]
61. Aghaei, P.; Visconti, C.G.; Groppo, G.; Tronconi, E. Development of a heat transport model for open-cell metal foams with high cell densities. *Chem. Eng. J.* **2017**, *321*, 432–446. [CrossRef]
62. De Jaeger, P.; T'Joen, C.; Huisseune, H.; Ameel, B.; De Paepe, M. An experimentally validated and parameterized periodic unit-cell reconstruction of open-cell foams. *J. Appl. Phys.* **2011**, *109*. [CrossRef]
63. Perrot, C.; Pannenton, R.; Olny, X. Periodic unit cell reconstruction of porous media: Application to open-cell aluminum foam. *J. Appl. Phys.* **2007**, *101*. [CrossRef]
64. Boomsma, K.; Poulikakos, D. The effect of compression and pore size variations on the liquids flow characteristics in metal foam. *J. Fluids Eng.* **2002**, *124*, 263–272. [CrossRef]
65. ERG Materials and Aerospace. Available online: http://ergaerospace.com/ (accessed on 24 March 2021).
66. Kececioglu, I.; Jiang, Y. Flow through porous media of packed spheres saturated with water. *J. Fluids Eng.* **1994**, *116*, 164–170. [CrossRef]
67. Dukhan, N.; Chen, K.C. Heat transfer measurements in metal foam subjected to constant heat flux. *Exp. Therm. Fluid Sci.* **2007**, *32*, 624–631. [CrossRef]

Article

Various Trade-Off Scenarios in Thermo-Hydrodynamic Performance of Metal Foams Due to Variations in Their Thickness and Structural Conditions

Trilok G [1], N Gnanasekaran [1,*] and Moghtada Mobedi [2,*]

[1] Department of Mechanical Engineering, National Institute of Technology Karnataka, Surathkal, Mangalore 575025, India; trilokg.197me506@nitk.edu.in

[2] Mechanical Engineering Department, Faculty of Engineering, Shizuoka University, 3-5-1 Johoku, Naka-ku, Hamamatsu-shi 432-8561, Japan

* Correspondence: gnanasekaran@nitk.edu.in (N.G.); moghtada.mobedi@shizuoka.ac.jp (M.M.)

Abstract: The long standing issue of increased heat transfer, always accompanied by increased pressure drop using metal foams, is addressed in the present work. Heat transfer and pressure drop, both of various magnitudes, can be observed in respect to various flow and heat transfer influencing aspects of considered metal foams. In this regard, for the first time, orderly varying pore density (characterized by visible pores per inch, i.e., PPI) and porosity (characterized by ratio of void volume to total volume) along with varied thickness are considered to comprehensively analyze variation in the trade-off scenario between flow resistance minimization and heat transfer augmentation behavior of metal foams with the help of numerical simulations and TOPSIS (Technique for Order of Preference by Similarity to Ideal Solution) which is a multi-criteria decision-making tool to address the considered multi-objective problem. A numerical domain of vertical channel is modelled with zone of metal foam porous media at the channel center by invoking LTNE and Darcy–Forchheimer models. Metal foams of four thickness ratios are considered (1, 0.75, 0.5 and 0.25), along with varied pore density (5, 10, 15, 20 and 25 PPI), each at various porosity conditions of 0.8, 0.85, 0.9 and 0.95 porosity. Numerically obtained pressure and temperature field data are critically analyzed for various trade-off scenarios exhibited under the abovementioned variable conditions. A type of metal foam based on its morphological (pore density and porosity) and configurational (thickness) aspects, which can participate in a desired trade-off scenario between flow resistance and heat transfer, is illustrated.

Keywords: metal foams; thickness ratio; pore density; porosity; heat transfer; pressure drop; trade-off; TOPSIS

Citation: G, T.; Gnanasekaran, N.; Mobedi, M. Various Trade-Off Scenarios in Thermo-Hydrodynamic Performance of Metal Foams Due to Variations in Their Thickness and Structural Conditions. *Energies* **2021**, *14*, 8343. https://doi.org/10.3390/en14248343

Academic Editor: Dmitry Eskin

Received: 26 October 2021
Accepted: 6 December 2021
Published: 10 December 2021

Publisher's Note: MDPI stays neutral with regard to jurisdictional claims in published maps and institutional affiliations.

Copyright: © 2021 by the authors. Licensee MDPI, Basel, Switzerland. This article is an open access article distributed under the terms and conditions of the Creative Commons Attribution (CC BY) license (https://creativecommons.org/licenses/by/4.0/).

1. Introduction

Metal foams have been widely researched for their excellent thermal transfer augmenting potential due to their extraordinary heat conducting properties, high surface area density and other desired properties such as high strength and low density. Applications where metal foams have been found to be advantageous include hotspot removal in electronic equipment [1,2] thermal energy storage [3,4], photo voltaic panel [5], thermal management in batteries [6,7], boiling heat transfer [8], solar collectors [9,10], etc. However, the benefit of augmented heat transfer with the use of metal foams in heat exchanging applications is always compromised by increased pressure drop due to high flow obstruction offered by such materials. Hence, it is crucial to analyze the thermo-hydraulic behavior of metal foams with both flow resistance and heat transfer enhancement. Several factors that influence flow and heat transfer through metal foam include material aspects (thermal conductivity [11]), structural aspects (porosity and pore density [12]), configurational aspects (thickness [11,13] and partial filling [14,15]), the foam geometrical aspect (shape [16]), and the arrangement aspect (foam multi-layer [17]), etc. It is interesting to note that different combinations of these aspects result in varied pressure drop and heat transfer, thus allowing a better trade-off between enhanced heat transfer with the accompanied increased pressure drop. In this regard, heat transfer and flow influencing, structural (pore density and porosity) and configurational (thickness) aspects that can be altered resulting in varied magnitudes of both heat transfer as well as flow characteristics of a thermal system involving metal foam are analyzed in this study.

In the previous studies [12,18], the influence of structural aspects (PPI and porosity) on varied flow resistance and thermal transmission behaviors was illustrated, highlighting the optimum selection of foam samples based on their structural aspects for a desired thermo-hydraulic performance. However, it is interesting to analyze the thermo-hydraulic performance of metal foams subjected to further accompanied variable conditions such as varied thickness scenarios. Many works involving investigation of optimum performance of high flow resistance inducing metal foams include either partial filling scenarios (varying thickness) or varying structural aspects (pore density and porosity). As per our best knowledge, there is no work in the literature that reflects the combined effect of varying thickness along with variations in structural aspects to describe varied magnitudes of flow resistance and thermal transmission in a system with metal foams.

Zuo et al. [19] numerically investigated optimal design of partially filled metal foams on the improved performance of a latent thermal storage component. Various filling thicknesses and filling angles were considered. Sardari et al. [20] performed a numerical study to investigate the behavior of a heat storage system with copper foam and phase change material. They considered foam structural variables such as porosity and pore density with different heater locations. Higher rates of heat transfer were observed with the addition of metal foam. The effect of porosity and PPI on temperature and liquid fraction in the porous-PCM heat storage unit was illustrated in this study. The significance of metal foams in enhancing heat transfer in heat storage units can also be seen in a study by Mohammed et al. [21]. The authors took the effect of pore density into consideration and observed that using metal foams of higher PPI, the circulation of generated heat in the domain gets more uniform, aiding the melting process. Recent studies also focused on numerical modeling of porous media with high Prandtl number fluids to analyze various naturally occurring phenomena [22]. Li et al. [23] experimentally investigated the enhancement of heat transfer using a partially filled gas tube with metal foams. The study revealed that for a considered thickness and porosity condition, an increase in PPI increased the heat transfer. Bianco et al. [24] made an attempt to understand the trade-off between pressure drop and heat transfer in a heat sink with metal foam and metal foam fins. They provided optimization procedures to enhance heat transfer at a given pumping power with the help of Pareto plots. Siavashi et al. [25] considered gradient and multilayered metal foams to understand flow and heat transfer behavior with nanofluid as the working fluid. Optimized properties of metal foams and various types of arrangements to enhance heat transfer with reduced pumping power were analyzed. Anuar et al. [26] investigated the effects of pore density and height of metal foam on pressure drop characteristics. The study reported an enhancement of pressure drop with an increase in blockage ratio and pore density. Lai et al. [27] studied the effect of pore density of coated hydrophilic foams on heat transfer and pressure drop when subjected to wet air flow. Variation of pore density was considered for the single porosity condition in this study, limiting the study to a single structural variable and coating type (hydrophobic or hydrophilic). A pore scale simulation by Sun et al. [28] also focuses on the augmentation of thermal transmission with an upsurge in porosity as well as with an increase in pore-density conditions. The study considered 0.9, 0.87 and 0.82 porosity foam samples of 40, 20 and 10 pores per inch, respectively. The comprehensive heat removal enhancement index was evaluated in the study and found that it was better exhibited by a foam sample of 0.87 porosity with 40 PPI pore density condition. Mancin et al. [29] investigated pressure drop and heat transfer through metal foams of 20 PPI with various heights. The study revealed that although the pressure drop was the same for both height conditions (20 and 40 mm), a significant change could be observed in heat transfer.

Jadhav et al. [15] investigated the performance of metal foams under the partially filled scenario of various configurations. Metal foams of altered combinations of PPI and porosity were included in the analysis. The study reported the variation in heat transfer corresponding to the metal foam sample and type of partial filling configuration. Singh et al. [13] investigated the influence of thickness on heat transfer through jet array impingement in foams to arrive at a scenario with high heat transfer and reduced pressure drop conditions. For this, 5, 10 and 20 PPI pore density foam samples of 19, 12.7 and 6.35 mm thicknesses were considered for the analysis. Optimum thermo-hydraulic performance for a considered pore density foam sample was reported to have been observed at the intermediate foam thickness.

Our literature survey showed that metal foam works in literature mostly consider the effect of individual variable conditions or set of constant variable conditions to assess flow resistance and heat transfer in a thermal system with metal foams. However, when both pressure drop and heat transfer vary individually as well as with combined effects of the given variable conditions, it becomes crucial to understand the flow and heat transfer behavior through metal foams corresponding to their orderly varied structural aspects (pore density and porosity) along with simultaneously varying thickness condition. Pressure drop (representing flow resistance), wall heat transfer co-efficient and wall Nusselt number (both representing heat transfer enhancement) are comprehensively analyzed using TOPSIS a multi-objective multi-criteria decision-making tool to understand the trade-off scenario between enhanced heat transfer and increased pressure drop for all the considered scenarios. This makes possible to understand various trade-off scenarios between pressure drop minimization and heat transfer enhancement associated with metal foams subjected to combined variable conditions, which is the goal of the current investigation.

2. Problem Statement

Kamath et al.'s [30] effort to experimentally analyze the thermo-hydraulic performances of metal foams in a flow channel (vertical) is numerically modelled in the present study. Figure 1 depicts the sketch of the experimental set up.

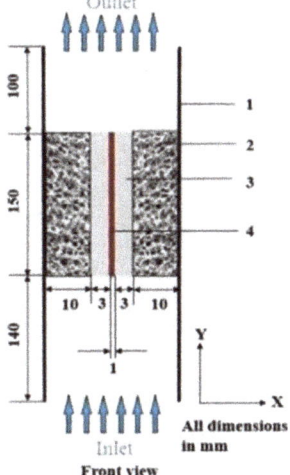

Figure 1. Sketch depicting the front view of the experimental set up used in [9] which has a depth of 250 mm. It is mainly constituted by the following: (1) vertical channel, (2) metal foam, (3) aluminum plate and (4) heater.

It is constituted by a 27 × 250 × 390 (mm) rectangular channel vertically placed, having a heater providing heat flux (constant) of 266.67 W/m² in the middle of the channel passage adjacent to two aluminum plates of 3 mm thickness. Metal foams are accommodated within the channel. LTNE and Darcy–Forchheimer models are implemented to model the thermo-hydrodynamic phenomenon through the metal foam. Aluminum metal foams of 0.95, 0.9, 0.85 and 0.8 porosity, each with a pore density of 25, 20, 15, 10 and 5 PPI are analyzed by housing in the mid-way of the channel adjacent to a thin aluminum plate subjected to constant heat flux condition. An air flow of 0.4 m/s is allowed into the channel consisting of heater-plate (Al)-porous assembly. Under a steady-state condition, the temperature field across the aluminum plate is obtained in terms of average wall temperature, in every case considering 20 various types of metal foams (of various structural aspects, i.e., pore density and porosity) each at 4 different thicknesses scenarios, totaling 80 different cases for simulations.

Using each structural (pore density and porosity combination) and configurational (various thickness) information, the average wall temperature of the aluminum plate is computed along with pressure drop incurred through the foam filled domain. Further, pressure drop (representing flow resistance), wall heat transfer co-efficient and wall Nusselt number (both representing heat transfer enhancement) are comprehensively analyzed using TOPSIS, a multi-objective multi-criteria decision-making tool to comprehend the trade-off scenario between enhanced heat transfer and increased pressure drop for all the considered scenarios.

3. Governing Equations

The conservative equations considered by the solver to exhibit flow and temperature fields in the non-foam region are provided in Equations (1)–(3).

Continuity equation:

$$\frac{\partial \left(\rho_f u_i\right)}{\partial x_i} = 0, \tag{1}$$

Momentum equation:

$$\frac{\partial \left(\rho_f u_i u_j\right)}{\partial x_j} = -\frac{\partial p}{\partial x_j} + \frac{\partial}{\partial x_j}\left(\mu_f\right)\left(\frac{\partial u_i}{\partial x_j} + \frac{\partial u_j}{\partial x_i}\right), \tag{2}$$

Energy equation (for fluid):

$$\frac{\partial \left(\rho_f C_{pf} u_j T\right)}{\partial x_j} = \frac{\partial}{\partial x_j}\left(\lambda_f \frac{\partial T_f}{\partial x_j}\right). \tag{3}$$

Governing equations corresponding to regions with foam are given in Equations (4)–(8). Heat transfer and flow influencing parameters such as thermal conductivity, pore density (pores per inch), porosity (ratio of void volume comprising fluid to total volume in a fluid saturated porous medium), surface area density (surface area to volume ratio of inter structures of the porous medium), interfacial heat transfer coefficient (corresponding to heat transmission across the structures of the porous medium), etc., are incorporated in the conservative equations. The source term appearing in the momentum equation given in Equation (9) rightly considers both viscous (as a result of interaction within the fluid) as well as inertial effects (as a result of interaction of fluid with solid structures) occurring in cases of flow through metal foam similar porous media.

Continuity equation:

$$\frac{\partial \left(\rho_f \varepsilon u_i\right)}{\partial x_i} = 0 \tag{4}$$

Momentum equation:

$$\frac{\partial(\rho_f u_i u_j)}{\partial x_j} = -\varepsilon\frac{\partial p}{\partial x_j} + \frac{\partial}{\partial x_j}\left(\mu_f\left(\frac{\partial u_i}{\partial x_i}+\frac{\partial u_j}{\partial x_j}\right)\right) - \varepsilon\left(\frac{\mu_{eff}}{K}u_i + \rho_f C|u|u_i\right), \quad (5)$$

where, K is permeability, and C is the inertia coefficient.

The energy equation for solid and fluid phases for a LTNE condition is for fluid,

$$\varepsilon\frac{\partial(\rho_f C_{pf} u_j T)}{\partial x_j} = \lambda_{fe}\varepsilon\frac{\partial}{\partial x_j}\left(\frac{\partial T_f}{\partial x_j}\right) + h_{sf} a_{sf}(T_s - T_f), \quad (6)$$

and for solid,

$$\lambda_{se}(1-\varepsilon)\frac{\partial}{\partial x_j}\left(\frac{\partial T_s}{\partial x_j}\right) = h_{sf} a_{sf}(T_s - T_f), \quad (7)$$

where

$$\lambda_{fe} = \lambda_f \cdot \varepsilon \text{ and } \lambda_{se} = \lambda_s \cdot (1-\varepsilon). \quad (8)$$

Boundary conditions for the considered problem are clearly shown in Figure 2. The uniform velocity boundary condition is given at the inlet of the flow passage, and the pressure outlet boundary condition is assigned at the outlet of the channel passage. The heater plate is given with constant heat flux boundary. The walls of the channel passage are assigned with the insulated boundary condition. The working fluid considered is air with 0.4 m/s velocity. It can be noted that the Reynolds number based on hydraulic diameter is found to be 1226, signifying a forced convection dominant regime.

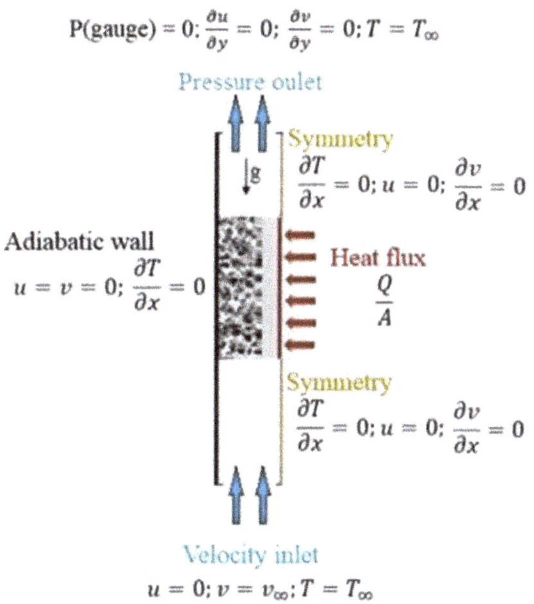

Figure 2. Sketch of complete thickness (100% foam filled) numerical domain with assigned boundary conditions.

The studied 4 length of the metal foam is also shown in Figure 3. The problem is solved for four cases, 100%, 75%, 50% and 25% length of the metal foam.

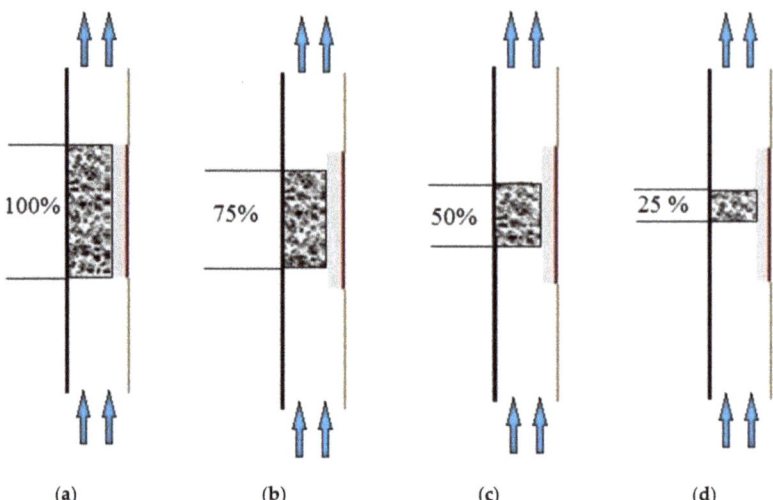

Figure 3. Sketch depicting the various thickness configurations considered in the present study: (**a**) 100% thickness, (**b**) 75% thickness, (**c**) 50% thickness and (**d**) 25% thickness.

The definitions of heat transfer coefficient and Nusselt number in this study are given below.

Heat transfer coefficient:

$$h_{wall} = \frac{Q}{A \Delta T_w}, \tag{9}$$

where

$$\Delta T_w = \overline{[T - T_a]}. \tag{10}$$

Nusselt number:

$$Nu_{wall} = \frac{h_{wall} D_h}{\lambda_{eff}}, \tag{11}$$

where

$$\lambda_{effective} = \left(\lambda_f^{\varepsilon} \times \lambda_s^{1-\varepsilon} \right). \tag{12}$$

Correlations provided by [31,32] to evaluate the interfacial parameters appearing in the energy equation, surface area density and heat transfer co-efficient are shown in Equations (13) and (14). The following section deals with the above-mentioned aspects.

$$a_{sf} = \frac{3\pi d_f (1 - exp - ((1 - \varepsilon)/0.04))}{(0.59 d_p)^2}, \tag{13}$$

$$\frac{h_{sf} d_f (1 - exp - ((1 - \varepsilon)/0.04))}{\lambda_f} = \begin{cases} 0.76 Re_{df}^{0.4} Pr^{0.37}, & \left(1 \leq Re_{df} \leq 40\right), \\ 0.52 Re_{df}^{0.5} Pr^{0.37}, & \left(40 \leq Re_{df} \leq 10^3\right), \\ 0.26 Re_{df}^{0.6} Pr^{0.37}, & \left(10^3 \leq Re_{df} \leq 2 \times 10^5\right), \end{cases} \tag{14}$$

$$d_p = \frac{0.0254}{PPI}, \tag{15}$$

$$\frac{d_f}{d_p} = 1.18 \sqrt{\frac{1-\varepsilon}{3\pi}} \times \frac{1}{1 - exp - ((1-\varepsilon)/0.04)}, \tag{16}$$

4. Numerical Details and Grid Independent Study

Sketch of two-dimensional numerical domains for computation is depicted in Figure 2 with described boundary conditions required for numerical modeling of the domain using commercially available ANSYS FLUENT software (Ansys, India). Contact regions of porous, solid and fluid regions are associated with the corresponding interfaces. The experimental domain is symmetric in nature about its axis along the flow direction as can be seen in Figure 1, and therefore, just one part of the symmetrical domain is considered for the numerical modeling that results in decreased computational effort by appropriately incorporating a symmetry boundary condition at the center line axis separating two symmetrical portions. The obtained numerical domain mainly constitutes a heater close to a solid plate of aluminum that faces inward into the channel passage accommodating the foam sample. Evaluation of the required flow and heat transfer influencing properties appearing in the governing equations are estimated as provided in [12,18].

The pressure-based coupled algorithm is used to solve continuity and momentum equations in order to achieve numerically modeled flow domain. The convergence determined by residuals of the solutions in the iterative steps is set as 10^{-5} for momentum and continuity and 10^{-10} for energy. A grid independence study was carried out for the numerical domain in the present study comparing the domain with 26,130; 56,700 and 88,400 cells as shown in Table 1. Having the highest cell domain as the base line, it was observed that the percentage deviation of pressure drop and wall temperature are found to be at least 0.07% and 0.42%, respectively, for the meshed domain of 56,700 cells. However, with a lower mesh size of 26,130 cells, the respective percentage deviations of pressure drop and wall temperature were found to be 0.22% and 0.94%, and any further decrease in the number of cells in the meshed domain would increase the deviation of the mentioned heat transfer and flow parameters. Therefore, in the current study for optimum computational effort, a mesh domain with 56,700 elements is chosen, and subsequent simulations were carried out in the same meshed domain throughout the present work.

Table 1. Grid independence study.

Mesh Type	No. of Elements	Pressure Drop ΔP, N/m^2	Temperature Difference ΔT °C	Deviations ΔP, %	Deviations ΔT, %
1	26,130	27.60	7.74	0.22	0.94
2	56,700	27.56	7.70	0.07	0.42
3	88,400	27.54	7.66	Base line	

5. Validation of the Numerical Solution

The numerical procedure involved in the present study is validated against the experimental work of [30]. Variations of numerically predicted thermal parameters, such as wall temperature and wall heat transfer coefficient, with those of Kamath et al.'s [30] experimental data are shown in Figure 4a for 20 PPI and 0.9 porosity porous foam. Similarly, the variation of the numerically simulated and experimentally obtained flow parameter is shown in Figure 4b in terms of variation in pressure drop for a foam sample of 0.9 porosity and 20 PPI. Numerically predicted data of thermal and flow parameters are observed to be in good agreement with that of the experimental work. Maximum percentage deviation of numerically predicted temperature data was found to be only 2.75% from that of experimental value and a 0.048% of minimum deviation was observed between the numerically predicted and experimentally obtained data. Similarly, with good agreement in trend of pressure drop variation, numerically predicted data showed an average percentage deviation of 12%. This confirms the aptness of the current numerical procedure and adopted models to predict flow and temperature field in a domain consisting of metal based porous media.

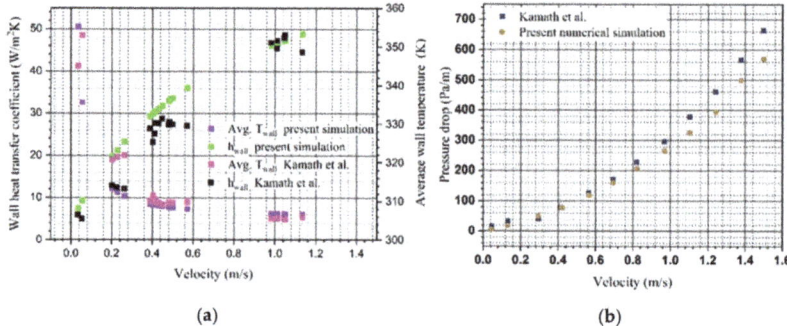

Figure 4. (a) Assessment of closeness of numerically predicted thermal parameters (wall heat transfer coefficient and wall temperature) for metal foam of 0.9 porosity and 20 PPI with that of experimental data. (b) Assessment of closeness of numerically predicted flow parameters (pressure drop) for metal foam of 0.9 porosity and 20 PPI with that of experimental data.

6. TOPSIS Technique and Various Trade-off Scenarios between Pressure Drop Minimization and Heat Transfer Augmentation Behavior

The TOPSIS technique allows multi-objective optimization to be employed where one objective is to minimize a parameter and, contrarily, another is to maximize the other parameter (in this study, flow resistance and heat transfer are the parameters to be minimized and maximized, respectively) to understand and distinguish the given variable conditions (under various pore density, porosity and thickness conditions) based on various trade-offs between pressure drop and heat transfer. It also makes it possible to know in what trade-off scenario a given variable condition is participating. The knowledge of the trade-off in which a given variable condition would result becomes crucial for designing a heat transfer device where maximization of heat transfer and minimization of pressure drop are of major interest. In that direction, the present work demonstrates various trade-offs (in terms of Criteria I to V) involving variable conditions such as pore density, porosity and thickness of the metal foam that can alter the trade-off between heat transfer and pressure drop.

Various steps involved in implementing the TOPSIS method are as follows [12,33]:

- Step 1: Obtain matrix of normalized columns.

Produce a matrix of two columns each comprising values of wall heat transfer coefficient (h_{i1}) and values of pressure drop (p_{i2}), respectively. Produce another set of two columns each comprising normalized magnitudes of heat transfer coefficient and pressure drop ($h_{\_i3}$) and ($p_{\_i4}$), respectively, using Equations (17) and (18).

$$\overline{h_{i3}} = \frac{h_{i1}}{\sqrt{\sum_{i=1}^{m} h_{i1}^2}}, \qquad (17)$$

$$\overline{p_{i4}} = \frac{h_{i2}}{\sqrt{\sum_{i=1}^{m} p_{i2}^2}}. \qquad (18)$$

where 'i' indices ($i = 1, 2, \ldots, m$) indicate the rows of the matrix, and 'm' represents the total number of metal foam models (pertaining to all variable conditions).

- Step 2: Obtain matrix of weight assigned normalized columns.

Additionally, the matrix is extended with weighted normalized columns, using Equations (19) and (20); the weight of unity is distributed (that varies from 0 to 1) to the formerly computed values of normalized magnitudes of heat transfer coefficient and pressure drop. The allotted weight entirely depends on the particular interest of the design.

$$V_{i5} = \overline{h_{i3}}.W_h, \tag{19}$$

$$V_{i6} = \overline{p_{i4}}.W_p. \tag{20}$$

where W_h and W_p are 0 and 1, respectively, for criteria I, whereas it is 0.25 and 0.75, 0.5 and 0.5, 0.75 and 0.25 and 1 and 0 for criteria II, III, IV and V, respectively. Various criteria and their significance are elaborated in the later discussion.

- Step 3: Obtain the ideal best $V+$ and ideal worst $V-$ values.

The ideal worst and best values are achieved from the formerly computed weighted normalized columns of pressure drop and the heat transfer coefficient values. Ideal best value ($V+$) is the highest value in the set of beneficial parameters and the lowest value in the set of unfavorable parameters. Similarly, ideal worst value ($V-$) is the lowest value in the set of favorable parameters and the highest value in the set of unfavorable parameters. In the present study, since pressure drop is considered as the unfavorable parameter while average heat transfer coefficient is considered as the favorable parameter, the highest value among the weighted normalized columns of pressure drop is regarded as the ideal worst, and the lowest value amongst the weighted normalized column of pressure drop is chosen as the ideal best value. On the other hand, the maximum value among the weighted normalized column of heat transfer coefficient is chosen as the ideal best value, and the lowest value among the weighted normalized column of heat transfer coefficient is considered as the ideal worst value. Identification of the ideal best value $V+$ and the ideal worst value $V-$ is expressed using the Equations (21)–(24).

$$V_h^+ = \max(V_{i5}), \tag{21}$$

$$V_p^+ = \min(V_{i6}), \tag{22}$$

$$V_h^- = \min(V_{i5}), \tag{23}$$

$$V_p^- = \max(V_{i6}). \tag{24}$$

- Step 4: Obtain the Euclidean distance.

Euclidean distance refers to the relative distance of each weighted normalized value from the obtained (best or worst) ideal values. Positive Euclidean distance is the measure of distance of each value in the weighted normalized column from the ideal best value and is evaluated as given in Equation (25). Similarly, negative Euclidean distance is the measure of distance of each value in the weighted normalized column from the ideal worst value and is evaluated as given in Equation (26).

$$S_i^+ = \left[(V_{i5} - V_h^+)^2 + (V_{i6} - V_p^+)^2\right]^{0.5}, \tag{25}$$

$$S_i^\pm = \left[(V_{i5} - V_h^-)^2 + (V_{i6} - V_p^-)^2\right]^{0.5}. \tag{26}$$

- Step 5: Evaluate the performance score.

Performance scores of metal foams of all considered variable conditions are evaluated using Equation (27). It ranks the metal foams subjected to variable conditions (pore density, porosity and thickness) based on the values of the heat transfer coefficient and pressure drop exhibited, under the restriction of how close is its performance in meeting the given weighted criteria.

$$P_i = \frac{S_i^-}{S_i^+ + S_i^-}. \tag{27}$$

7. Results and Discussion
7.1. Pressure Drop and Heat Transfer Characteristics

Pressure drop variation is shown in Figure 5 for varying thickness scenarios, each for foam materials of various PPI and porosity blends. It can be observed that for any given thickness of the foam sample considered, pressure drop increases with an increase in pore density for any given porosity condition due to the increased resistance to flow. This is primarily due to the increase in number of metal fibers with the increase in pore density condition, which results in increased flow obstruction leading to an increased pressure drop. Contrarily, the pressure drop intensifies with a decrease in porosity for any given pore density condition. It has to be noted that varied porosity for a given pore density condition is achieved as a result of a compromise in the fiber diameter of the foam samples. In this regard, for a given pore density situation, the porosity of a foam sample is increased as a result of a decrease in the fiber diameter allowing lesser flow obstruction to the flow and, consequently, a lowered pressure drop. For a given pore density condition, porosity decreases with an increase in fiber diameter, resulting in an increased flow resistance and, consequently, increasing the pressure drop. When the thickness of the metal foams is reduced, as an obvious behavior the pressure drop is observed to decrease. The interesting observation can be made from Figure 5 that, although the pressure drop continuously decreases or increases for a given PPI and porosity combination with variation in thickness, a certain pressure drop values can be observed to be close for foam samples of different foam structural properties (pore density and porosity) under varied thickness scenario. Since heat transfer characteristics also vary with the mentioned variables, it would be interesting to analyze what combinations of PPI and porosity of a foam material, under which thickness scenario, are participating in a desired trade-off scenario between curtailing pressure drop and enhancing heat transfer.

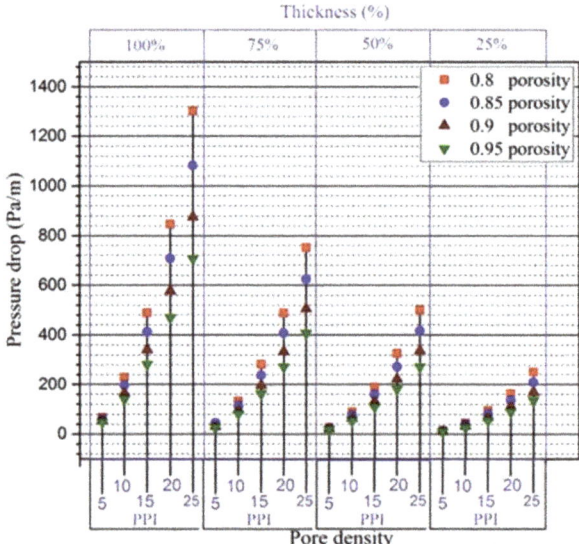

Figure 5. Variation of pressure drop for varying thickness scenario at various pore density and porosity conditions.

Heat transfer features of the metal foams of considered pore density and porosity combinations with varying thickness condition are shown in Figure 6, in terms of average wall heat transfer coefficient, and in Figure 7, in terms of wall average wall Nusselt number. It can be observed that for any given thickness scenario, the average wall heat transfer coefficient increases with upsurge in pore density for a considered porosity condition as a result of augmented interfacial area density participating in improved heat transfer [12]. Similarly, for a given pore density condition in any thickness scenario considered, the wall heat transfer coefficient is observed to increase with an increase in porosity due to enhanced interfacial heat transfer coefficient as a result of increased fluid volume . This is primarily due to the increased fluid to solid ratio with the increase in porosity enabling more fluid to participate in the heat transfer [12]. It can be seen that the wall heat transfer coefficient increases greatly with an increase in pore density for a given porosity condition compared to the increase in the same coefficient with an upsurge in porosity for a given PPI condition. Interfacial heat transfer is a key feature of heat transfer through a porous medium that contributes to the overall heat transfer in a medium filled with metal foam as porous medium. It is characterized by an interfacial specific surface area (a_{sf}) and interfacial heat transfer co-efficient (h_{sf}). The product of these parameters $a_{sf} \times h_{sf}$ can be seen appearing in the energy equation (Equations (6) and (7)). An increase in pore density with constant porosity increases both a_{sf} and h_{sf}, whereas an increase in porosity at constant pore density decreases the interfacial specific surface area and slightly increases interfacial heat transfer coefficient as a result of increased fluid volume in the domain. Owing to these effects, heat transfer characterized by a wall heat transfer co-efficient can be observed to greatly increase with pore density for a given porosity condition compared to its increase with an increase in porosity for a given pore density condition. Considering the variation of wall Nusselt number under the considered variable conditions, a similar trend in relative deviation of wall Nusselt number similar to that of wall heat transfer coefficient is observed. However, Nusselt number is observed to greatly increase with porosity rather than with changes in pore density. Unlike wall heat transfer coefficient behavior, Nusselt number accounts the heat transfer through convection compared to heat transfer that would have resulted through conduction. In this regard, an increase in porosity resulted in a higher Nusselt number with greater magnitude than it would be with an increase in pore density, corresponding to effective thermal conductivity value as described in Equation (12). From the variation of both average wall heat transfer coefficient and average wall Nusselt number, the heat transfer can be perceived to increase with an increase in pore density and as well as with porosity. In addition to this, an increase in heat transfer can be seen with an increase in thickness as an obvious result.

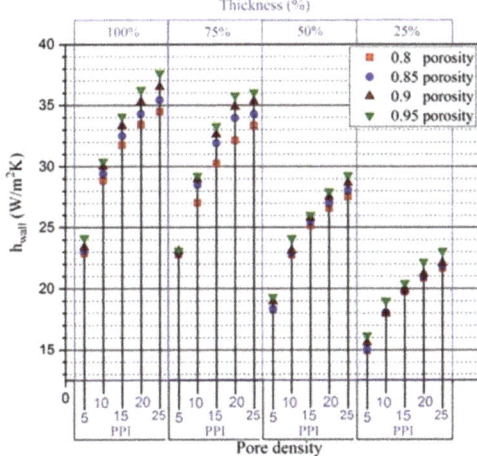

Figure 6. Variation of wall heat transfer coefficient for varying thickness scenarios at various pore density and porosity conditions.

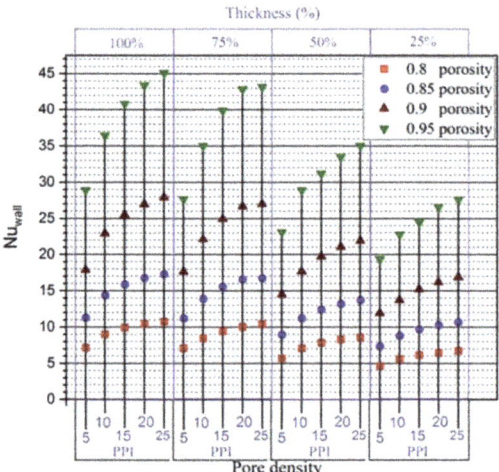

Figure 7. Variation of wall Nusselt number for varying thickness scenarios at various pore density and porosity conditions.

It can be observed that the pressure drop and heat transfer characteristics of metal foams are similar to those of other metal foams subjected to variation in pore density, porosity as well as thickness. For comparison, the pressure contours of 100% thickness and 50% thickness of 0.95 porosity 10 PPI foam are shown in Figure 8a,b. Similarly, temperature contours of 100% thickness and 50% thickness of 0.95 porosity 10 PPI foam are also shown in Figure 9a,b.

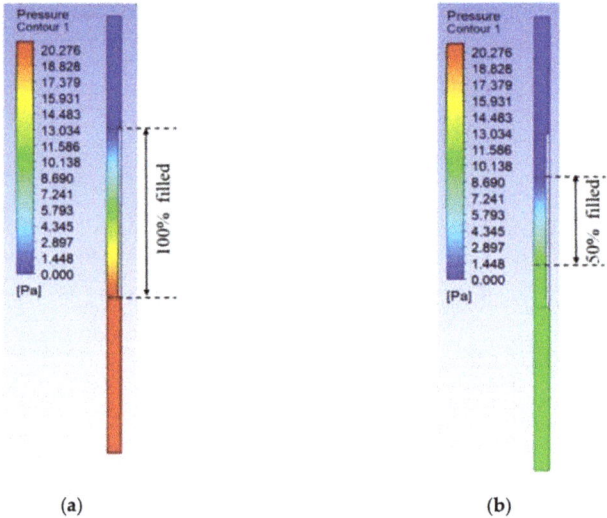

Figure 8. (a) Pressure contours of 100% thickness of 0.95 porosity 10 PPI foam. (b) Pressure contours of 50% thickness of 0.95 porosity 10 PPI foam.

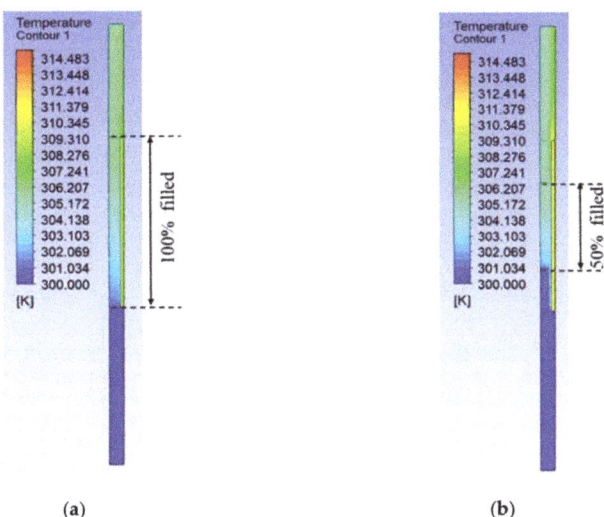

Figure 9. (**a**) Temperature contours of 100% thickness of 0.95 porosity 10 PPI foam. (**b**) Temperature contours of 50% thickness of 0.95 porosity 10 PPI foam.

Each metal foam of a given pore density and porosity combination under the considered thickness exhibits unique heat transfer and pressure drop characteristics. This raises the interesting question of by what magnitude the pressure drop is compensated to exhibit the corresponding heat transfer behavior. Noting that this trade-off scenario varies with the structural (pore density and porosity) and as well as with configurational (thickness) aspects, a comprehensive analysis attempting to understand the tradeoff mechanism between minimizing pressure drop abilities and maximizing heat transfer characteristics of a metal foam under a given variable condition would be very beneficial in the field of heat transfer involving metal foams.

7.2. TOPSIS Analysis

As a well-known fact, the benefit of an increase in heat transfer is always supplemented by an undesired increased pressure drop with the use of metal foams in heat exchanging applications. It is a long-standing issue which requires attention to arrive at an optimum performance. There are several aspects that determine the exhibited trade-off process in the heat transfer enhancement and pressure drop minimization phenomenon; among them, varying thickness scenario and varying PPI and porosity of foams are the best choice of parameters to include in analyzing the tradeoff process as they can be easily varied. In the present study, varying thickness scenarios of four kinds are considered simultaneously with altering pore densities of five kinds (5, 10, 15, 20 and 25 PPI), each with a combination of various porosities such as 0.8, 0.85, 0.9 and 0.95 porosity. The flow resistance, characterized by pressure drop, and the heat transfer, characterized by average wall heat transfer coefficient and Nusselt number exhibited by the metal foams of the above-mentioned variable conditions, are shown in Figures 5–7, respectively. However, none of the mentioned graphs gives information on what is the trade-off in the flow resistance minimization behavior of metal foams in order to exhibit a given heat transfer characteristic. It is here that TOPSIS (technique for order of preference by similarity to ideal solution), a multi-objective, multi-criteria decision-making tool, plays a key role in analyzing which metal foams of given variable condition perform best in meeting a desired trade-off scenario. In other words, the TOPSIS methodology makes it possible to understand how the flow resistance minimization behavior of metal foams is compromised in order to exhibit a given heat transfer behavior subjected to variable conditions such as pore density and porosity of the metal foam and the thickness considered. In this section, the results of TOPSIS analysis on the present problem are presented for various trade-off scenarios (in terms of Criteria I to V) involving variable conditions such as pore density, porosity and thickness of the metal foam that can alter the trade-off between pressure drop and heat transfer.

Criteria 1 (h_{max}:P_{min}::0:1 and Nu_{max}:P_{min}::0:1) represents the trade-off scenario between pressure drop and heat transfer where complete emphasis is given to minimization of pressure drop with no attention given to enhancement of heat transfer. Various performances of metal foams under the considered variable conditions in order to meet the specific weighted objective pertaining to criteria 1 are shown in Figure 10a,b, considering trade-off between pressure drop with wall heat transfer coefficient and Nusselt number, respectively. Under the circumstances of criteria 1, where all the emphasis is focused on minimization of pressure drop (by placing complete weight of '1' on minimizing pressure drop objective) and zero emphasis is given on maximizing heat transfer (by placing weight of '0' on maximizing heat transfer objective), as an obvious interpretation, it can be noted that for a given thickness scenario, metal foams of lower PPI that offer less obstruction to flow are scored best for a given porosity condition. Similarly, for a given thickness scenario, it can be witnessed that for a foam material of a considered pore density, the performance of higher porosity foams is better in relative to lesser porosity foams that offer higher flow resistance. Comparing this performance with varying thickness condition, it can be noted that for any foam material of a given PPI and porosity condition, a better performance is achieved with decrease in thickness as an obvious result of reduced flow resistance accompanying reduced thickness scenario of the metal foams. It is interesting to note that the performance of metal foam of particular pore density and porosity conditions in a given thickness scenario can be closer to the performance of metal foam of a different pore density and porosity combination under a different thickness scenario. This kind of performance charts help choose the desired variable conditions, for instance, thickness of metal foams, where there is restricted variations in pore density and porosity combinations of the metal foams.

Criteria 2 (h_{max}:P_{min}::0.25:0.75 and Nu_{max}:P_{min}::0.25:0.75) representing the case where slight emphasis is given to heat transfer maximization capabilities of foams (with 25% distributed weight on maximizing heat transfer objective) and still a larger emphasis given to minimizing the pressure drop objective (with 75% distributed weight on this objective) are shown in Figure 11a,b. It can be observed that a 5 PPI foam sample of 25% thickness that performed best in meeting the objectives of criteria 1, performs the worst while subjected to the objectives of criteria 2. However, for other pore density scenarios apart from 5 PPI, the relative deviation of performance scores of foam materials is observed to get closer to that of criteria 1 particularly with an upsurge in the pore density (PPI) condition.

Criteria 3 (h_{max}:P_{min}::0.5:0.5 and Nu_{max}:P_{min}::0.5:0.5) represents the case where equal emphasis is given to the heat transfer maximization capabilities of foams (with 50% distributed weight on maximizing heat transfer objective) and to minimizing the pressure drop objective (with 50% distributed weight on this objective); the results are shown in Figure 12a,b. It can be perceived that foam samples of 10 PPI pore density with a 75% thickness condition perform the finest in meeting the criteria with an increase in porosity; 50% thickness foam samples of 15 PPI pore density can be observed to outperform 25% thickness foam samples of 15 PPI with a higher porosity condition, unlike the situation in criteria 1 and 2 where 25% thickness foam samples still performed better than foam samples of 50% thickness. The tendency of 50% thickness foam samples of 20 PPI pore density to outperform the 25% thickness scenario can also be observed with a higher porosity condition. With a 5 PPI pore density condition, the 75% and 100% thickness conditions are observed to perform very close to each other. However, in scenarios of higher pore density, such as 25 PPI, the relative deviation of the performance scores of foam samples of considered variable conditions is similar to that of criteria 1 and 2 where a foam material of a given pore density and porosity condition performed best with a 25% thickness condition followed by 50, 75 and 100 percent thickness conditions.

Criteria 4 ($h_{max}:P_{min}::0.75:0.25$ and $Nu_{max}:P_{min}::0.75:0.25$), representing the case where more emphasis is given to the heat transfer maximization capabilities of foams (with 75% distributed weight on maximizing heat transfer objective) and comparatively lesser importance given to minimizing the pressure drop objective (with 25% distributed weight on this objective), are shown in Figure 13a,b. Foam samples of 75% and 100% thickness are observed to closely perform (subject to the objectives of these criteria) under all porosity conditions under 5 PPI and 10 PPI conditions, with the 100% thickness condition showing inclination to perform better than the 75% thickness condition in the highest porosity case. However, this behavior is clearly exhibited under higher pore density conditions such as 15, 20 and 25 PPI where foam samples of 75% thickness can be seen dominating all other thickness scenarios under the highest porosity condition. Variation performance inclinations (either increasing or decreasing behavior) can be seen under varying porosity conditions due to changes in the trade-off scenarios between pressure drop and heat transfer. Moreover, a slight inconsistency in the relative deviation of the performance scores can be observed when analyzing the heat transfer coefficient and Nusselt number as heat transfer parameters. This is because of dissimilar variations in magnitude of the heat transfer coefficient and Nusselt number as shown in Figures 6 and 7, which results in slight variations in the trade-off scenarios with pressure drop (pressure drop being an unaltered parameter both in the comparison with heat transfer coefficient as well as with the Nusselt number).

Criteria 5 ($h_{max}:P_{min}::1:0$ and $Nu_{max}:P_{min}::1:0$), representing the case where the highest emphasis is given to the heat transfer maximization capabilities of foams (with 100% distributed weight on maximizing heat transfer objective) and the least emphasis is given to the objective of minimizing pressure drop (with 25% distributed weight on this objective), are shown in Figure 14a,b. In other words, criteria 5 rank the foam samples subjected to different variable conditions such as variable pore density, porosity and thickness based on complete attention given to maximizing the heat transfer abilities of the foam samples with no attention given to minimization of the pressure drop. Hence, subjected to these criteria, those foams samples of given variable conditions are scored best that exhibit the highest heat transfer irrespective of pressure drop. As it can be observed from Figure 14a,b, for any foam material of a given PPI and porosity, 100% thickness foams perform best in meeting this criterion, followed by 75, 50 and 25% thickness foams. For a given thickness and pore density condition, the performance scores (subjected to the objectives of criteria 5) of foam samples are observed to increase with porosity. In each pore density condition, 100% filled cases with the highest porosity can be seen performing best in meeting these criteria. The highest pore density case with the highest porosity and 100% thickness exhibits the highest heat transfer which can be seen ranked as the best in Figure 14a,b. It can be noted that although the pressure drop is relatively larger in a higher pore density case and a complete thickness case, the case of a completely filled 25 PPI foam is ranked best, considering its highest heat transfer enhancement behavior irrespective of its flow resistance behavior as decided by criteria 5.

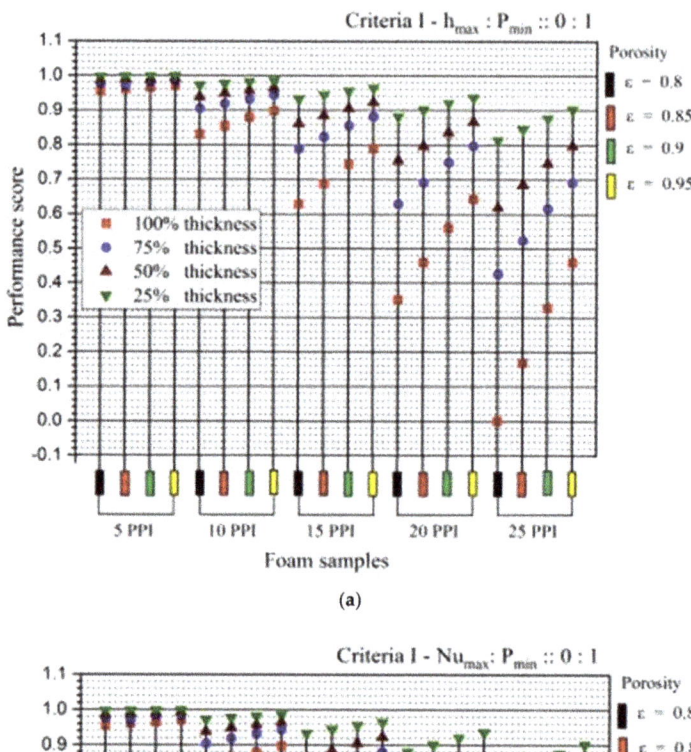

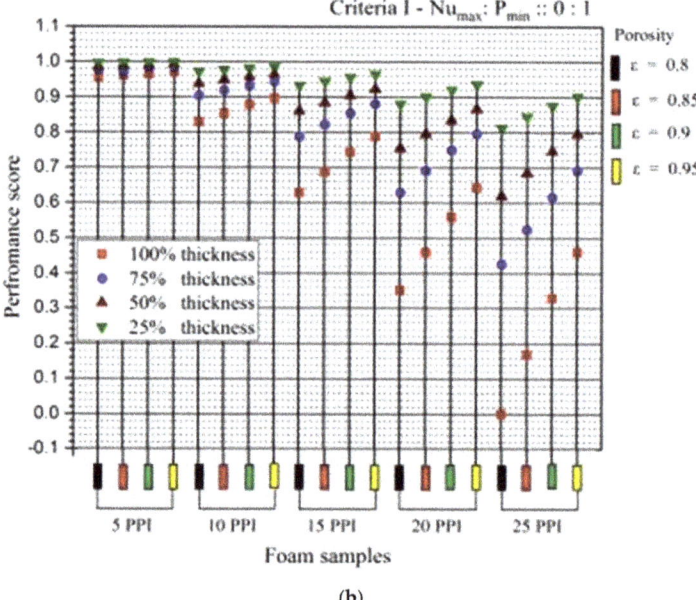

Figure 10. (a) Criteria I ($h_{max}:P_{min}$). (b) Criteria I ($Nu_{max}:P_{min}$).

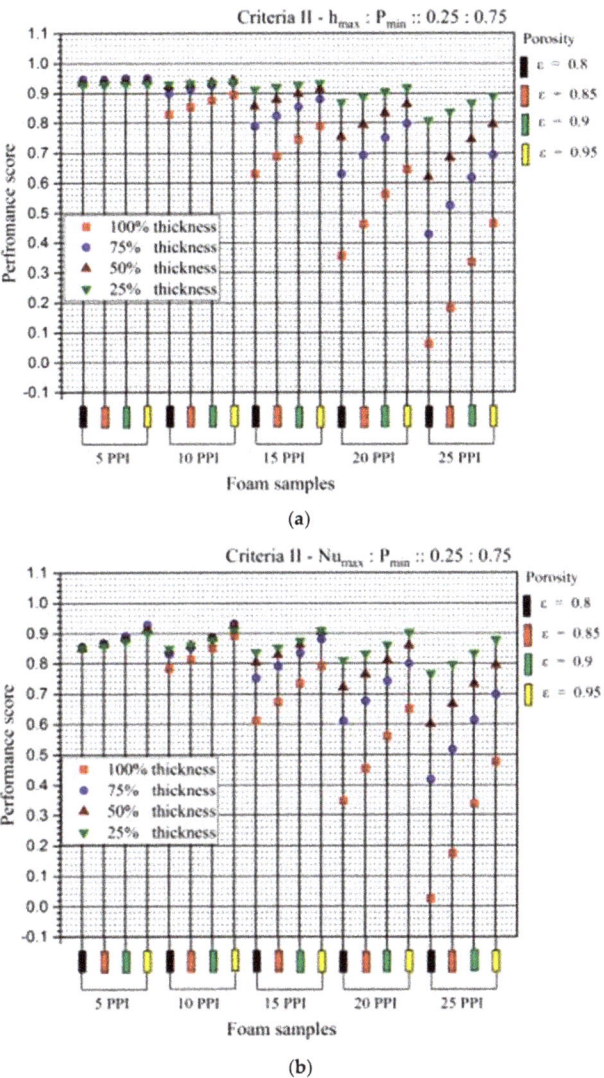

Figure 11. (**a**) Criteria II ($h_{max}:P_{min}$). (**b**) Criteria II ($Nu_{max}:P_{min}$).

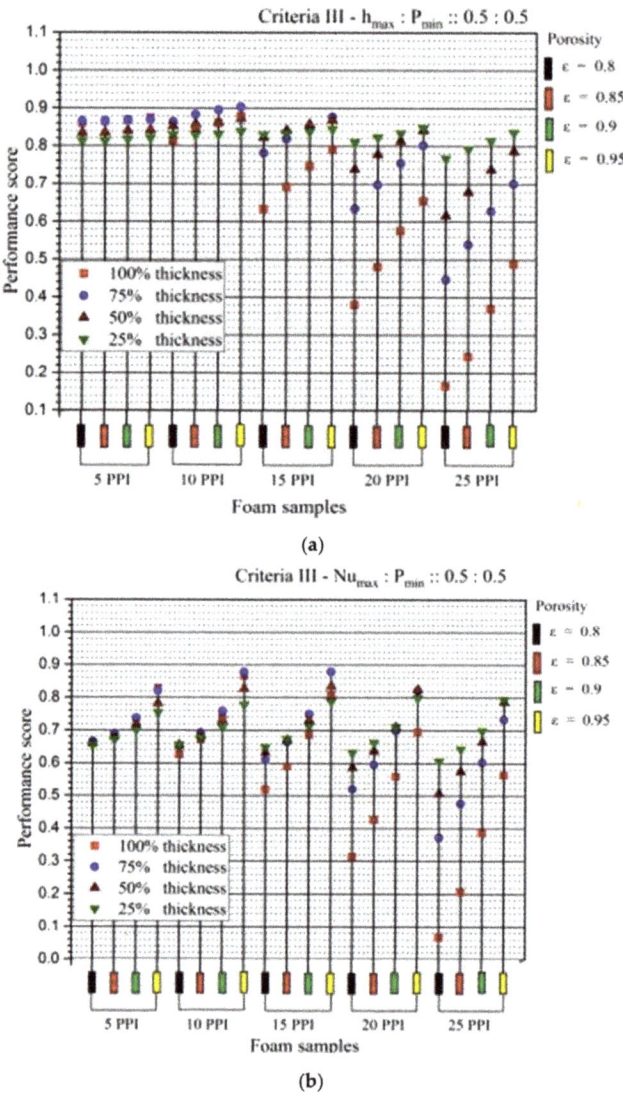

Figure 12. (a) Criteria III (h_{max}:P_{min}). (b) Criteria III (Nu_{max}:P_{min}).

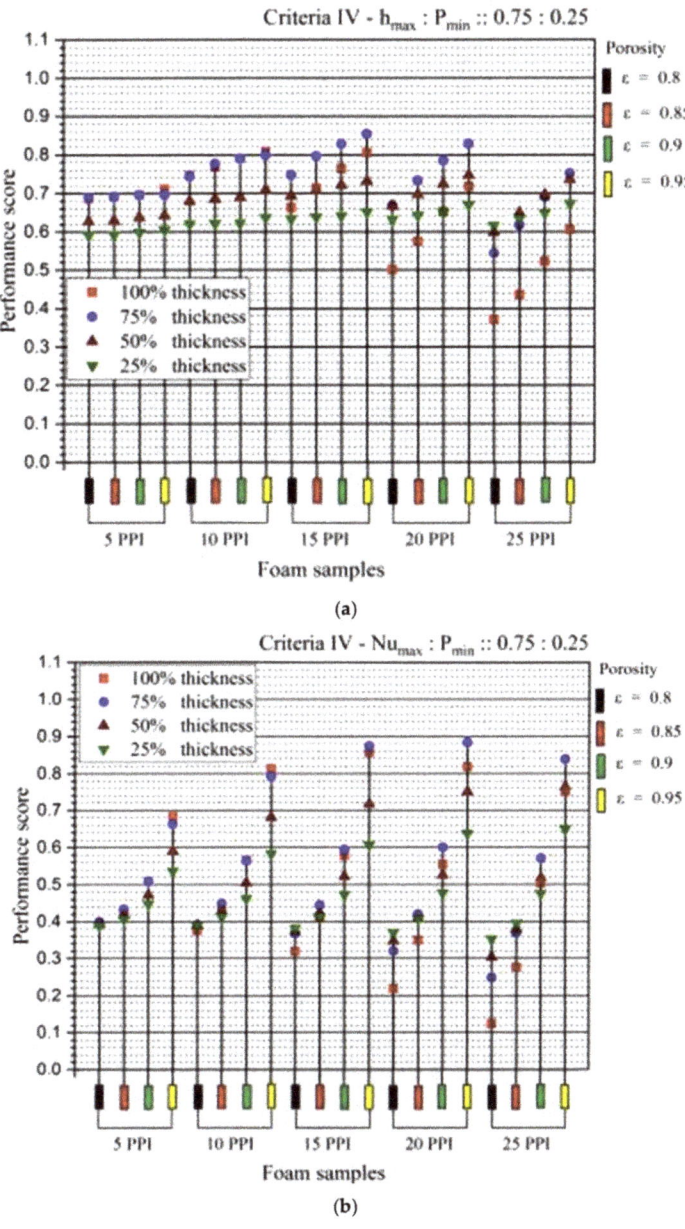

Figure 13. (**a**) Criteria IV ($h_{max}:P_{min}$). (**b**) Criteria IV ($Nu_{max}:P_{min}$).

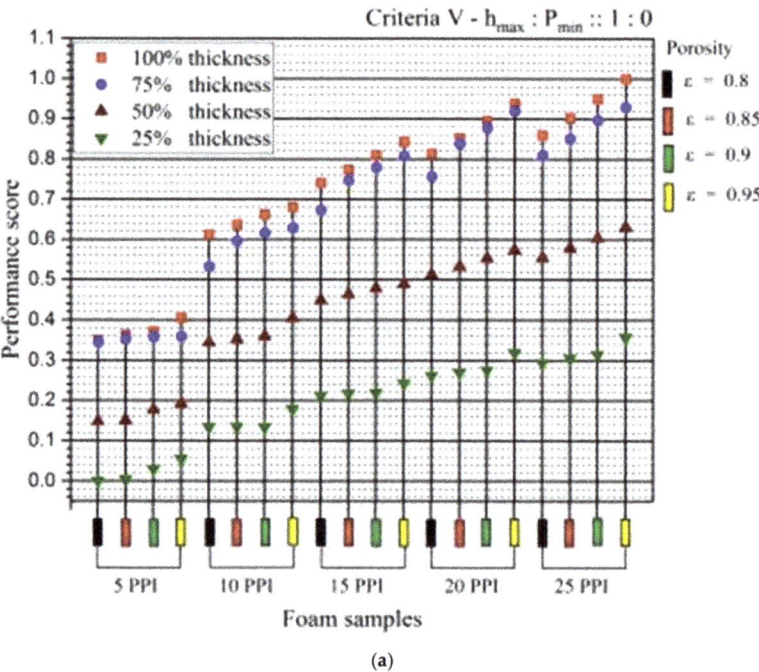

(a)

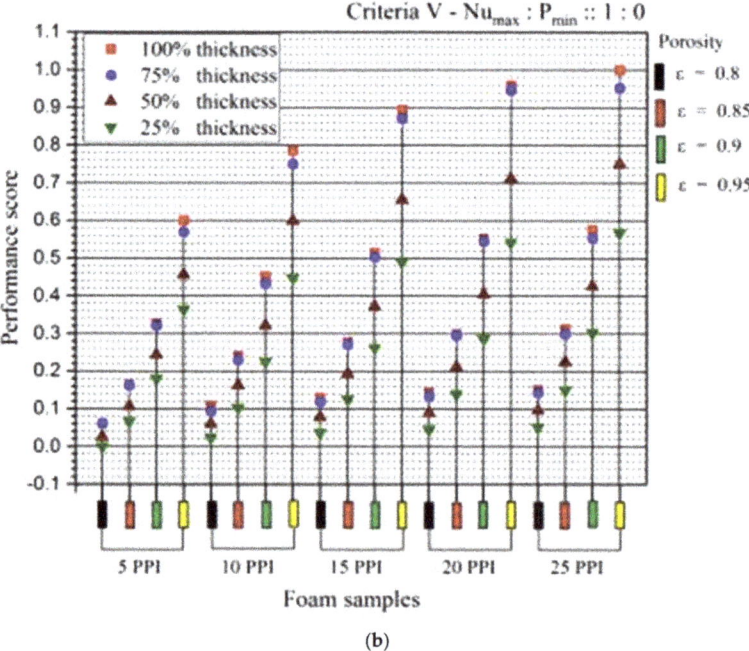

(b)

Figure 14. (a) Criteria V (h_{max}:P_{min}). (b) Criteria V (Nu_{max}:P_{min}).

From Figure 10a to Figure 14b, it can be noted that every foam sample of given variable conditions (pore density, porosity and thickness) shows a different tendency towards meeting the various criteria (subjected to different objectives of various magnitudes of importance). The variation in relative performance of foams of particular variable condition with changes in criteria shows how differently does a foam sample of particular variable conditions participates in the trade-off between heat transfer and flow resistance. Such analysis not only helps decide which variable conditions to consider for given or desired heat transfer and pressure drop characteristics but also helps judge how well a given condition is participating in the trade-off between heat transfer and pressure drop, which is a crucial point in designing a heat exchanging device especially dealing with metal foam-like materials to enhance heat transfer that always comes with a penalty of increased pressure drop in a thermal system.

8. Conclusions

A critical analysis of various trade-off scenarios between heat transfer and pressure drop involving metal foams has been accomplished in the present study. For this purpose, metal foams of three prime variable conditions including pore density (5, 10, 15, 20 and 25 PPI), porosity (0.8, 0.85, 0.9 and 0.95 porosity) and thickness (100, 75, 50 and 25 percent) were considered in a vertical channel subjected to a constant heat flux condition. Individually analysis showed that heat transfer characteristics increased with pore density, porosity and thickness of the foam sample as expected. Moreover, the pressure drop was observed to decrease with a decrease in the pore density and thickness conditions and with an increase in porosity. As per the authors' best knowledge, this is the first time that both flow resistance and heat transfer characteristics are simultaneously analyzed for foam samples of three orderly varying variable conditions, namely, pore density, porosity and thickness, under different objectives (one is to maximize heat transfer and another to minimize pressure drop) subjected to various magnitudes of importance, to understand various trade-off scenarios in which a foam sample of given variable conditions would participate. Interesting trade-off scenarios between the enhancement of heat transfer and the reduction of flow resistance behavior were demonstrated using TOPSIS, a multi-objective multi-criteria decision-making tool that comprehensively illustrates various potentials of every foam sample of considered variable conditions in meeting desired trade-off conditions.

The criteria of $h_{max}:P_{min}$ and $Nu_{max}:P_{min}$ is used to represent the trade-off scenario between pressure drop and heat transfer. For the criteria of $h_{max}:P_{min}::0:1$ and $Nu_{max}:P_{min}::0:1$, metal foam with 5 PPI, 25% thickness yields the best score. For the criteria of $h_{max}:P_{min}::0.5:0.5$ and $Nu_{max}:P_{min}::0.5:0.5$, the best score belongs to the metal foam with 10 PPI, 75% thickness with porosity of 0.95. The metal foam with 25 PPI, 100% thickness with porosity of 0.95 received the best score for the criteria of $h_{max}:P_{min}::1.0:0.0$ and $Nu_{max}:P_{min}::1.0:0.0$.

The selection of the best pore density, thickness and porosity depends on the criteria representing the trade-off between the pressure drop and heat transfer. A score given by the TOPSIS method can deduce the best configurational and structural parameters for a specified application.

Author Contributions: Conceptualization, T.G. and N.G.; methodology, T.G., N.G. and M.M.; software, T.G. and N.G.; validation, T.G. and N.G.; formal analysis, T.G. and N.G.; investigation, T.G., N.G. and M.M.; resources, T.G. and N.G.; data curation, T.G.; writing—original draft preparation, T.G.; writing—review and editing, T.G., N.G. and M.M.; visualization, T.G.; supervision, N.G. and M.M.; project administration, T.G. and N.G.; funding acquisition, N.G. All authors have read and agreed to the published version of the manuscript.

Funding: This research was funded by Science and Engineering Research Board (SERB), Department of Science and Technology, India. Grant number EEQ/2018/000322.

Institutional Review Board Statement: Not applicable.

Informed Consent Statement: Not applicable.

Data Availability Statement: Not applicable.

Acknowledgments: Science and Engineering Research Board (SERB), DST No: EEQ/2018/000322, India, supported and funded this work.

Conflicts of Interest: The authors declare no conflict of interest. The funders had no role in the design of the study; in the collection, analyses or interpretation of data; in the writing of the manuscript or in the decision to publish the results.

Nomenclature

a_{sf}	Area density (m^{-1})
C	Inertial resistance coefficient (m^{-1})
C_p	Specific heat (J/kgK)
d_p	Pore diameter (m)
d_f	Fiber diameter (m)
h	Heat transfer coefficient (W/m^2K)
h_{sf}	Interstitial heat transfer coefficient (W/m^2K)
K	Permeability (m^2)
Nu	Nusselt number
P	Pressure (N/m^2)
Pi	Performance index
Pr	Prandtl number of fluid
Re	Reynolds number
S	Euclidean distance
T	Temperature (K)
u	Velocity vector (m/s)
v	Velocity at inlet (m/s)

Greek symbols

ε	Porosity
ω	pore density
λ	Thermal conductivity (W/mK)
μ	Dynamic viscosity (N-s/m^2)
ν	Kinematic viscosity (m^2/s)
ρ	Density (kg/m^3)

Subscript

f	Fluid
fe	Fluid effective
s	Solid
se	Solid effective

References

1. Brahim, T.; Jemni, A. CFD analysis of hotspots copper metal foam flat heat pipe for electronic cooling applications. *Int. J. Therm. Sci.* **2021**, *159*, 106583. [CrossRef]
2. Mahjoob, S.; Kashkuli, S. Thermal Transport Analysis of Injected Flow through Combined Rib and Metal Foam in Converging Channels with Application in Electronics Hotspot Removal. *Int. J. Heat Mass Transf.* **2021**, *177*, 121223. [CrossRef]
3. Qureshi, Z.A.; Al-Omari, S.A.B.; Elnajjar, E.; Al-Ketan, O.; Al-Rub, R.A. Using triply periodic minimal surfaces (TPMS)-based metal foams structures as skeleton for metal-foam-PCM composites for thermal energy storage and energy management applications. *Int. Commun. Heat Mass Transf.* **2021**, *124*, 105265. [CrossRef]
4. Talebizadehsardari, P.; Mohammed, H.I.; Mahdi, J.M.; Gillott, M.; Walker, G.S.; Grant, D.; Giddings, D. Effect of airflow channel arrangement on the discharge of a composite metal foam-phase change material heat exchanger. *Int. J. Energy Res.* **2021**, *45*, 2593–2609. [CrossRef]
5. Mahdi, J.M.; Pal Singh, R.; Taqi Al-Najjar, H.M.; Singh, S.; Nsofor, E.C. Efficient thermal management of the photovoltaic/phase change material system with innovative exterior metal-foam layer. *Sol. Energy* **2021**, *216*, 411–427. [CrossRef]
6. Heyhat, M.M.; Mousavi, S.; Siavashi, M. Battery thermal management with thermal energy storage composites of PCM, metal foam, fin and nanoparticle. *J. Energy Storage* **2020**, *28*, 101235. [CrossRef]
7. Liu, H.; Ahmad, S.; Shi, Y.; Zhao, J. A parametric study of a hybrid battery thermal management system that couples PCM/copper foam composite with helical liquid channel cooling. *Energy* **2021**, *231*, 120869. [CrossRef]
8. Mohammed, H.I.; Sardari, P.T.; Giddings, D. Multiphase flow and boiling heat transfer modelling of nanofluids in horizontal tubes embedded in a metal foam. *Int. J. Therm. Sci.* **2019**, *146*, 106099. [CrossRef]

9. Heyhat, M.M.; Valizade, M.; Abdolahzade, S.; Maerefat, M. Thermal efficiency enhancement of direct absorption parabolic trough solar collector (DAPTSC) by using nanofluid and metal foam. *Energy* **2020**, *192*, 116662. [CrossRef]
10. Saedodin, S.; Zamzamian, S.A.H.; Nimvari, M.E.; Wongwises, S.; Jouybari, H.J. Performance evaluation of a flat-plate solar collector filled with porous metal foam: Experimental and numerical analysis. *Energy Convers. Manag.* **2017**, *153*, 278–287. [CrossRef]
11. Kotresha, B.; Gnanasekaran, N. Effect of thickness and thermal conductivity of metal foams filled in a vertical channel—A numerical study. *Int. J. Numer. Methods Heat Fluid Flow* **2019**, *29*, 184–203. [CrossRef]
12. Trilok, G.; Gnanasekaran, N. Numerical study on maximizing heat transfer and minimizing flow resistance behavior of metal foams owing to their structural properties. *Int. J. Therm. Sci.* **2021**, *159*, 106617. [CrossRef]
13. Singh, P.; Nithyanandam, K.; Zhang, M.; Mahajan, R.L. The effect of metal foam thickness on jet array impingement heat transfer in high-porosity aluminum foams. *J. Heat Transf.* **2020**, *142*, 052301. [CrossRef]
14. Kotresha, B.; Gnanasekaran, N. Investigation of Mixed Convection Heat Transfer Through Metal Foams Partially Filled in a Vertical Channel by Using Computational Fluid Dynamics. *J. Heat Transfer* **2018**, *140*, 112501. [CrossRef]
15. Jadhav, P.H.; Gnanasekaran, N.; Perumal, D.A.; Mobedi, M. Performance evaluation of partially filled high porosity metal foam configurations in a pipe. *Appl. Therm. Eng.* **2021**, *194*, 117081. [CrossRef]
16. Ambrosio, G.; Bianco, N.; Chiu, W.K.S.; Iasiello, M.; Naso, V.; Oliviero, M. The effect of open-cell metal foams strut shape on convection heat transfer and pressure drop. *Appl. Therm. Eng.* **2016**, *103*, 333–343. [CrossRef]
17. Donmus, S.; Mobedi, M.; Kuwahara, F. Double-layer metal foams for further heat transfer enhancement in a channel: An analytical study. *Energies* **2021**, *14*, 672. [CrossRef]
18. Trilok, G.; Kumar, K.K.; Gnanasekaran, N.; Mobedi, M. Numerical assessment of thermal characteristics of metal foams of orderly varied pore density and porosity under different convection regimes. *Int. J. Therm. Sci.* **2022**, *172*, 107288. [CrossRef]
19. Zuo, H.; Wu, M.; Zeng, K.; Zhou, Y.; Kong, J.; Qiu, Y.; Lin, M.; Flamant, G. Numerical investigation and optimal design of partially filled sectorial metal foam configuration in horizontal latent heat storage unit. *Energy* **2021**, *237*, 121640. [CrossRef]
20. Sardari, P.T.; Mohammed, H.I.; Giddings, D.; Walker, G.S.; Gillott, M.; Grant, D. Numerical study of a multiple-segment metal foam-PCM latent heat storage unit: Effect of porosity, pore density and location of heat source. *Energy* **2019**, *189*, 116108. [CrossRef]
21. Mohammed, H.I.; Talebizadehsardari, P.; Mahdi, J.M.; Arshad, A.; Sciacovelli, A.; Giddings, D. Improved melting of latent heat storage via porous medium and uniform Joule heat generation. *J. Energy Storage* **2020**, *31*, 101747. [CrossRef]
22. Chen, S.; Li, W.; Mohammed, H.I. Heat transfer of large Prandtl number fluids in porous media by a new lattice Boltzmann model. *Int. Commun. Heat Mass Transf.* **2021**, *122*, 105129. [CrossRef]
23. Li, Y.; Wang, S.; Zhao, Y. Experimental study on heat transfer enhancement of gas tube partially filled with metal foam. *Exp. Therm. Fluid Sci.* **2018**, *97*, 408–416. [CrossRef]
24. Bianco, N.; Iasiello, M.; Mauro, G.M.; Pagano, L. Multi-objective optimization of finned metal foam heat sinks: Tradeoff between heat transfer and pressure drop. *Appl. Therm. Eng.* **2021**, *182*, 116058. [CrossRef]
25. Siavashi, M.; Talesh Bahrami, H.R.; Aminian, E. Optimization of heat transfer enhancement and pumping power of a heat exchanger tube using nanofluid with gradient and multi-layered porous foams. *Appl. Therm. Eng.* **2018**, *138*, 465–474. [CrossRef]
26. Shikh Anuar, F.; Ashtiani Abdi, I.; Odabaee, M.; Hooman, K. Experimental study of fluid flow behaviour and pressure drop in channels partially filled with metal foams. *Exp. Therm. Fluid Sci.* **2018**, *99*, 117–128. [CrossRef]
27. Lai, Z.; Hu, H.; Ding, G. Influence of pore density on heat transfer and pressure drop characteristics of wet air in hydrophilic metal foams. *Appl. Therm. Eng.* **2019**, *159*, 113897. [CrossRef]
28. Sun, M.; Hu, C.; Zha, L.; Xie, Z.; Yang, L.; Tang, D.; Song, Y.; Zhao, J. Pore-scale simulation of forced convection heat transfer under turbulent conditions in open-cell metal foam. *Chem. Eng. J.* **2020**, *389*, 124427. [CrossRef]
29. Mancin, S.; Zilio, C.; Rossetto, L.; Cavallini, A. Foam height effects on heat transfer performance of 20 ppi aluminum foams. *Appl. Therm. Eng.* **2012**, *49*, 55–60. [CrossRef]
30. Kamath, P.M.; Balaji, C.; Venkateshan, S.P. Experimental investigation of flow assisted mixed convection in high porosity foams in vertical channels. *Int. J. Heat Mass Transf.* **2011**, *54*, 5231–5241. [CrossRef]
31. Calmidi, V.V.; Mahajan, R.L. Forced convection in high porosity metal foams. *J. Heat Transfer* **2000**, *122*, 557–565. [CrossRef]
32. Zukauskas, A. Convective heat transfer in cross flow. In *Handbook of Single-Phase Convective Heat Transfer*; John Wiley & Sons: Hoboken, NJ, USA, 1987.
33. Jadhav, P.H.; Trilok, G.; Gnanasekaran, N.; Mobedi, M. Performance score based multi-objective optimization for thermal design of partially filled high porosity metal foam pipes under forced convection. *Int. J. Heat Mass Transf.* **2022**, *182*, 121911. [CrossRef]

Article

Heat Transfer Potential of Unidirectional Porous Tubes for Gas Cooling under High Heat Flux Conditions

Kazuhisa Yuki [1,*], Risako Kibushi [1], Ryohei Kubota [1], Noriyuki Unno [1], Shigeru Tanaka [2] and Kazuyuki Hokamoto [2]

[1] Department of Mechanical Engineering, Tokyo University of Science, Yamaguchi 1-1-1 Daigakudori, Sanyo-Onoda 756-0884, Japan; kibushi@rs.socu.ac.jp (R.K.); f118025@ed.socu.ac.jp (R.K.); unno@rs.socu.ac.jp (N.U.)
[2] Institute of Industrial Nanomaterials, Kumamoto University, 2-39-1 Kurokami, Chuo-ku, Kumamoto 860-8555, Japan; tanaka@mech.kumamoto-u.ac.jp (S.T.); hokamoto@mech.kumamoto-u.ac.jp (K.H.)
* Correspondence: kyuki@rs.socu.ac.jp

Abstract: To discuss a suitable porous structure for helium gas cooling under high heat flux conditions of a nuclear fusion divertor, we first evaluate effective thermal conductivity of sintered copper-particles in a simple cubic lattice by direct numerical heat-conduction simulation. The simulation reveals that the effective thermal conductivity of the sintered copper-particle highly depends on the contacting state of each particle, which leads to the difficulty for the thermal design. To cope with this difficulty, we newly propose utilization of a unidirectional porous tube formed by explosive compression technology. Quantitative prediction of its cooling potential using the heat transfer correlation equation demonstrates that the heat transfer coefficient of the helium gas cooling at the pressure of 10 MPa exceeds 30,000 W/m^2/K at the inlet flow velocity of 25 m/s, which verifies that the unidirectional porous copper tubes can be a candidate for the gas-cooled divertor concept.

Keywords: unidirectional porous tube; gas cooling; high heat flux condition; fusion reactor; divertor; effective thermal conductivity; sintered particles; porous media

1. Introduction

The surrounding of the core plasma in a fusion reactor contains structures and equipment exposed to extremely severe heat loads, such as the first wall and the divertor. These must withstand the radiation from the plasma and the load of high energy particles while continuously maintaining these functions. For example, the divertor, to which α-particles flow directly, is subjected to a localized, but steady, heat load of about 10 MW/m^2. It goes without saying that in such an extreme thermal load environment, a sufficient cooling margin is essential but, at the same time, the establishment of a thermal design with excellent economic efficiency, soundness, and maintainability is the key for the realization of nuclear fusion power reactors.

Based on the above background, the development of the divertor cooling technology is one of the most important issues of the reactor engineering. In ITER (International Thermonuclear Experimental Reactor), which is currently under construction in Cadarache, France, a water-cooled system is adopted for the divertor cooling because of the emphasis on self-ignition and demonstration of the reactor engineering technologies. To enhance the cooling performance, a swirl tube with an inserted twisted tape is applied [1,2]. As the fluid flows in the swirl tube while turning spirally, a secondary flow is formed by centrifugal force. In this way, fluid mixing is promoted, resulting in thinning the temperature boundary layer and improving the cooling performance, as well as the increase in the critical heat flux. On the other hand, the application of screw tubes with a threaded structure on the inner surface of the tube has been studied in the divertor of JT-60SA and the prototype reactor [3,4]. The cooling performance of the screw tube proved higher than that of the swirl tube with the same pumping power because of the effect of increased heat transfer area and fluid stirring in the vicinity of the tube wall. As another water-cooling technique for the divertor, Toda has proposed an ultra-low flowrate-type evaporative heat transfer device, EVAPORON (Evaporated Fluid Porous Thermodevice), which utilizes the latent heat of vaporization of a cooling liquid in a metal porous medium [5]. Furthermore, the author has upgraded it to EVAPORON-2, which is equipped with subchannels for enhancement of vapor discharge and has demonstrated cooling performance exceeding 20 MW/m^2 at extremely low flow rates [6,7]. So far, the authors have also proposed EVAPORON-3 [8], adaptable to large divertor surfaces, and EVAPORON-4 [9], which applies a unidirectional porous medium.

In recent years, however, the helium gas cooling has been reconsidered as a highly safe cooling method for divertors [10–12]. In particular, international joint research had been conducted as one of the tasks of PHOENIX, the Japan/U.S. fusion research collaboration project. To apply helium, which has a low heat capacity, as a coolant, the HEMJ (Helium-cooled Multi-Jet) cooling method has been studied, in which helium is compressed up to about 10 MPa and injected as an impinging jet flow into a narrow channel equipped with a number of nozzles [13]. Although a heat transfer coefficient exceeding 30,000 W/m^2/K has been obtained at the pressure of 10 MPa, deterioration of the heat transfer performance due to re-laminarization has also been pointed out [14].

To complement the low heat transfer performance of the gas cooling, it is necessary to introduce a cooling technology that encompasses all the technologies related to heat transfer enhancement, such as (1) increasing the heat transfer area, (2) promoting turbulent heat transfer, and (3) utilizing micro- and mini-channels. We can know various kinds of heat transfer promoters for the gas flow from Tao's review article [15]. From the viewpoint of high heat flux conditions that needs extremely high heat transfer coefficient, the cooling technology using porous media is the one that satisfies all of the above criteria. In the existing studies for the divertor cooling, Hermsmeyer et al. theoretically demonstrated that the heat transfer coefficient of the helium gas flow in pin fin porous arrays exceeds 60,000 W/m^2/K at the inlet flow velocity of 20 m/s (averaged local velocity: 120 m/s) and the pressure of 10 MPa [16]. Sharafat et al. focused on the cooling technique that uses metal foam and his CFD simulation verified that the heat transfer coefficient is in the range from 12,500 to 25,000 W/m^2/K for the modeled flow at the inlet flow velocity of 150 m/s and the pressure of 4 MPa [17,18]. Yuki et al. studied the cooling performance using sintered particles and the heat transfer coefficient of N_2 gas impinging jet flow with the porous medium is much higher than that of common impinging jet flow without the porous medium from the viewpoint of not only flow velocity, but also pumping power [19]. Up to now, however, the issue of a porous medium suitable for cooling the high heat flux of the divertor has not been sufficiently addressed. Toward the optimal control of the porous structure for the high heat flux removal, 3D-printed porous metals fabricated by selective laser metal melting (SLMM) technology, such as a porous lattice [20], are also candidates. Various kinds of the 3D-printed porous metal [21–25] can be expected to bring out the heat transfer potential because we can easily control the porosity and pore size distribution against the heat load and cooling conditions. However, the fabrication of a 3D-printed porous "copper" with micro/mini channels seems to be considerably difficult in the conventional 3D-printed technology.

Against these backgrounds, in this study, we first focus on again the pros and cons of utilizing sintered copper-particles porous media with excellent thermal conductivity and vast heat transfer surface from the viewpoint of the high heat flux removal. After that, we newly propose the introduction of a unidirectional porous tube formed by explosive compression technology, which was developed by Hokamoto et al. [26], and demonstrate its cooling potential by the heat transfer correlation equation constructed by the author's heat transfer experiments.

2. Prediction of Effective Thermal Conductivity of Sintered Copper-Particles by Direct Numerical Simulation of Heat Conduction

2.1. Procedure for Evaluating Effective Thermal Conductivity of Porous Medium

A general formula for estimating the effective thermal conductivity k_{eff} [W/m/K] of porous media is shown as follows:

$$k_{\text{eff}} = \varepsilon k_f + (1 - \varepsilon) k_s \tag{1}$$

where, ε is the porosity, and k_f [W/m/K] and k_s [W/m/K] are the thermal conductivities of the fluid and the porous solid, respectively. Although Equation (1) indicates that the effective thermal conductivity is solely based on the porosity of the porous media, a highly versatile correlation equation for the effective thermal conductivity should be taken into account using the porosity, the pore size, the structure, and other parameters that affect the effective thermal conductivity (e.g., thermal contact resistance between particles, etc.).

For that purpose, in this study, the effective thermal conductivity is directly evaluated using a three-dimensional numerical simulation of heat conduction inside the porous medium. Here, we focus on sintered copper-particles as the porous media because the sintered particles have higher thermal conductivity compared to other porous media.

Figure 1 shows the simulation model to evaluate the effective thermal conductivity. The porous medium is a structure in which the sphere particles are placed in a simple cubic lattice, which is assumed to have the highest effective thermal conductivity among all means of placing the particles. The particles, 1.0 mm in diameter, are packed by placing five pieces along the horizontal, vertical, and lateral sides, making the size of the formed porous medium 5 × 5 × 5 mm (see Figure 1 on the right). The porosity ε for this structure is 0.48. Figure 1 on the right shows the method for jointing the sphere particles. To reproduce a porous bed in which the sphere particles are sintered or point-contacted, the particles are joined by a cylinder having a diameter d [mm], which is determined by the central angle θ [°] of the particles (hereafter, contact angle). Square rods, 5.0 mm wide and 50 mm long, are attached to the upper and lower parts of the porous medium. A square plate of 1.0 mm in thickness is attached to the upper surface of the rod 2 to set a constant temperature during the simulation. The finite element method is used for solving the 3-dimentional equation of the heat conduction. We utilized a commercial software "CreoParametric ver.6.0" for the simulation. Pure copper with thermal conductivity k_s = 398 W/m/K is set for all solid parts of the simulation model. The void portion of the porous bed is filled with a static air (thermal conductivity k_f = 0.0256 W/m/K). As for the boundary conditions, all the side walls are defined as adiabatic conditions, the heat flux of 500,000 W/m² is uniformly applied to the bottom surface of the rod 1, and the top surface of the rod 2 is set to 100 °C. The most important parameter in this simulation is the contact angle θ formed between the particles. There are six patterns: θ = 5°, in which the particles are close to point contact; and θ = 10°, 20°, 30°, 40°, and 50°, close to the state in a sintered particle. Before the simulation, the suitable mesh size and structure were sufficiently evaluated and discussed many times especially in the contacting region of the particles. The effective thermal conductivity k_{eff} [W/m/K] of the sintered copper-particles is estimated on the basis of the temperature difference T_1–T_2 obtained by the simulation, as follows:

$$k_{\text{eff}} = \frac{q \times \Delta y}{T_1 - T_2} \tag{2}$$

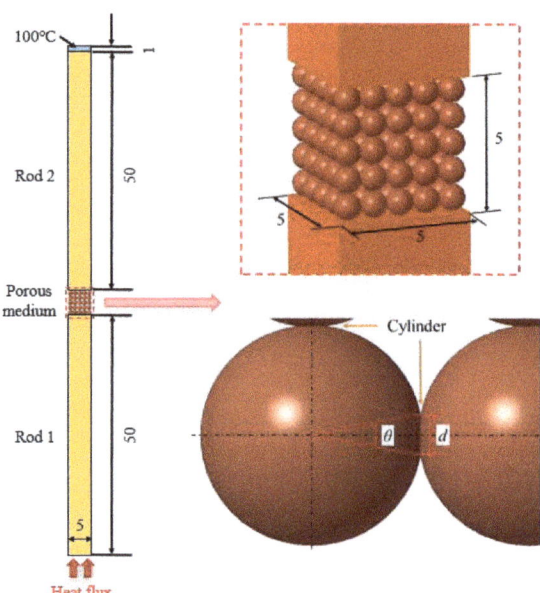

Figure 1. Simulation model for evaluating effective thermal conductivity.

In this equation, q [W/m^2] is the heat flux, Δy [m] is the length of the porous bed in the y-direction, and T_1 [K] and T_2 [K] represent the temperatures of the joint surface between the rod 1 or rod 2 and the porous bed, respectively. These temperatures are determined by approximating the temperature distribution in the rods as a linear function using the least squares method, where the temperature distribution in the axial direction can be regarded as one-dimensional profile. This evaluation method of the thermal conductivity is well-known as the steady method (experimental method).

2.2. Effective Thermal Conductivity of Sintered Copper-Particles

Figure 2 shows the effective thermal conductivity for each contact angle. The value (%) in the figure is the ratio of the effective thermal conductivity to the thermal conductivity of pure copper. According to the results of the simulation, the effective thermal conductivities are 15.1, 27.5, 49.8, 72.0, 93.6, and 114.6 W/m/K for the contact angles $\theta = 5°$, 10°, 20°, 30°, 40°, and 50° respectively. This clearly verifies that the effective thermal conductivity increases with increasing contact area (i.e., the degree of sintering). Compared to the thermal conductivity of pure copper, if the contact area between the particles is small (i.e., the contact angle $\theta = 5°$, which is a state close to point contact), the effective thermal conductivity is reduced to 3.8% of that of pure copper, whereas even at $\theta = 50°$, simulating a sufficiently sintered state, the thermal conductivity of copper is reduced to approximately 29%. These results indicate that less than 30% of the thermal conductivity of pure copper can be utilized even with a simple cubic lattice, which has the highest effective thermal conductivity among different particle placements. Furthermore, it is worth mentioning that the effective thermal conductivity estimated by the porosity weighting Equation (1) is significantly different from the simulation results and overestimates the effective thermal conductivity when the thermal conductivity of the porous solid, k_s, is given as that of pure copper. This demonstrates that the porosity information is not sufficient for calculating the effective thermal conductivity and that a correlation equation that reflects the tortuosity of the porous media, i.e., packing structure of the particles and the degree of sintering, is required.

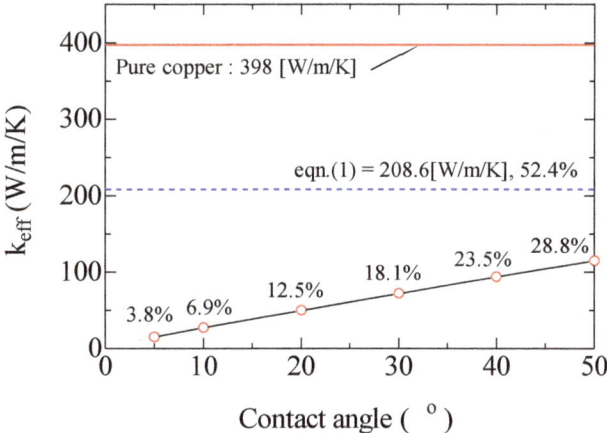

Figure 2. Effective thermal conductivity of copper sintered particle.

Furthermore, these results verify that, even in the case of placing the particles in a certain cubic lattice, the optimal design for the heat transfer enhancement must take into account the actual effective thermal conductivity of the sintered copper-particle, as well as the thickness of the porous layer to increase the fin efficiency, etc. In addition, the contacting state between the particles, as well as between the particle-sintered porous medium and the heat transfer surface, highly affects the fin effect inside the porous medium, which makes it more difficult to precisely predict the cooling performance of forced convective heat transfer using the sintered copper-particle. Especially under high heat flux conditions such as nuclear fusion divertors, precise evaluation of the effective thermal conductivity highly affects the prediction of the surface temperature of an armor material of the divertor that faces the core plasma.

3. Heat Transfer Potential of Unidirectional Porous Copper Tube of Gas Flow

Heat Transfer Correlation of Gas Flow in Unidirectional Porous Tubes Fabricated by Explosive Compression Technology

To cope with the difficulty of precise thermal design using the sintered copper-particle porous media for the divertor cooling, we focus on completely new porous copper tubes with uniformly-distributed pore holes fabricated by explosive compression technique that was developed by Hokamoto et al. [26]. This unique unidirectional porous tube (hereafter, porous tube) is fabricated by compressing a bundle of thin copper tubes by a gunpowder explosion that is set around the bundle (see Figure 3). One of the features of the porous tube is high thermal conductivity, even in the radial direction. For instance, the effective thermal conductivities of the porous tube in the axial and radial directions ($k_{eff//}$ and $k_{eff\perp}$) are approximately 210 W/m/K and 110 W/m/K at the porosity of 50 %, respectively, which enables active utilization of the vast heat transfer surface of the porous tube. Here, $k_{eff//}$ can be estimated by Equation (1) and $k_{eff\perp}$ is based on the Ogushi's equation [27].

$$k_{eff\perp} = \frac{(\beta+1)+\varepsilon(\beta-1)}{(\beta+1)-\varepsilon(\beta-1)} k_s \qquad (3)$$

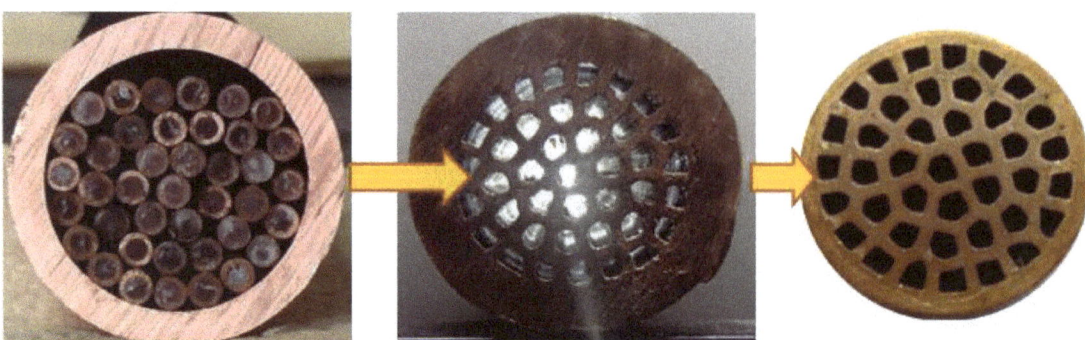

Figure 3. Unidirectional porous copper tube #1 (before and after explosive compaction).

Figure 4 shows a high fin efficiency of the unidirectional porous plate fins (1.0 mm × 10 mm × 10 mm), compared to that of a sintered copper-particle fin (Each porosity is 50%). In addition, the pore diameter and the porosity can be optionally adjustable by changing the thin copper tube with different inner diameter and thickness. Utilizing the porous tube as a heat transfer promoter also makes it possible to reduce the pressure loss in comparison with other porous media such as a sintered particle because of its unidirectional pore structure. As the result, the porous tube also enables to reduce the pumping power.

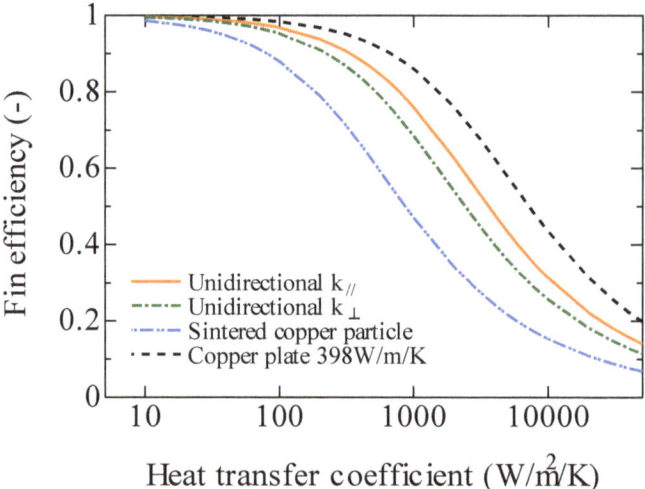

Figure 4. Fin efficiency of sintered copper-particle and unidirectional porous copper.

In past studies regarding the unidirectional porous tube, Fiedler et al. evaluated the mechanical and thermal properties [28]. Regarding the heat transfer characteristics of the unidirectional porous copper tube, the author firstly demonstrated its extremely high heat transfer performance for gas flow [29,30]. In addition, Kibushi et al. [31] evaluated the heat transfer characteristics of many kinds of the porous tubes as shown in Table 1. The heat transfer experiments were performed using a double-tube heat exchanger, where the porous tube is set as the inner tube. Hot water flows inside the annular channel between the outer tube and the porous tube, and air driven by the compressor flows through the porous tube. As to the details of the heat transfer experiments, please see reference [31]. The experimental results clarified that the heat transfer performance of the porous tubes is

7.4 times higher than that of the circular tube flow at the maximum and superior compared with other heat transfer promoter tubes such as the swirl tube and artificially roughened ducts (see Figure 5). Furthermore, we constructed a heat transfer correlation equation.

Table 1. Specifications of porous tubes.

	#1	#2	#3	#4	#5	#6
Length (mm)	470	455	340	475	450	440
Number of pore	40	40	40	21	13	9
Pore size (mm)	1.69	2.16	2.71	2.60	3.51	4.61
Porosity (%)	35.5	57.7	81.3	44.2	49.5	59.1

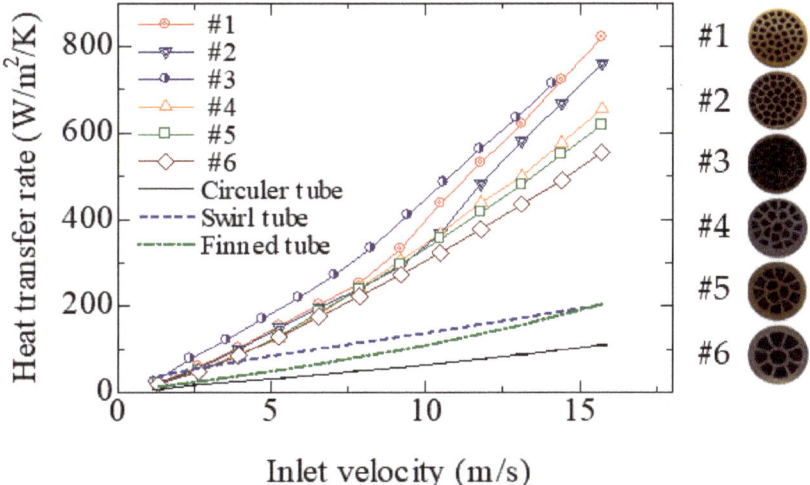

Figure 5. Heat transfer performance of porous tubes.

Of course, the demonstration experiment of the porous tube toward the heat transfer coefficient of 30,000 W/m²/K at least, should be carried out under the divertor cooling conditions using helium flow, but the experimental demonstration is considerably difficult because the pressure is extremely high, approximately 10 MPa, and the temperature of the helium flow is higher than 300 °C. In that sense, the heat transfer potential should be evaluated instead, as the first step, by applying the following heat transfer correlation equation of the porous tube flow we constructed in reference [31].

$$Nu = 0.286\, Re^{0.8} Pr^{0.4} \left(\frac{d_{pore}}{D}\right)^{-0.388} \left(\frac{k_{eff}}{k_{gas}}\right)^{-0.176} \qquad (4)$$

Here, $Nu = hD/k_{gas}$ (k_{gas}) is thermal conductivity of gas, and D is inside diameter of the porous tube [m]). The porous copper tube has long and unidirectional pore structure, therefore, the heat transfer correlation equation is constructed based on the Dittus-Boelter correlation equation, which is a well-known correlation for turbulent convective heat transfer in a circular tube. In addition, the correlation equation for the porous tubes takes into account the pore structure as the ratio of the pore size d_{pore} to the tube diameter D, and the effective thermal conductivity as the ratio of the effective thermal conductivity of the porous tube k_{eff} to the thermal conductivity of the gas k_{gas}. Here, the effective thermal conductivity of the porous copper tubes k_{eff} is calculated using the Equation (3).

Here, the heat transfer coefficient of the helium gas flow in the porous copper tube is predicted using the physical property of the helium gas at the temperature of 300 °C and the pressure of 10 MPa. The porous tube #1 in Table 1, which showed the highest heat transfer performance in Figure 5, is used for the potential evaluation, because the porous tube with thicker solid wall is more suitable for achieving the high heat transfer coefficient from the viewpoint of the fin theory. Figure 6 shows the predicted heat transfer performance of the helium gas flow. The heat transfer coefficient exceeds 30,000 W/(m²·K) at the inlet flow velocity of 25 m/s (average velocity in each pore is 70.4 m/s). In addition, the heat transfer coefficient of over 30,000 W/m²/K could easily be possible by optimizing the pore size and the porosity of the porous tube, which indicates that the porous copper tubes can be one of candidates for the gas-cooled divertor concept. As to the pressure loss, we can apply the Darcy–Weisbach equation, which is commonly utilized for the pressure loss prediction of a circular tube flow, depending on the actual piping geometry for the divertor cooling, because the unidirectional porous tube is an assembly of circular tubes (if there is no deformation of the inner thin tube).

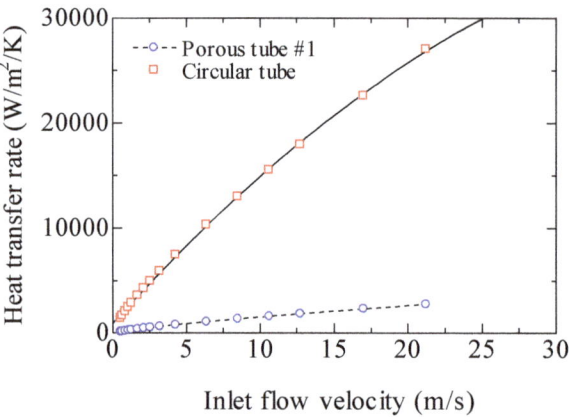

Figure 6. Heat transfer performance of helium gas flow in unidirectional porous tube.

4. Conclusions

To discuss suitable porous structure for gas-cooled divertor concept under high heat flux conditions of 10 MW/m², we first evaluated effective thermal conductivity of sintered copper-particle in a simple cubic lattice by direct numerical heat-conduction simulation. The simulation revealed that the effective thermal conductivity of the sintered copper-particle is in the range from 15.1 to 114.6 W/m/K (pure copper: 398 W/m/K) and highly depends on the contacting state of the particles, which makes it difficult to predict the exact temperature of divertor armor material. To cope with this difficulty, we newly proposed the utilization of a unidirectional porous copper tube formed by explosive compression technology. As the experimental demonstration is considerably difficult under the divertor cooling conditions, quantitative prediction of its cooling potential using the heat transfer correlation equation proved that the heat transfer coefficient of the porous tube with the porosity of 35.5% and the averaged pore size of 1.69 mm exceeds 30,000 W/m²/K at the inlet flow velocity of 25 m/s and the pressure of 10 MPa, which verifies that the porous copper tubes can be one of candidates for the gas-cooled divertor concept.

Author Contributions: Conceptualization and methodology, K.Y.; validation and analysis, R.K. (Risako Kibushi), R.K. (Ryohei Kubota) and N.U.; writing—original draft preparation, K.Y.; writing—review and editing, R.K. (Risako Kibushi) and N.U.; supervision, K.H. and S.T.; project administration, K.Y. and K.H. All authors have read and agreed to the published version of the manuscript.

Funding: This research was funded by AMADA Foundation [AF-2019005-A3].

Institutional Review Board Statement: Not applicable.

Informed Consent Statement: Not applicable.

Data Availability Statement: Not applicable.

Acknowledgments: This work was joint research between Kumamoto University and our group and supported by Institute of pulsed power science (Institute of industrial nanomaterials, and the authors acknowledge the financial support by AMADA Foundation [AF-2019005-A3]). Additionally, we deeply appreciate effort by Yoshiaki Sato and Yoshiaki Miyamoto, who are graduates of Yuki laboratory, and Akira Ogushi of Ogushi Manufactory (Miyagi, Japan) and his sophisticated machining skill.

Conflicts of Interest: The authors declare that they have no known competing financial interests or personal relationships that could have appeared to influence the work reported in this paper.

References

1. Raffray, A.R.; Schlosser, J.; Akiba, M.; Araki, M.; Chiocchio, S.; Driemeyer, D.; Escourbiac, F.; Grigoriev, S.; Merola, M.; Tivey, R. Critical heat flux analysis and R&D for the design of the ITER divertor. *Fusion Eng. Des.* **1999**, *45*, 377–407.
2. Araki, M.; Sato, K.; Suzuki, S.; Akiba, M. Critical-heat-flux experiment on the screw tube under one-sided-heating conditions. *Fusion Technol.* **1996**, *29*, 519–528. [CrossRef]
3. Boscary, J.; Araki, M.; Suzuki, S.; Ezato, K.; Akiba, M. Critical heat flux in subcooled water flow of one-side-heated screw tubes. *Fusion Technol.* **1999**, *35*, 289–296. [CrossRef]
4. Ezato, K.; Suzuki, S.; Sato, K.; Akiba, M. Thermal fatigue experiment of screw cooling tube under one-sided heating condition. *J. Nucl. Mater.* **2004**, *329*, 820–824. [CrossRef]
5. Toda, S.; Ebara, S.; Hashizume, H. Development of an advanced cooling device using porous media with active boiling flow counter to high heat flux. Fundamental study on high heat flux removal system using evaporated fluid in metal porous media. In Proceedings of the 11th International Heat Transfer Conferences (IHTC11), Kyongju, Korea, 23–28 August 1998; pp. 503–508.
6. Yuki, K.; Hashizume, H.; Toda, S. Sub-channels-inserted porous evaporator for efficient divertor cooling. *Fusion Sci. Technol.* **2011**, *60*, 238–242. [CrossRef]
7. Yuki, K.; Hashizume, H.; Toda, S.; Sagara, A. Divertor cooling with sub-channels-inserted metal porous media. *Fusion Sci. Technol.* **2013**, *64*, 325–330. [CrossRef]
8. Takai, K.; Yuki, K.; Sagara, A. Heat transfer performance of EVAPORON-3 developed for an enlarged heat transfer surface of divertor. *Plasma Fusion Res.* **2017**, *12*, 1405015. [CrossRef]
9. Takai, K.; Yuki, K.; Yuki, K.; Kibushi, R.; Unno, N. Heat transfer performance of an energy-saving heat removal device with uni-directional porous copper for divertor cooling. *Fusion Eng. Des.* **2018**, *136 Pt A*, 518–521. [CrossRef]
10. Raffray, A.R.; Malang, S.; Wang, X. Optimizing the overall configuration of a He-cooled W-alloy divertor for a power plant. *Fusion Eng. Des.* **2009**, *84*, 1553–1557. [CrossRef]
11. Yoda, M.; Abdel-Khalik, S.I.; Sadowski, D.L.; Mills, B.H.; Rader, J.D. Experimental evaluation of the thermal-hydraulics of helium-cooled divertors. *Fusion Sci. Technol.* **2015**, *67*, 142–157. [CrossRef]
12. Yoda, M.; Abdel-Khalik, S.I. Overview of thermal hydraulics of helium-cooled solid divertors. *Fusion Sci. Technol.* **2017**, *72*, 285–293. [CrossRef]
13. Zhao, B.B.B.; Musa, S.; Abdel-Khalik, S.; Yoda, M. Experimental and numerical studies of helium-cooled modular divertors with multiple jets. *Fusion Eng. Des.* **2018**, *136*, 67–71. [CrossRef]
14. Yokomine, T.; Oohara, K.; Kunugi, T. Experimental investigation on heat transfer of HEMJ type divertor with narrow gap between nozzle and impingement surface. *Fusion Eng. Des.* **2016**, *109*, 1543–1548. [CrossRef]
15. Ji, W.-T.; Jacobi, A.M.; He, Y.-L.; Tao, W.-Q. Summary and evaluation on the heat transfer enhancement techniques of gas laminar and turbulent pipe flow. *Int. J. Heat Mass Transf.* **2017**, *111*, 467–483. [CrossRef]
16. Hermsmeyer, S.; Malang, S. Gas-cooled high performance divertor for a power plant. *Fusion Eng. Des.* **2002**, *61–62*, 197–202. [CrossRef]
17. Sharafat, S.; Mills, A.; Youchison, D.; Nygren, R.; Williams, B.; Ghoniem, N. Ultra low pressure-drop helium-cooled porous-tungsten PFC. *Fusion Sci. Technol.* **2007**, *52*, 559–565. [CrossRef]
18. Sharafat, S.; Aoyama, A.T.; Ghoniem, N.; Williams, B. Design and fabrication of a rectangular he-cooled refractory foam hx-channel for divertor applications. *Fusion Sci. Technol.* **2011**, *60*, 208–212. [CrossRef]
19. Yuki, K.; Kawamoto, M.; Hattori, M.; Suzuki, K.; Sagara, A. Gas-cooled divertor concept with high thermal conductivity porous media. *Fusion Sci. Technol.* **2015**, *61*, 197–202. [CrossRef]
20. Takarazawa, S.; Ushijima, K.; Fleischhauer, R.; Kato, J.; Terada, K.; Cantwell, W.J.; Kaliske, M.; Kagaya, S.; Hasumoto, S. Heat-transfer and pressure drop characteristics of micro-lattice materials fabricated by selective laser metal melting technology. *Heat Mass Transf.* **2022**, *58*, 125–141. [CrossRef]
21. Trevizoli, P.V.; Teyber, R.; da Silveira, P.S.; Scharf, F.; Schillo, S.M.; Niknia, I.; Govindappa, P.; Christiaanse, T.V.; Rowe, A. Thermal-hydraulic evaluation of 3D printed microstructures. *Appl. Therm. Eng.* **2019**, *160*, 113990. [CrossRef]
22. Elkholy, A.; Kempers, R. Enhancement of pool boiling heat transfer using 3D-printed polymer fixtures. *Exp. Therm. Fluid Sci.* **2020**, *114*, 110056. [CrossRef]

23. Guo, Y.; Yang, H.; Lin, G.; Jin, H.; Shen, X.; He, J.; Miao, J. Thermal performance of a 3D printed lattice-structure heat sink packaging phase change material. *Chin. J. Aeronaut.* **2021**, *34*, 373–385. [CrossRef]
24. Hu, X.; Gong, X. Experimental study on the thermal response of PCM-based heat sink using structured porous material fabricated by 3D printing. *Case Stud. Therm. Eng.* **2021**, *24*, 100844. [CrossRef]
25. Liang, D.; He, G.; Chen, W.; Chen, Y.; Chyu, M.K. Fluid flow and heat transfer performance for micro-lattice structures fabricated by Selective Laser Melting. *Int. J. Therm. Sci.* **2022**, *172*, 107312. [CrossRef]
26. Hokamoto, K.; Vesenjak, M.; Ren, Z. Fabrication of cylindrical uni-directional porous metal with explosive compaction. *Mater. Lett.* **2014**, *137*, 323–327. [CrossRef]
27. Ogushi, T.; Chiba, H.; Nakajima, H.; Ikeda, T. Measurement and analysis of effective thermal conductivities of lotus-type porous copper. *J. Appl. Phys.* **2004**, *95*, 5843–5847. [CrossRef]
28. Fiedler, T.; Borovinšek, M.; Hokamoto, K.; Vesenja, M. High-performance thermal capacitors made by explosion forming. *Int. J. Heat Mass Transf.* **2014**, *83*, 366–371. [CrossRef]
29. Yuki, K.; Sato, Y.; Kibushi, R.; Unno, N.; Suzuki, K.; Tomimura, T.; Hokamoto, K. Heat transfer performance of porous copper pipe with uniformly-distributed holes fabricated by explosive welding technique. In Proceedings of the 27th International Symposium on Transport Phenomena (ISTP27), Honolulu, HI, USA, 20–23 September 2016. ISTP27-126.
30. Sato, Y.; Yuki, K.; Abe, Y.; Kibushi, R.; Unno, N.; Hokamoto, K.; Tanaka, S.; Tomimura, T. Heat transfer characteristics of a gas flow in uni-directional porous copper pipes. In Proceedings of the 16th International Heat Transfer Conference (IHTC16), Beijing, China, 10–15 August 2018; pp. 8189–8193.
31. Kibushi, R.; Yuki, K.; Unno, N.; Tanaka, S.; Hokamoto, K. Heat transfer and pressure drop correlations for a gas flow in unidirectional porous copper tubes fabricated by explosive compaction. *Int. J. Heat Mass Transf.* **2022**. submitted.

Article

Thermal Cloaking in Nanoscale Porous Silicon Structure by Molecular Dynamics

Jian Zhang, Haochun Zhang *, Yiyi Li, Qi Wang and Wenbo Sun

School of Energy and Engineering, Harbin Institute of Technology, Harbin 150001, China;
20b902063@stu.hit.edu.cn (J.Z.); 18b902056@stu.hit.edu.cn (Y.L.); 21s002039@stu.hit.edu.cn (Q.W.);
21s102109@stu.hit.edu.cn (W.S.)
* Correspondence: hczhang@hit.edu.cn

Abstract: Nanoscale thermal cloaks have great potential in the thermal protection of microelectronic devices, for example, thermal shielding of thermal components close to the heat source. Researchers have used graphene, crystalline silicon film, and silicon carbide to design a variety of thermal cloaks in different ways. In our previous research, we found that the porous structure has lower thermal conductivity compared to bulk silicon; thus, so we tried to use the porous structure to construct the functional region to control the heat flux. We first calculated the thermal conductivity of crystalline silicon and porous silicon films by means of nonequilibrium molecular dynamics, proving that the porous structure satisfied the conditions for building a thermal cloak. A rectangular cloak with a porous structure was constructed, and a crystalline silicon film was used as a reference to evaluate its performance by the index of the ratio of thermal cloaking. We found that the thermal cloak built with a porous structure could produce an excellent cloaking effect. Lastly, we explain the mechanism of the cloaking phenomenon produced by a porous structure with the help of phonon localization theory. Porous structures have increased porosity compared to bulk silicon and are not conducive to phonon transport, thus producing strong phonon localization and reducing thermal conductivity. Our research expands the construction methods of nanocloaks, expands the application of porous structure materials, and provides a reference for the design of other nanodevices.

Keywords: porous silicon; thermal cloak; phonon localization; molecular dynamics; nanoscale

Citation: Zhang, J.; Zhang, H.; Li, Y.; Wang, Q.; Sun, W. Thermal Cloaking in Nanoscale Porous Silicon Structure by Molecular Dynamics. *Energies* **2022**, *15*, 1827. https://doi.org/10.3390/en15051827

Academic Editors: Moghtada Mobedi and Kamel Hooman

Received: 6 February 2022
Accepted: 25 February 2022
Published: 2 March 2022

Publisher's Note: MDPI stays neutral with regard to jurisdictional claims in published maps and institutional affiliations.

Copyright: © 2022 by the authors. Licensee MDPI, Basel, Switzerland. This article is an open access article distributed under the terms and conditions of the Creative Commons Attribution (CC BY) license (https://creativecommons.org/licenses/by/4.0/).

1. Introduction

The manipulation and regulation of heat flux has always been a research hotspot, and the transformation thermotics theory [1] and thermal metamaterials have further promoted its development. Researchers have realized the regulation of heat flux. For example, the thermal cloak [2–11] is a thermal function device that protects the objects in the functional region from the outside temperature without disturbing the background temperature field. Hu et al. [12] put forward the concept of "inverse heat cloak", whereby the heat flux is successfully concentrated to the functional region to achieve local heating, also known as the thermal concentrator [13]. In addition, researchers have designed a thermal rotator [14], thermal camouflage [15–21], thermal illusion [22–25], encrypted thermal printing [26], etc. and conducted experimental verification. So far, the regulation of heat flux has been achieved in various situations on the macroscale. This has also greatly promoted the development of thermodynamics and put forward different thermodynamic evaluation indicators, such as local entropy production rate [27], and response entropy [28], thus also better evaluating the effect of heat flux regulation and providing a powerful tool for performance optimization of heat flux regulation devices. To advance the application of heat flux control devices in industrial production, Li et al. [29] evaluated the cloaking performance and environmental response of a 2D thermal cloak based on a dynamic environment in the form of sinusoids, using variables such as ambient amplitude and temperature difference. Considering the heat dissipation and energy loss of the whole process, the local entropy generation rate was introduced to analyze the influence of environmental parameters on the cloaking system. The results showed that, in the functional region, the thermal dissipation and irreversible energy loss of the system were caused by the difference in thermal conductivity between adjacent layers, and the heat dissipation capacity of the thermal cloak increased with the increase in the ambient amplitude and temperature difference. According to the distribution of entropy yield, it was found that thermal cloaking has a corresponding application range for different environmental variables. Considering the requirements for different energy types and functions in practical applications, research on metamaterial structures that simultaneously realize complex tuning functions in multiphysics is particularly important. Xu et al. [30] solved the complex regulation problem in multiphysics by designing a discrete array of "sources" using the "rotation–linear" mapping method. Discrete arrays of "sources" were designed to reconfigure each physical field produced by a single excitation source and form a "multiple source" array field distribution. This research provided a new way to efficiently manipulate and distribute the energy generated by excitation sources, which could lead to the development of solar cells and thermoelectric devices. Li et al. [31] designed and built an intelligent flux transfer regulator with uniformly distributed conduction parameters and efficient operation in both thermal and DC electric fields. In addition, with the help of the discretization method, effective medium theory, and other control devices, the effectiveness of the established device was verified experimentally. This work not only further develops the diversity and applicability of energy regulation methods, but also provides a reference for studying energy transport phenomena from different disciplines. However, there are still great challenges to achieve complete thermal cloaking in practical experiments.

In recent years, due to the miniaturization of electronic equipment, the heat flux density of electronic devices has become larger and larger, and nanoscale heat flux control has become particularly important. Phonons, as micro/nanoscale heat transfer carriers, have unique characteristics of wave–particle duality. Researchers have designed thermal diodes [32,33], the phonon Hall effect [34–36], thermal rectification [37–40], etc. with the help of molecular dynamics and other means. The commonly used means for thermal protection of electronic devices is to use physical isolation for thermal shielding of thermal components close to the heat source. The research on the nanoscale thermal cloak is very important. Ye et al. [41] first used a graphene film to construct a nanocloak of chemical functionalization. The partial chemical functionalization on graphene was used to form heat flux channels that avoid specific regions. Due to the phonon localization, the heat flux automatically avoided the transition region, and the central part was protected from the heat flux. Considering the popularity of graphene, this study not only has important application value for thermal protection of graphene-based devices, but also has far-reaching implications for the development of other nanoelectronic devices. Liu et al. [42] used a crystalline silicon film with the "melting–quenching" technique to build a nanocloak. The results showed that the thermal cloaking effect was proportional to the width of the cloak ring. By exploring the underlying mechanism of nanoscale thermal cloaking by computing PDOS and MPR, it was found that phonon localization occurred in the nanocloak ring region and was responsible for the reduced thermal conductivity of amorphous silicon. The proposed method to engineer nanocloaks by in situ tuning of the material's atomic lattice structure is more practical and efficient, and it may open the way for nanoscale thermal functions and thermal management in nanophotonics and nanoelectronics. Choe et al. [43] designed a particle irradiation platform and observed the phenomenon of thermal cloaking experimentally. The ability to control microscale heat flux using this platform provides an opportunity to explore new ideas for microscopic thermal management. Furthermore, the demonstrated ion-writing microcalorics has the potential to be a versatile platform for controlling heat flow at the microscale, similar to what nanofluids do to fluids. In addition, we built a nanocloak by perforating a crystalline silicon film [44–46], while the effects of the number, size, and arrangement of perforations on its performance were explored. The results showed that the nanocloak designed by the perforation method could also show a good cloaking effect, and the cloaking effect was positively correlated with the number and size of the holes. When the number and size of holes were fixed, the best arrangement was circular. The nanocloak performance was optimized using the response surface method, and the fitting equations of multiple influencing factors were obtained. The best parameter selection interval was obtained by analyzing the interaction between two different parameters. By selecting the parameters in the best interval, the best cloaking effect could be obtained. The dynamic response and the effect of concave depth on its cloaking performance were studied, and the functional relationships of response temperature and the ratio of thermal cloaking with concave depth were obtained. The results showed that, under the condition of ensuring the stability of the structure, a greater depth led to better cloaking performance. Xiao et al. [47] used nanoporous thin films to realize the thermal cloaking. This work presented periodic square nanoholes and rectangular nanogrooves as efficient methods to locally tune the thermal conductivity and thermal anisotropy of thin films and potentially atomically thick materials. Thermal cloaking efficiencies were calculated and compared with theoretical predictions of continuous thermal conductivity distributions. This new approach enables two-dimensional nanocloaking of thin-film devices by eliminating the challenges of fabricating nanocomposites with well-controlled thermal properties. The periodic arrangement of the porous structure can be equivalent to a phononic crystal. According to the research of Chen et al. [48], the porosity has a great influence on the thermal conductivity of the phononic crystal. A greater porosity leads to a lower thermal conductivity, because a larger porosity corresponds to larger nanocubic pores, thereby narrowing the path for phonon transmission and increasing the surface-to-body ratio of the system, which leads to stronger structural reflections and, ultimately, lower thermal conductivity. No thermal cloak based on a porous structure has been found. Therefore, we used a porous structure to build a thermal cloak in this research.

In this paper, on the basis of nonequilibrium dynamics (NEMD), we calculate the thermal conductivity of porous structures and crystalline silicon films, and we use the porous structure to build a rectangular nanocloak. According to the index of the ratio of thermal cloaking, we find that the porous structure can be used to build a nanocloak to produce an excellent cloaking effect. Furthermore, the heat flux and temperature distribution are drawn to intuitively show the cloaking phenomenon. In addition, we explain the mechanism of the cloaking phenomenon produced using a porous structure with the help of phonon localization theory. Porous structures have increased porosity compared to bulk silicon and are not conducive to phonon transport, thus producing strong phonon localization and reducing the thermal conductivity. Our research can promote the application of porous materials and expand the construction methods of nanocloaks. Section 2 introduces the model and the basic theory of numerical simulation, Section 3 analyzes the simulation results, and Section 4 introduces the main conclusions of this paper. Figure 1 shows the overall calculation flow of this paper.

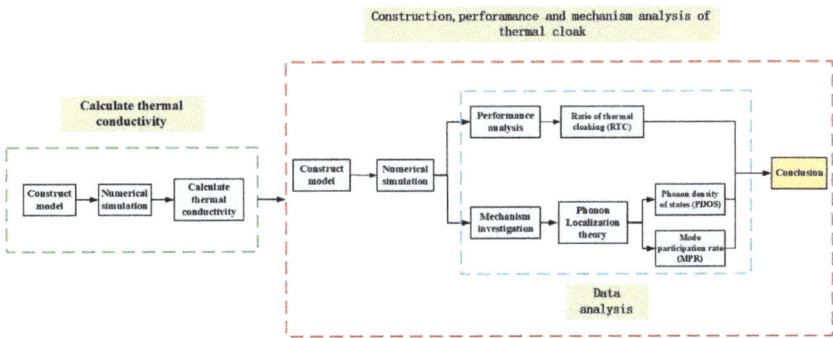

Figure 1. The detailed calculation process of this paper.

2. Model and Methodology

2.1. Construction of Computational Models

In this paper, the material we chose was crystalline silicon with a lattice constant of 5.431 Å. The basic unit constituted a unit cell (UC). We used LAMMPS [49] to construct the model and conduct the simulation. The built-in modeling commands of LAMMPS were used for modeling. First, a silicon film plate model was constructed as a calculation reference, and then a porous structure was built on the plate. Since the number of regions is limited in LAMMPS, we created it in two steps. The first step was to delete the plate x- and y-direction atoms, as shown in Figure 2b, while the second step was to delete the z-direction atoms. The thermal conductivity calculation model is shown in Figure 2. The size of the simulation box was 50 × 12 × 4 UC.

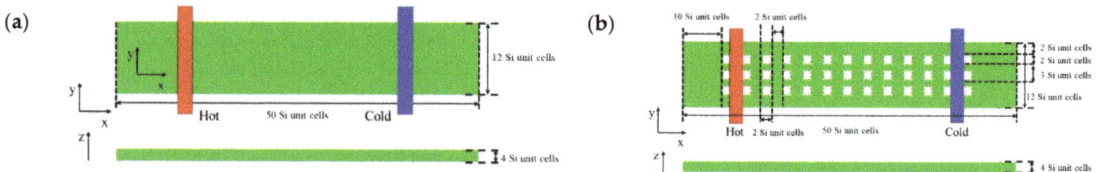

Figure 2. The thermal conductivity calculation model: (**a**) crystalline film; (**b**) porous structure film.

Since the porous structure selected in this paper was a cubic nanoporous structure, it was not easy to build a classic circular nanocloak. Therefore, according to our previous research experience, we selected a more reasonable scale to build a porous rectangular thermal cloak. The calculation model of the nanocloak is shown in Figure 3. The model size was 60 × 40 × 4 UC; the functional region was constructed with a porous structure, while the remainder was crystalline silicon. The two ends were fixed regions to prevent the whole model from moving, and the length was 2 UC. The thermostat region was divided into cold and hot baths, both of which were 8 UC in length. The remaining regions were background regions.

2.2. Methodology

In this research, the Verlet algorithm [50] was used to solve Newton's equations of motion, as it is the most accurate velocity calculation method. The canonical ensemble (NVT) and the microcanonical ensemble (NVE) were used in the simulation. The correctness of the molecular dynamics simulation results depends on the accuracy of the potential function. According to our previous research, we chose the Tersoff potential [51].

$$E = \sum_i E_i = \frac{1}{2}\sum_{i \neq j} f_C(r_{ij})\left[f_R(r_{ij}) + b_{ij}f_A(r_{ij})\right], \tag{1}$$

where E is the total energy, i and j are atom labels, r is the distance between atoms, f is the potential function, C is the cutoff function, R and A are repulsive and attractive pairs, respectively, and b is the bond order between atoms.

2.2.1. Thermal Conductivity Calculation

The key to design a nanocloak is to construct a functional region with low thermal conductivity; thus, calculating the thermal conductivity is crucial for successful construction. There are two commonly used methods, EMD and NEMD. We chose the NEMD method, which is also known as the "nonequilibrium" method. The so-called nonequilibrium state is a state opposite to the equilibrium state. In most molecular dynamics simulations, relaxation is required. Relaxation is also called the "running equilibrium", whereby the shape of the material in the equilibrium state is obtained, while the temperature and energy of the system remain basically unchanged. The nonequilibrium state is the opposite, whereby the temperature is not constant, but maintains a certain gradient. We can calculate the lattice thermal conductivity as follows:

$$J = -\kappa \nabla T, \quad (2)$$

where J is the given heat flux, ∇T is the temperature gradient, and κ is a tensor [52–55]. Our calculated thermal conductivity was κ_{xx}. Since the simulation was mainly in the xy plane, κ_{zz} was ignored, periodic boundary conditions were used in the simulation process, and the model was symmetrically distributed; hence, in the xy plane, thermal conductivity was isotropic, $\kappa_{xx} = \kappa_{yy}$.

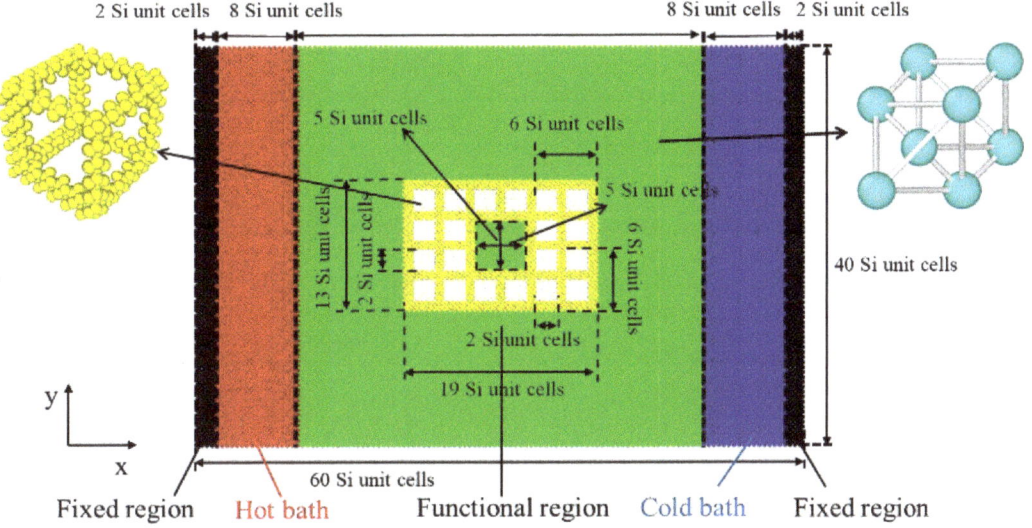

Figure 3. Schematic diagram of each region of the nanocloak.

2.2.2. Heat Flux and Temperature Calculation

In order to evaluate the cloaking performance of the nanocloak and visually display the cloaking phenomenon, referring to [56], we can calculate the single-atom heat flux as follows:

$$\mathbf{J} = \frac{1}{V}\left[\sum_i e_i \mathbf{v}_i - \sum_i \mathbf{S}_i \mathbf{v}_i\right], \quad (3)$$

where $\mathbf{J}$ is the heat flux, V is the volume, i is an atomic label, e is the total energy, which includes kinetic and potential energy, $\mathbf{v}$ is the velocity vector, and $\mathbf{S}$ is the pressure tensor. In all three directions, the heat flux can be described as follows:

$$J_x = \frac{1}{V}\left[\sum_i e_i v_{xi} - \sum_i (S_{ixx} v_{ix} + S_{ixy} v_{iy} + S_{ixz} v_{iz})\right] \quad (4)$$

$$J_y = \frac{1}{V}\left[\sum_i e_i v_{yi} - \sum_i (S_{iyx} v_{ix} + S_{iyy} v_{iy} + S_{iyz} v_{iz})\right] \quad (5)$$

$$J_z = \frac{1}{V}\left[\sum_i e_i v_{zi} - \sum_i (S_{izx}v_{ix} + S_{izy}v_{iy} + S_{izz}v_{iz})\right] \tag{6}$$

According to two regions A and B in Figure 3, the ratio of thermal cloaking (RTC) can be calculated as follows:

$$RTC = \frac{J_A}{J_B}, \tag{7}$$

where J is the average heat flux. In addition, we can also calculate the single-atom temperature during the simulation as follows:

$$\sum_i \varepsilon_i = \frac{dim}{2}k_B NT, \tag{8}$$

where ε is the kinetic energy, i is an atomic label, dim is the dimensionality ($dim = 3$), k_B is the Boltzmann constant, and N is the total number of atoms.

2.2.3. Phonon Localization Theory

Phonon localization theory has important applications in various fields. In a related study on graphene [57–59], with the help of phonon localization theory, it was found that phonon localization mainly affects long-wave and high-frequency modes, and it was found that the strong localization of low-frequency phonons caused by reverse short thermal contacts leads to a clear temperature transition. In the study of heterostructures [60,61], with the help of phonon localization theory, the internal mechanism of the abnormal interfacial thermal conductance (ITC) change at the interface of Gr and h-BN was explored, aiming at the topological defect-based semi-defective graphene/hexagonal boron nitride. With respect to the thermal conductivity of the in-plane heterostructure interface, it was found that the phonon coupling on both sides of the interface and the phonon localization effect of the heterostructure were the two keys determining the heterostructure ITC factor. In other fields [62], the effect of anti-substitution on thermal conductivity was investigated using nonequilibrium molecular dynamics simulations, and it was found that, when the defect concentration was low, localization was the main reason for the decrease in thermal conductivity, whereas, when the defect concentration was high, the main reason was phonon defect scattering in all phonon modes.

With the help of the phonon localization theory, we can explain the cloaking mechanism by calculating the phonon density of states (PDOS). To make the results meaningful, we selected the same region in the cloak and the crystalline film. Since the cloaking phenomenon is caused by the existence of the functional region, we chose the whole functional region as the calculation region. PDOS can be described as

$$\text{PDOS}(\omega) = \frac{1}{N\sqrt{2\pi}} \int e^{-i\omega t} \left\langle \sum_{j=1}^{N} \mathbf{v}_j(0)\mathbf{v}_j(t) \right\rangle dt, \tag{9}$$

where N is the total number of atoms, ω is the phonon frequency, $\mathbf{v}$ is the velocity vector, and j is an atomic label.

To better understand the PDOS, the mode participation rate (MPR) can be calculated as follows [63]:

$$\text{MPR}(\omega) = \frac{1}{N}\frac{\left[\sum_i \text{PDOS}_i(\omega)^2\right]^2}{\sum_i \text{PDOS}_i(\omega)^4}, \tag{10}$$

where N is the total number of atoms, and $\text{PDOS}_i(\omega)$ is local density of states based on Equation (9).

2.3. Simulation Process

We used NEMD to calculate the thermal conductivity of the constructed model. This method requires a heat source and a heat sink to be applied to the model, both of which were 0.0005 eV·Å/ps in this paper. The initial temperature of the whole system was fixed at 300 K, and the simulation was run for 10 ns.

We then simulated the thermal cloak. The timestep was 1 fs. Periodic boundary conditions were used in all three directions during the simulation. The whole process was divided into two stages. In the preparation stage, the energy of the whole system was first minimized, and all atoms except the fixed region were given an initial velocity satisfying a Gaussian distribution at a temperature of 300 K (achieved by specifying a random number in LAMMPS; the random number in this paper was 4,928,459). In the simulation stage, the selected calculated system was first placed in the NVT ensemble, simulated for 100 ps at 300 K for full relaxation, and then placed in the NVE ensemble and simulated for 1000 ps. The hot bath and cold bath were 350 K and 250 K, respectively, and the thermostat used was Nose–Hoover [64].

$$\frac{d}{dt}p_i = F_i - \gamma p_i, \tag{11}$$

$$\frac{d}{dt}\gamma = \frac{1}{\tau^2}\left[\frac{T(t)}{T_0} - 1\right], \tag{12}$$

$$T(t) = \frac{2}{3Nk_B}\sum_i \frac{p_i^2}{2m_i}, \tag{13}$$

where i is an atomic label, p is the momentum, F is the force, γ is a dynamic parameter, τ is the relaxation time, m is the mass, k_B is the Boltzmann constant, and N is the number of atoms in the thermostats.

3. Results and Discussion

3.1. The Thermal Conductivity

We obtained the temperature change curve through simulation and fit it to obtain the slope of the curve, as shown in Figure 4. The slope was ∇T. Using Equation (2), we calculated the thermal conductivity. The results prove that the thermal conductivity of the porous structure was lower than that of bulk silicon; hence, it could be used to construct the nanocloak.

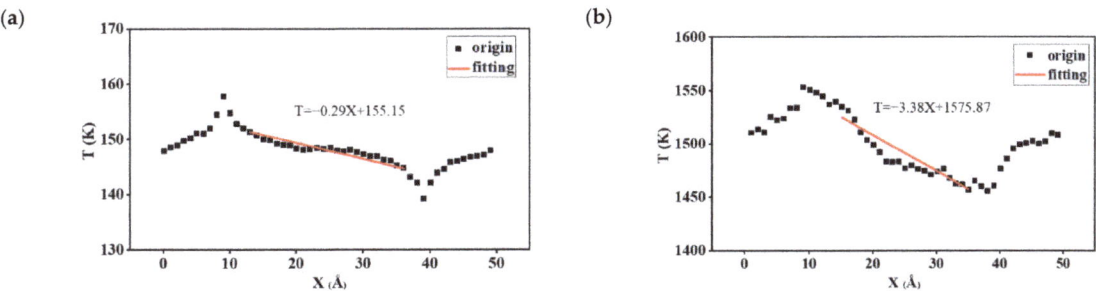

Figure 4. Temperature of the calculated model: (**a**) crystalline silicon; (**b**) porous silicon structure.

3.2. The Calculation of Thermal Cloak Performance

Figure 5 shows the RTC of the thermal cloak and crystalline silicon film at 600 ps. The RTC of the thermal cloak was much higher than that of the crystalline silicon film, confirming the cloaking phenomenon, and the cloaking effect was obvious. Compared with bulk silicon, the porous silicon structure had a large porosity, which was not conducive to the transmission of phonons, resulting in low thermal conductivity and producing the cloaking phenomenon. Increasing the porosity of the porous structure could produce excellent cloaking effects.

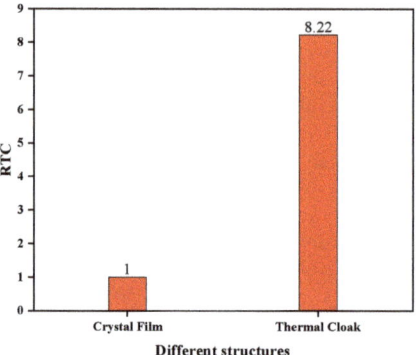

Figure 5. The RTC with different structures at 600 ps.

To prove the occurrence of the cloaking phenomenon, the heat flux and temperature distribution of the crystalline film and the nanocloak at 600 ps are plotted in Figure 6. In the simulation process, we calculated the heat flux and temperature of a single atom; however, we do not care about the state of a single atom but the state of the whole system. Therefore, we first divided the whole system into 40 small blocks, and then calculated the average heat flux and average temperature in each small block. By analyzing the heat flux distribution, we can clearly see the cloaking phenomenon. The heat flux of the whole functional region of the thermal cloak was kept extremely low, and the heat flux was spread around the functional region. However, when observing the temperature distribution, the cloaking phenomenon was not obvious, because the nanocloak is a typical heat flux control device, rather than a temperature control device.

3.3. Analysis of Cloaking Mechanism of Nanocloak

Figure 7 shows the PDOS of the crystalline silicon film and the thermal cloak. For the crystalline silicon film, consistent with a previous study [65], a weak and a strong peak appeared at 5 THz and 17 THz, respectively. Phonons are generated due to the excitation of atomic vibration. For the same atom, the vibration frequency was the same, and the phonons excited by the vibration were roughly the same. For the thermal cloak, due to the reduction in the number of atoms in the functional region, both the low- and the high-frequency peaks shifted to the left, and the PDOS was larger than the crystalline silicon film in the range of 0–5 THz, while the PDOS was smaller than the crystalline silicon film in the range of 15.5–18.5 THz. In addition, a large number of phonon modes appeared in the range of 5–15 THz.

Figure 8 shows the MPR in the calculated region of the different structures. For the crystalline silicon film, the MPR was uniformly distributed, basically greater than 0.6, indicating that these phonon modes were decentralized. For the thermal cloak, all MPRs were lower than 0.6, and the distribution was uneven, proving the occurrence of the phonon localization phenomenon [66]. The main reason is that the number of atoms in the functional region of the thermal cloak was reduced, while there were fewer phonons excited by vibration in the same region, and the porous structure was not conducive to the propagation of phonons, which in turn affected the PDOS. The PDOS at low frequency increased, and the peak shifted to the left, while the PDOS at high frequency decreased, and the peak shifted to the left, resulting in a strong phonon localization, which in turn reduced the thermal conductivity of the functional region, resulting in the cloaking phenomenon.

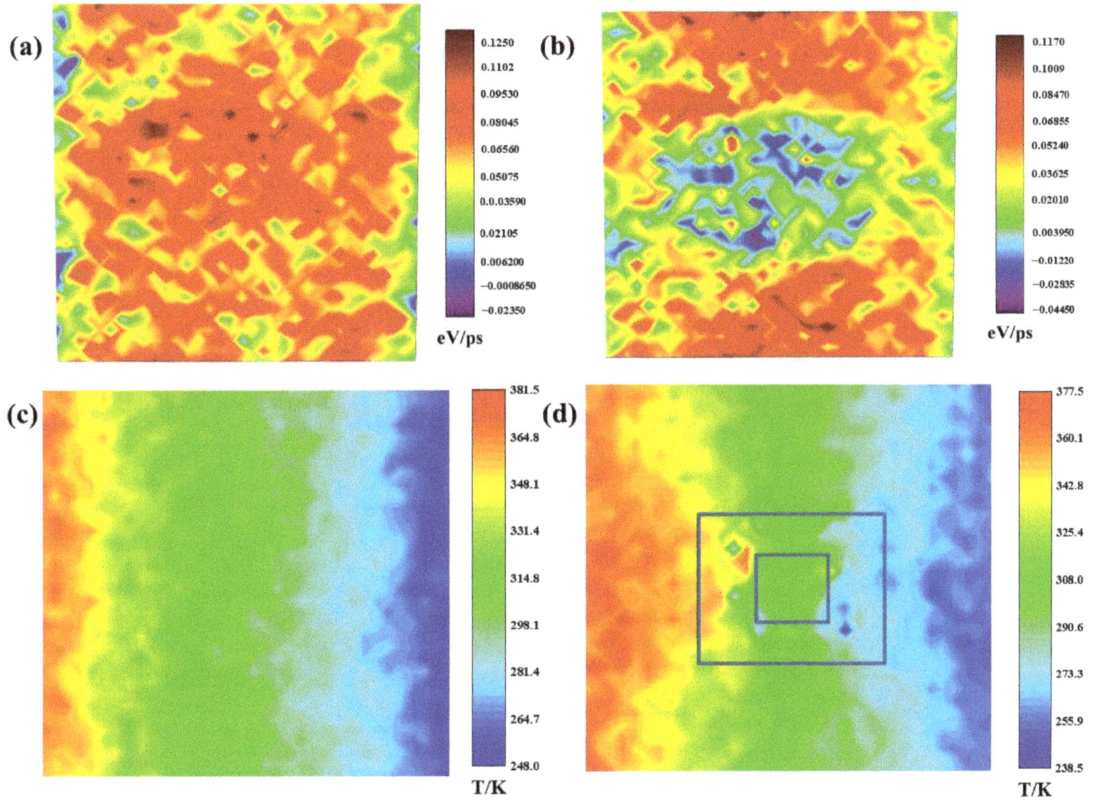

Figure 6. (**a**,**b**) The heat flux distribution at 600 ps, where (**a**) is the crystalline film, and (**b**) is the nanocloak; (**c**,**d**) the temperature distribution at 600 ps, where (**c**) is the crystalline film, and (**d**) is the nanocloak.

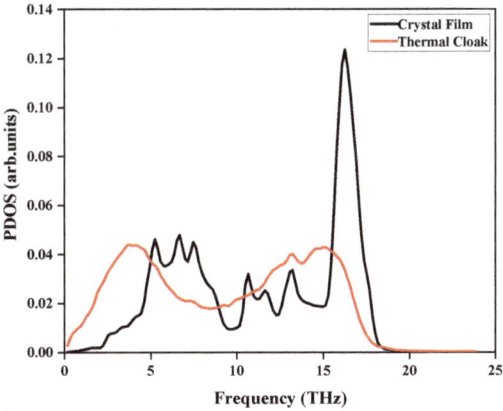

Figure 7. PDOS in the calculated region of the different structures.

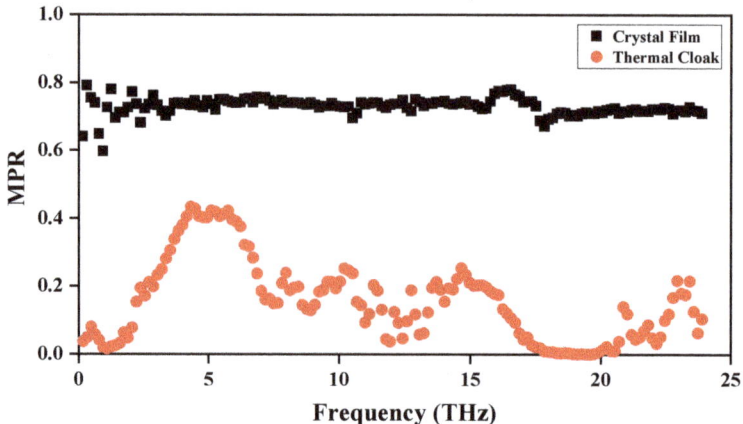

Figure 8. MPR in the calculated region of the different structures.

4. Conclusions

In this paper, we used LAMMPS software for modeling and simulation, and we constructed thermal conductivity and thermal cloak calculation models. The RTC was used to evaluate the cloaking performance, and the phonon localization theory was used to explain its cloaking mechanism. This research has important applications in the thermal isolation of microelectronic devices, expands the construction methods of thermal cloaks, and promotes the application of porous materials. The main conclusions are summarized as follows:

(1) Through NEMD simulation, we found that the thermal conductivity of the porous structure was significantly lower than that of the crystalline film, satisfying the conditions for building the thermal cloak;

(2) The RTC was used to evaluate the cloaking performance, proving that the nanocloak built with a porous structure could produce excellent cloaking effects, which was verified by drawing the heat flux distribution;

(3) The cloaking mechanism of the nanocloak was analyzed by phonon localization theory; we calculated the PDOS and MPR, and we found that due to the reduction in the number of atoms in the functional region of the thermal cloak, the phonons generated by vibration were reduced, while the porous structure was not conducive to the propagation of phonons, which in turn affected the PDOS. The PDOS at low frequency increased, and the peak shifted to the left, while the PDOS at high frequency decreased, and the peak shifted to the left, resulting in a strong phonon localization. As a result, the thermal conductivity of the functional region of the thermal cloak was reduced, and the phenomenon of cloaking occurred.

In this paper, it was pointed out that an increase in porosity of the porous structure led to a decrease in thermal conductivity, but the effect of porosity on the cloaking performance of the nanocloak was not explored, as this paper only explored the cloaking performance of the porous cloak within a constant temperature boundary. In order to facilitate engineering applications, the dynamic environment response is also critical, and we will focus on addressing these issues in the future to advance the application of porous structures in the field of heat flow regulation.

Author Contributions: Conceptualization, J.Z. and H.Z.; methodology, J.Z.; software, J.Z.; validation, J.Z., Y.L. and Q.W.; data curation, W.S. and Y.L.; writing—original draft preparation, J.Z.; visualization, Q.W. and W.S.; supervision, H.Z. All authors have read and agreed to the published version of the manuscript.

Funding: This research was funded by the National Natural Science Foundation of China, grant number 51776050.

Institutional Review Board Statement: Not applicable.

Informed Consent Statement: Not applicable.

Data Availability Statement: The data presented in this study are available in this article.

Conflicts of Interest: The authors declare no conflict of interest.

References

1. Fan, C.Z.; Gao, Y.; Huang, J.P. Shaped graded materials with an apparent negative thermal conductivity. *Appl. Phys. Lett.* **2008**, *92*, 251907. [CrossRef]
2. Han, T.C.; Zhao, J.J.; Yuan, T.; Lei, D.Y.; Li, B.W.; Qiu, C.W. Theoretical realization of an ultra-efficient thermal-energy harvesting cell made of natural materials. *Energy Environ. Sci.* **2013**, *6*, 3537–3541. [CrossRef]
3. Yang, T.Z.; Huang, L.J.; Chen, F.; Xu, W.K. Heat flux and temperature field cloaks for arbitrarily shaped objects. *J. Phys. D Appl. Phys.* **2013**, *46*, 305102. [CrossRef]
4. Han, T.C.; Bai, X.; Gao, D.L.; Thong, J.T.L.; Li, B.W.; Qiu, C.W. Experimental Demonstration of a Bilayer Thermal Cloak. *Phys. Rev. Lett.* **2014**, *112*, 054302. [CrossRef] [PubMed]
5. Nguyen, D.M.; Xu, H.G.; Zhang, Y.M.; Zhang, B.L. Active thermal cloak. *Appl. Phys. Lett.* **2015**, *107*, 121901. [CrossRef]
6. Han, T.C.; Bai, X.; Liu, D.; Gao, D.L.; Li, B.W.; Thong, J.T.L.; Qiu, C.W. Manipulating Steady Heat Conduction by Sensu-shaped Thermal Metamaterials. *Sci. Rep.* **2015**, *5*, 10242. [CrossRef] [PubMed]
7. Yang, T.Z.; Bai, X.; Gao, D.L.; Wu, L.Z.; Li, B.W.; Thong, J.T.L.; Qiu, C.W. Invisible Sensors: Simultaneous Sensing and Camouflaging in Multiphysical Fields. *Adv. Mater.* **2015**, *27*, 7752–7758. [CrossRef]
8. Li, T.H.; Zhu, D.L.; Mao, F.C.; Huang, M.; Yang, J.J.; Li, S.B. Design of diamond-shaped transient thermal cloaks with homogeneous isotropic materials. *Front. Phys.* **2016**, *11*, 110503. [CrossRef]
9. Shen, X.Y.; Li, Y.; Jiang, C.R.; Ni, Y.S.; Huang, J.P. Thermal cloak-concentrator. *Appl. Phys. Lett.* **2016**, *109*, 031907. [CrossRef]
10. Zhou, L.L.; Huang, S.Y.; Wang, M.; Hu, R.; Luo, X.B. While rotating while cloaking. *Phys. Lett. A* **2019**, *383*, 759–763. [CrossRef]
11. Zhu, Z.; Ren, X.C.; Sha, W.; Xiao, M.; Hu, R.; Luo, X.B. Inverse design of rotating metadevice for adaptive thermal cloaking. *Int. J. Heat Mass Transf.* **2021**, *176*, 121417. [CrossRef]
12. Hu, R.; Wei, X.L.; Hu, J.Y.; Luo, X.B. Local heating realization by reverse thermal cloak. *Sci. Rep.* **2014**, *4*, 3600. [CrossRef]
13. Chen, F.; Lei, D.Y. Experimental Realization of Extreme Heat Flux Concentration with Easy-to-Make Thermal Metamaterials. *Sci. Rep.* **2015**, *5*, 11552. [CrossRef]
14. Guenneau, S.; Amra, C. Anisotropic conductivity rotates heat fluxes in transient regimes. *Opt. Express* **2013**, *21*, 6578–6583. [CrossRef]
15. Han, T.C.; Bai, X.; Thong, J.T.L.; Li, B.W.; Qiu, C.W. Full Control and Manipulation of Heat Signatures: Cloaking, Camouflage and Thermal Metamaterials. *Adv. Mater.* **2014**, *26*, 1731–1734. [CrossRef] [PubMed]
16. Yang, T.Z.; Su, Y.S.; Xu, W.K.; Yang, X.D. Transient thermal camouflage and heat signature control. *Appl. Phys. Lett.* **2016**, *109*, 121905. [CrossRef]
17. Wang, J.; Bi, Y.Q.; Hou, Q.W. Three-dimensional illusion thermal device for location camouflage. *Sci. Rep.* **2017**, *7*, 7541. [CrossRef]
18. Li, Y.; Bai, X.; Yang, T.Z.; Luo, H.L.; Qiu, C.W. Structured thermal surface for radiative camouflage. *Nat. Commun.* **2018**, *9*, 273. [CrossRef]
19. Qu, Y.R.; Li, Q.; Cai, L.; Pan, M.Y.; Ghosh, P.; Du, K.K.; Qiu, M. Thermal camouflage based on the phase-changing material GST. *Light Sci. Appl.* **2018**, *7*, 26. [CrossRef] [PubMed]
20. Hu, R.; Xi, W.; Liu, Y.D.; Tang, K.C.; Song, J.L.; Luo, X.B.; Wu, J.Q.; Qiu, C.W. Thermal camouflaging metamaterials. *Mater. Today* **2020**, *45*, 120–141. [CrossRef]
21. Liu, Y.D.; Song, J.L.; Zhao, W.X.; Ren, X.C.; Cheng, Q.; Luo, X.B.; Fang, N.X.; Hu, R. Dynamic thermal camouflage via a liquid-crystal-based radiative metasurface. *Nanophotonics* **2020**, *9*, 855–863. [CrossRef]
22. Zhu, N.Q.; Shen, X.Y.; Huang, J.P. Converting the patterns of local heat flux via thermal illusion device. *AIP Adv.* **2015**, *5*, 053401. [CrossRef]
23. Alwakil, A.; Zerrad, M.; Bellieud, M.; Amra, C. Inverse heat mimicking of given objects. *Sci. Rep.* **2017**, *7*, 43288. [CrossRef]
24. Hu, R.; Zhou, S.L.; Li, Y.; Lei, D.Y.; Luo, X.B.; Qiu, C.W. Illusion Thermotics. *Adv. Mater.* **2018**, *30*, 1707237. [CrossRef]
25. Zhou, S.L.; Hu, R.; Luo, X.B. Thermal illusion with twinborn-like heat signatures. *Int. J. Heat Mass Transf.* **2018**, *127*, 607–613. [CrossRef]
26. Hu, R.; Huang, S.Y.; Wang, M.; Luo, X.B.; Shiomi, J.; Qiu, C.W. Encrypted Thermal Printing with Regionalization Transformation. *Adv. Mater.* **2019**, *31*, 1807849. [CrossRef] [PubMed]
27. Xu, G.Q.; Zhang, H.C.; Zhang, X.; Jin, Y. Investigating the thermodynamic performances of TO-based metamaterial tunable cells with an entropy generation approach. *Entropy* **2017**, *19*, 538. [CrossRef]
28. Xu, G.Q.; Zhang, H.C.; Zou, Q.; Jin, Y. Predicting and analyzing interaction of the thermal cloaking performance through response surface method. *Int. J. Heat Mass Transf.* **2017**, *109*, 746–754. [CrossRef]

29. Li, Y.Y.; Zhang, H.C.; Sun, M.Y.; Zhang, Z.H.; Zhang, H.M. Environmental Response of 2D Thermal Cloak under Dynamic External Temperature Field. *Entropy* **2020**, *22*, 461. [CrossRef]
30. Xu, G.Q.; Zhou, X.; Zhang, H.C.; Tan, H.P. Creating illusion of discrete source array by simultaneously allocating thermal and DC fields with homogeneous media. *Energy Convers. Manag.* **2019**, *187*, 546–553. [CrossRef]
31. Li, Y.Y.; Zhang, H.C.; Chen, Y.J.; Zhang, D.; Huang, Z.L.; Wang, H.M. Realization and analysis of an Intelligent flux transfer regulator by allocating thermal and DC electric fields. *Int. J. Heat Mass Transf.* **2021**, *179*, 121677. [CrossRef]
32. Liu, Y.Z.; Xu, Y.; Zhang, S.C.; Duan, W.H. Model for topological phononics and phonon diode. *Phys. Rev. B* **2017**, *96*, 064106. [CrossRef]
33. Wang, S.; Cottrill, A.L.; Kunai, Y.; Toland, A.R.; Liu, P.W.; Wang, W.J.; Strano, M.S. Microscale solid-state thermal diodes enabling ambient temperature thermal circuits for energy applications. *Phys. Chem. Chem. Phys.* **2017**, *19*, 13172–13181. [CrossRef]
34. Zhang, L.F.; Ren, J.; Wang, J.S.; Li, B.W. Topological Nature of the Phonon Hall Effect. *Phys. Rev. Lett.* **2010**, *105*, 225901. [CrossRef]
35. Qin, T.; Zhou, J.H.; Shi, J.R. Berry curvature and the phonon Hall effect. *Phys. Rev. B* **2012**, *86*, 104305. [CrossRef]
36. Li, N.B.; Ren, J.; Wang, L.; Zhang, G.; Haenggi, P.; Li, B.W. Colloquium: Phononics: Manipulating heat flow with electronic analogs and beyond. *Rev. Mod. Phys.* **2012**, *84*, 1045–1066. [CrossRef]
37. Chang, C.W.; Okawa, D.; Majumdar, A.; Zettl, A. Solid-state thermal rectifier. *Science* **2006**, *314*, 1121–1124. [CrossRef] [PubMed]
38. Hu, J.N.; Ruan, X.L.; Chen, Y.P. Molecular Dynamics Study of Thermal Rectification in Graphene Nanoribbons. *Int. J. Thermophys.* **2012**, *33*, 986–991. [CrossRef]
39. Tian, H.; Xie, D.; Yang, Y.; Ren, T.L.; Zhang, G.; Wang, Y.F.; Zhou, C.J.; Peng, P.G.; Wang, L.G.; Liu, L.T. A Novel Solid-State Thermal Rectifier Based on Reduced Graphene Oxide. *Sci. Rep.* **2012**, *2*, 523. [CrossRef]
40. Wang, Y.; Vallabhaneni, A.; Hu, J.N.; Qiu, B.; Chen, Y.P.; Ruan, X.L. Phonon Lateral Confinement Enables Thermal Rectification in Asymmetric Single-Material Nanostructures. *Nano Lett.* **2014**, *14*, 592–596. [CrossRef] [PubMed]
41. Ye, Z.Q.; Cao, B.Y. Nanoscale thermal cloaking in graphene via chemical functionalization. *Phys. Chem. Chem. Phys.* **2016**, *18*, 32952–32961. [CrossRef]
42. Liu, Y.D.; Cheng, Y.H.; Hu, R.; Luo, X.B. Nanoscale thermal cloaking by in-situ annealing silicon membrane. *Phys. Lett. A* **2019**, *383*, 2296–2301. [CrossRef]
43. Choe, H.S.; Prabhakar, R.; Wehmeyer, G.; Allen, F.; Lee, W.; Jin, L.; Li, Y.; Yang, P.D.; Qiu, C.W.; Dames, C.; et al. Ion Write Microthermotics: Programing Thermal Metamaterials at the Microscale. *Nano Lett.* **2019**, *19*, 3830–3837. [CrossRef]
44. Zhang, J.; Zhang, H.C.; Li, Y.Y.; Zhang, D.; Wang, H.M. Numerical analysis on nanoscale thermal cloak in three-dimensional silicon film with circular cavities. *Numer. Heat Transf. Part A Appl.* **2022**, *81*, 1–14. [CrossRef]
45. Zhang, J.; Zhang, H.C.; Sun, W.B.; Wang, Q. Mechanism analysis of double-layer nanoscale thermal cloak by silicon film. *Colloids Surf. A Physicochem. Eng. Asp.* **2022**, *634*, 128022. [CrossRef]
46. Zhang, J.; Zhang, H.; Wang, H.; Xu, C.; Wang, Q. Performance prediction of nanoscale thermal cloak by molecular dynamics. *Appl. Phys. A* **2021**, *127*, 790. [CrossRef]
47. Xiao, Y.; Chen, Q.Y.; Hao, Q. Inverse thermal design of nanoporous thin films for thermal cloaking. *Mater. Today Phys.* **2021**, *21*, 100477. [CrossRef]
48. Chen, G.; Dresselhaus, M.S.; Dresselhaus, G.; Fleutial, J.P.; Caillat, T. Recent developments in thermoelectric materials. *Int. Mater. Rev.* **2003**, *48*, 45–66. [CrossRef]
49. Plimpton, S. Fast Parallel Algorithms for Short-Range Molecular Dynamics. *J. Comput. Phys.* **1995**, *117*, 1–19. [CrossRef]
50. Zhang, Y.H.; Feller, S.E.; Brooks, B.R.; Pastor, R.W. Computer simulation of liquid/liquid interfaces. I. Theory and application to octane/water. *J. Chem. Phys.* **1995**, *103*, 10252–10266. [CrossRef]
51. Tersoff, J. Modeling solid-state chemistry: Interatomic potentials for multicomponent systems. *Phys. Rev. B* **1989**, *39*, 5566–5568. [CrossRef] [PubMed]
52. Liu, B.K.; Vu-Bac, N.; Rabczuk, T. A stochastic multiscale method for the prediction of the thermal conductivity of Polymer nanocomposites through hybrid machine learning algorithms. *Compos. Struct.* **2021**, *273*, 114269. [CrossRef]
53. Rabizadeh, E.; Bagherzadeh, A.S.; Anitescu, C.; Alajlan, N.; Rabczuk, T. Pointwise dual weighted residual based goal-oriented a posteriori error estimation and adaptive mesh refinement in 2D/3D thermo-mechanical multifield problems. *Comput. Methods Appl. Mech. Eng.* **2020**, *359*, 112666. [CrossRef]
54. Vu-Bac, N.; Duong, T.X.; Lahmer, T.; Zhang, X.; Sauer, R.A.; Park, H.S.; Rabczuk, T. A NURBS-based inverse analysis for reconstruction of nonlinear deformations of thin shell structures. *Comput. Methods Appl. Mech. Eng.* **2018**, *331*, 427–455. [CrossRef]
55. Rabczuk, T.; Ren, H.L.; Zhuang, X.Y. A Nonlocal Operator Method for Partial Differential Equations with Application to Electromagnetic Waveguide Problem. *Comput. Mater. Contin.* **2019**, *59*, 31–55. [CrossRef]
56. Chen, X.K.; Hu, J.W.; Wu, X.J.; Jia, P.; Peng, Z.H.; Chen, K.Q. Tunable Thermal Rectification in Graphene/Hexagonal Boron Nitride Hybrid Structures. *J. Phys. D* **2018**, *51*, 396–400. [CrossRef]
57. Loh, G.C.; Teo, E.H.T.; Tay, B.K. Phonon localization around vacancies in graphene nanoribbons. *Diam. Relat. Mater.* **2012**, *23*, 88–92. [CrossRef]
58. Jiang, P.F.; Hu, S.Q.; Ouyang, Y.L.; Ren, W.J.; Ren, C.Q.; Zhang, Z.W.; Chen, J. Remarkable thermal rectification in pristine and symmetric monolayer graphene enabled by asymmetric thermal contact. *J. Appl. Phys.* **2020**, *127*, 235101. [CrossRef]
59. Lu, S.; Ouyang, Y.L.; Yu, C.Q.; Jiang, P.F.; He, J.; Chen, J. Tunable phononic thermal transport in two-dimensional C6CaC6 via guest atom intercalation. *J. Appl. Phys.* **2021**, *129*, 225106. [CrossRef]

60. Liang, T.; Zhou, M.; Zhang, P.; Yuan, P.; Yang, D.G. Multilayer in-plane graphene/hexagonal boron nitride heterostructures: Insights into the interfacial thermal transport properties. *Int. J. Heat Mass Transf.* **2020**, *151*, 119395. [CrossRef]
61. Wu, X.; Han, Q. Thermal transport in pristine and defective two-dimensional polyaniline (C3N). *Int. J. Heat Mass Transf.* **2021**, *173*, 121235. [CrossRef]
62. Zhou, M.; Liang, T.; Wu, B.Y.; Liu, J.J.; Zhang, P. Phonon transport in antisite-substituted hexagonal boron nitride nanosheets: A molecular dynamics study. *J. Appl. Phys.* **2020**, *128*, 234304. [CrossRef]
63. Loh, G.C.; Teo, E.H.T.; Tay, B.K. Phononic and structural response to strain in wurtzite-gallium nitride nanowires. *J. Appl. Phys.* **2012**, *111*, 103506. [CrossRef]
64. Berendsen, H.J.C.; Postma, J.P.M.; van Gunsteren, W.F.; DiNola, A.; Haak, J.R. Molecular dynamics with coupling to an external bath. *J. Chem. Phys.* **1984**, *81*, 3684–3690. [CrossRef]
65. Li, H.P.; Zhang, R.Q. Anomalous effect of hydrogenation on phonon thermal conductivity in thin silicon nanowires. *Eur. Lett.* **2014**, *105*, 56003. [CrossRef]
66. Ma, D.K.; Ding, H.R.; Meng, H.; Feng, L.; Wu, Y.; Shiomi, J.; Yang, N. Nano-cross-junction effect on phonon transport in silicon nanowire cages. *Phys. Rev. B* **2016**, *94*, 165434. [CrossRef]

Article

Correlations and Numerical Modeling of Stacked Woven Wire-Mesh Porous Media for Heat Exchange Applications

Trilok G [1], Kurma Eshwar Sai Srinivas [1], Devika Harikrishnan [1], Gnanasekaran N [1,*] and Moghtada Mobedi [2,*]

[1] Department of Mechanical Engineering, National Institute of Technology Karnataka, Surathkal, Mangalore 575025, India; trilokg.197me506@nitk.edu.in or trilokg.tristar@gmail.com (T.G.); kessrinivas.181me141@nitk.edu.in or kessrinivas25@gmail.com (K.E.S.S.); devikaharikrishnan.191me226@nitk.edu.in or devihariikshnan98@gmail.com (D.H.)

[2] Mechanical Engineering Department, Faculty of Engineering, Shizuoka University, 3-5-1 Johoku, Naka-ku, Hamamatsu-shi 432-8561, Japan

* Correspondence: gnanasekaran@nitk.edu.in (G.N.); moghtada.mobedi@shizuoka.ac.jp (M.M.)

Abstract: Metal foams have gained attention due to their heat transfer augmenting capabilities. In the literature, correlations describing relations among their morphological characteristics have successfully been established and well discussed. However, collective expressions that categorize stacked wire mesh based on their morphology and thermo-hydraulic expressions required for numerical modeling are less explored in the literature. In the present study, cross relations among the morphological characteristics of stacked wire-mesh were arrived at based on mesh-size, wire diameter and stacking type, which are essential for describing the medium and determining key input parameters required for numerical modeling. Furthermore, correlation for specific surface area, a vital parameter that plays a major role in interstitial heat transfer, is provided. With the arrived correlations, properties of stacked wire-mesh samples of orderly varied mesh-size and porosity are obtained for various stacking scenarios, and corresponding thermo-hydraulic parameters appearing in the governing equations are evaluated. A vertical channel housing the categorized wire-mesh porous media is numerically modeled to analyze thermal and flow characteristics of such a medium. The proposed correlations can be used in confidence to evaluate thermo-hydraulic parameters appearing in governing equations in order to numerically model various samples of stacked wire-mesh types of porous media in a variety of heat transfer applications.

Keywords: porous media modeling; woven wire-mesh; pore density; porosity; LTNE; correlations; stacking types

Citation: G, T.; Srinivas, K.E.S.; Harikrishnan, D.; N, G.; Mobedi, M. Correlations and Numerical Modeling of Stacked Woven Wire-Mesh Porous Media for Heat Exchange Applications. *Energies* 2022, 15, 2371. https://doi.org/10.3390/en15072371

Academic Editor: Artur Blaszczuk

Received: 13 February 2022
Accepted: 21 March 2022
Published: 24 March 2022

Publisher's Note: MDPI stays neutral with regard to jurisdictional claims in published maps and institutional affiliations.

Copyright: © 2022 by the authors. Licensee MDPI, Basel, Switzerland. This article is an open access article distributed under the terms and conditions of the Creative Commons Attribution (CC BY) license (https://creativecommons.org/licenses/by/4.0/).

1. Introduction

Metal based porous media have been extensively researched for their ability to enhance heat transfer that benefits the function of many heat exchanging devices. Numerical modeling methodology, developed in order to mimic thermal and flow phenomena through such media has been a reason for abundant research on a variety of heat exchanging applications. In regard to the same, metal foams have been widely analyzed for their thermo-hydraulic behavior [1]. Some of the major applications include compact heat exchangers [2], fuel cells [3], thermal storage [4], geothermal applications [5], solar receivers [6], and heat sinks [7]. In a previous study [8] using the method of porous modeling, such metal foams have been analyzed for their varying behavior in maximizing heat transfer and minimizing flow resistance based on their structural aspects (porosity and pore density). Key thermo-hydraulic parameters such as the specific surface area and the interstitial heat transfer coefficient of the sample highly depend on the combination of porosity and pore density of the considered samples. However, in a porous medium such as metal foams, it is highly difficult to obtain the desired porosity and pore density combinations, as changing cell size and maintaining a desired fiber diameter in order to achieve a sample of desired porosity is not easy. However, porous media formed by stacking of woven wire mesh screens make it possible to achieve the desired porosity and specific surface area by obtaining screens of the desired pore density (dependent on mesh/cell size) and fiber diameter with most ease.

Stacked woven wire-mesh has been known to aid the thermo-hydraulic performance in many prominent engineering applications and has been used as the conventional heat transfer medium in regenerators of Stirling engines [9]. Recently, Zhao et al. [10] reviewed forced convection in porous structures emphasizing expressions describing morphological characteristics such as porosity, fiber diameter, pore diameter and surface area density and correlations depicting thermo-hydraulic behaviors of porous media of various types such as metal foam, lattice frame, packed bed and woven wire-mesh. The scarcity of comprehensive expressions (cross relations) for describing the morphological characteristics of woven wire-mesh types of porous media can be observed compared to the widely discussed and established cross relations for other kinds of porous media (especially open-celled type of porous media). Elaborate reviews of metallic woven wire structures can be found in [11], emphasizing mechanical, topological and thermal properties of various types of woven-wire porous structures.

Among the plethora of existing and possible metallic woven wire-mesh structures, textile-core, a multi-layered woven wire mesh proposed by Sypeck and Wadley [12], where the porous block of woven wire-mesh is formed by node to node (inline) stacking of the weaved wire mesh screens has received attention as variations in cell size, porosity, pore density can easily be achieved. Tian et al. [13] experimentally investigated heat transfer and flow resistance in textile-core woven wire-mesh porous structures. The study demonstrated the effect of porosity and cell topology on flow resistance. Heat transfer was attributed to surface area density, porosity and solid conductivity. The study highlighted that, at a given flow rate, surface area density that depends on pore density and porosity in turn depends on fiber diameter and cell size plays a major role in heat transfer augmentation. However, the thermo-hydraulic behavior of such porous samples of orderly varying pore density and porosity combinations was not addressed in the study.

Stacking of the woven wire-mesh screen can be accomplished in two major ways: aligned (inline), where there is node-to-node arrangement of the screens, and misaligned (staggered type), where there is misalignment in stacking that accommodates more void spaces with wire material. Costa et al. investigated pressure drop [14] and heat transfer [15] in stacked woven wire structures as an application to Stirling engine regenerators. Exact three dimensional geometries were modelled that considered aligned and misaligned stacking [14], and parallel and cross stacking [15]. Flow through the wire-mesh porous block was accomplished numerically and respective correlations for the pressure drop and Nusselt number were obtained in the study. Since the study involved exact geometry simulations, discussion related to parameters that are essential for obtaining heat transfer and flow models of such porous media was not emphasized, without which numerical modeling cannot be accomplished. Bussiere et al. [16] made an experimental analysis on flow resistance in stacked woven wire-mesh used in electrical safety. This study emphasized assessment of the drag coefficient and pressure drop of woven wire-mesh screens and their stacked arrangement in order to increase the efficiency of circuit breakers.

As a result of well-established and well-debated correlations pertaining to metal foams that provide expressions for describing/categorizing their thermo-hydraulic and morphological parameters, enormous numerical investigations [17–19] have been performed on porous media of metal foam type to evaluate and/or obtain optimum performance of metal foams involving thermal exchange phenomena. However, expressions for evaluating thermo-hydraulic and morphological parameters of wire-mesh type porous media have not been well established or debated in the open literature. Without this knowledge, it becomes difficult to perform a comprehensive numerical study using porous media modeling methodologies, considering that this type of porous media is subjected to variation in its morphological properties and stacking manner. Armour and Cannon [20] elaborately studied flow through woven wire-mesh layers of five various weave patterns. Expressions for screen thickness, specific area, porosity and pore diameter were provided for the considered woven wire-mesh screen layers. However, it has to be noted that these expressions are exclusive to single mesh screen layers and not for porous media formed from the stacks of wire-mesh screens, which is the conventional way of incorporating them in heat exchange applications.

In a recent study, Garg et al. [21] performed a numerical study of regenerators for Stirling cryocoolers using porous media modeling techniques to simulate flow and heat transfer through a stack of stainless steel wire-mesh screens. The part containing the stack of wire-mesh screens was modelled as a porous medium. The porosity, interstitial area density (specific area) and interstitial heat transfer coefficient were determined based on expressions following the works of [22] for porosity and [23] for the other two parameters. However, proper justifications for the use of expressions to obtain the mentioned parameters are not provided by the authors in the study and relevant discussion pertaining to the suitability of the used expressions are missing in the open literature. Recently [24] made a numerical analysis of the thermo-hydraulic behavior in woven wire mesh screens. They used expressions provided by [20] to describe the specific surface area and porosity of the mesh screens. Correlations for heat transfer and pressure drop were provided in this study; however, they were restricted to single mesh screen layers of various porosities and specific surface areas. Wang et al. [25] recently obtained an analytical model for the friction factor as an exclusive function of pore structure and quantified pressure drop using the wire mesh's morphological parameters. The study also defined the mesh screen's properties, such as porosity and specific surface area, using expressions provided by [20].

Of relevance to the current study, Xu et al. [26] made an effort to provide expressions for porosity and specific surface area of textile-core type stacked wire mesh porous media for heat exchange applications. However, the study simplified the expression by assuming the curved wires woven over the shute wires to be straight. Through this assumption, the authors were able to arrive at a simple expression to calculate the mentioned parameters. However, the possibility of incurring large errors with an increased number of screens in the stack or with increased fiber diameters of the screen wires is inevitable. Zhao et al. [27] provided new expressions to calculate these parameters by considering the stacking manner, fiber diameter, compactness factor, etc.; however, the complexity of the expressions and difficulty in appropriately obtaining the compactness factor defined in the study was high. All these factors make necessary simple and comprehensive expressions that can describe/classify such types of porous media with the help of information on only a few easily available parameters such as mesh-size/pore-density and fiber (wire) diameter that can differentiate a given stacked woven wire-mesh type of porous medium corresponding to various stacking scenarios. Kurian et al. [28] conducted an experimental study on the thermodynamic performance of brass woven wire-mesh of textile-core type that was observed to be misaligned in stacking in a vertical channel. This study showed that the overall performance factor could be achieved at wire-mesh porous samples of higher porosity; though the analysis was made for wire-mesh porous samples of three different porosities, no discussion relevant to pore density that affects interstitial area density was considered in the study. Similar analysis was performed in another study by the same authors [29], but for a stainless steel wire-mesh porous block. Kotresha and Gnanasekaran [30] made a numerical analysis by considering porous media modeling techniques for investigating fluid flow and heat transfer through stacked wire-mesh porous media. However, for modeling, expressions provided by [13,26] were used to obtain the specific surface area (one of the parameters that highly affects heat transfer) of a wire-mesh stack with the same characteristics as in [28]. Though the study successfully demonstrated modeling techniques for the stacked wire-mesh type of porous media, it did not emphasize the suitability of the used expressions.

It is quite evident that the discussion related to expressions required for modeling of the stacked wire-mesh type of porous medium is limited in the open literature, but it plays a major role in providing authenticity to the flow and heat transfer characteristics obtained from numerical studies carried out for this type of porous medium. In the present study, expressions that describe morphological characteristics of stacked woven wire-mesh types of porous media are arrived at that can categorize various samples of this type of porous medium by various characteristics such as porosity and specific surface area as a result of variation in parameters such as fiber diameter of the woven wires, pore-density/mesh-size and type of stacking. Furthermore, with the use of the obtained expressions, the categorized stacked wire-mesh porous samples of orderly varying porosity and mesh size are investigated for their performance in enhancing heat transfer in a vertical channel subjected to constant heat flux and laminar air flow. The flowchart of the present work methodology is shown in Figure 1.

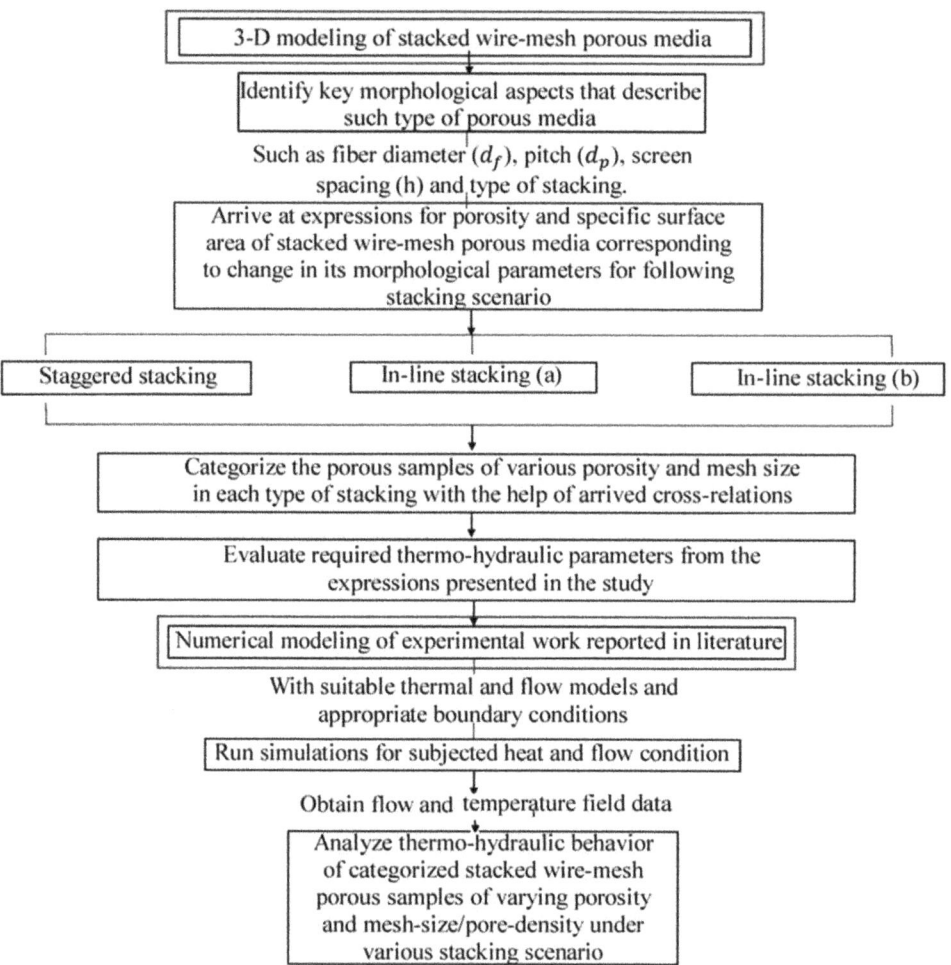

Figure 1. Chart describing the flow of the present work.

2. Problem Statement

With knowledge of the significance of morphological parameters of porous media in the numerical modeling of flow and heat transfer, cross relations for evaluating morphological parameters such as porosity, specific surface area of stacked brass wire-mesh porous media are arrived at for various stacking types with the help of easily available information such as fiber diameter (d_f) and pitch (d_p) of the woven mesh screen. Generalized expressions for screen spacing (h) for the closely packed scenario under various stacking scenarios are obtained and identified. With the aid of the arrived expressions, such stacked wire-mesh samples are categorized based on orderly varying porosity and pore-density/mesh-size combinations, and the respective heat transfer and flow influencing parameters are evaluated, thereby allowing the numerical modeling of the categorized porous samples for analyzing their thermal performance.

With the aid of the arrived expressions, stacked wire-mesh samples of 0.8, 0.85, 0.9 and 0.95 porosity each with pore densities 5, 10, 15, 20, 30, 40 and 45 PPI (of woven wire-mesh screens) are categorized under three different stacking scenarios, namely, staggered, inline type-a and inline type-b. Such categorized porous samples are evaluated for thermal performance in a vertical channel subjected to a constant heat input of 20 W under laminar flow achieved with 0.25 m/s air velocity under steady state conditions. The thermal phenomena identified by the wall heat transfer coefficient are analyzed for the categorized wire-mesh porous samples. Temperature information is numerically obtained; the excess temperature of the vertical plate is calculated using Equation (2) and is then used to compute the wall heat transfer coefficient using Equation (1).

The heat transfer coefficient is given by:

$$h_{wall} = \frac{Q}{A \Delta T_w} \tag{1}$$

where,

$$\Delta T_w = [T - T_\infty]_{avg} \tag{2}$$

3. Modeling and Arriving at Expressions for Stacked Woven Wire-Mesh Porous Samples

A cube of size $\Delta X \times \Delta \times \Delta$ is adopted within which the whole fluid saturated woven wire-mesh samples are geometrically modelled. The straight cylindrical wires are drawn in the Y direction and the curved woven wires are woven in the X direction along the length of the straight cylindrical wires. Woven wire mesh screens thus obtained are arranged in the Z direction to obtain a block of stacked woven wire mesh screens forming a fluid (empty spaces) saturated wire-mesh porous medium. Various views of the obtained geometry of the considering wire-mesh blocks are shown in Figure 2a–d. The front view of wire mesh layers of staggered, inline type-a and inline type-b stacking under the closely packed scenario are shown in Figures 3–5.

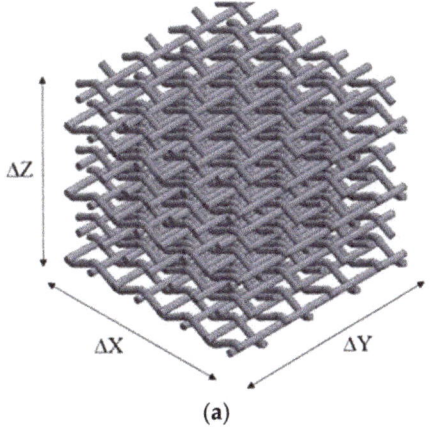

Figure 2. *Cont.*

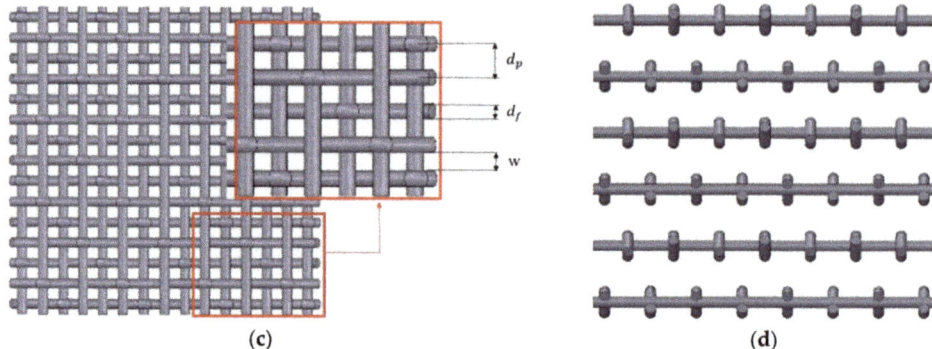

Figure 2. Corresponding to woven wire-mesh geometry with spacing between screens: (**a**) Isometric view. (**b**) top view. (**c**) front view and (**d**) side view.

Figure 3. Front view of staggered stacking for closely packed scenario (with h_{min} spacing).

Figure 4. Front view of inline stacking type-a for closely packed scenario (with h_{min} spacing).

Figure 5. Front view of inline stacking type-b for closely packed scenario (with h_{min} spacing).

3.1. Determination of Interstitial Area Density or Specific Area

By definition [31], the interstitial area density or specific area (a_{sf}) is determined as follows:

$$a_{sf} = \frac{Total\ surface\ area\ of\ wire - mesh}{al\ volume\ of\ the\ whole\ block\ (entire\ fluid\ saturated\ wire - mesh\ block)} \quad (3)$$

where, the dimension of the cube comprising fluid saturated woven wire-mesh porous media is represented as, $\Delta X \times \Delta \times \Delta$.

Referring to Figure 2a, the number of woven wire-mesh screens (n) can be expressed as shown in Equation (4), where 'h' is the distance between two consecutive mesh layers measured from center to center of straight running,

$$n = \frac{\Delta Z}{h} \tag{4}$$

It can be observed from Figure 2a that each layer of the wire-mesh block is formed by weaving cylindrical wires onto those cylindrical wires passing straight in the perpendicular direction. Therefore, the total surface area of the wires forming each layer of the wire-mesh block can be put in the following form.

The total surface area of each mesh layer (a_1) is the sum of the product of (a) the area of straight running cylindrical wire (a_s) and the total number of straight running cylindrical wires (N_s) and (b) the area of the weaving cylindrical wire (a_w) and the total number of weaving cylindrical wires (N_w), which can be expressed as:

$$a_1 = (a_s \times N_s) + (a_w \times N_w) \tag{5}$$

Observing Equations (4) and (5), the total area of the solid cylindrical wires forming the whole mesh block (A_{sm}) can be expressed as:

$$A_{sm} = ((a_s \times N_s) + (a_w \times N_w)) \times \frac{\Delta Z}{h} \tag{6}$$

Now, the parameters appearing in Equation (6) can individually be expressed as follows:

$$N_s = \frac{\Delta X}{d_p} \tag{7}$$

$$N_w = \frac{\Delta Y}{d_p} \tag{8}$$

$$a_s = \pi d_f \Delta Y + \frac{\pi}{2}(d_f)^2 \tag{9}$$

$$a_w = \pi d_f L + \frac{\pi}{2}(d_f)^2 \tag{10}$$

The parameter 'L' appearing in Equation (10) is the length of the weaving wire, where its value can be expressed from observing Figure 6a,b as:

$$L = OD \times number\ of\ pores\ (as\ there\ exists\ a\ curve\ 'OD'\ in\ every\ pore) \tag{11}$$

OD: as shown in Figure 6b.

The expression for the number of pores can be approximated as

$$N_p = N_s = \frac{\Delta X}{d_p} \tag{12}$$

In Equation (11), in order to find the expression for length 'L', the expression for 'OD' is derived and multiplied by the number of pores. For that, observing Figure 6a, angle $AO'B$ can be expressed as:

$$Cos\theta = \frac{O'A}{O'B} = \frac{2d_f}{d_p} \tag{13}$$

$$\theta = cos^{-1}\left(\frac{2d_f}{d_p}\right) \tag{14}$$

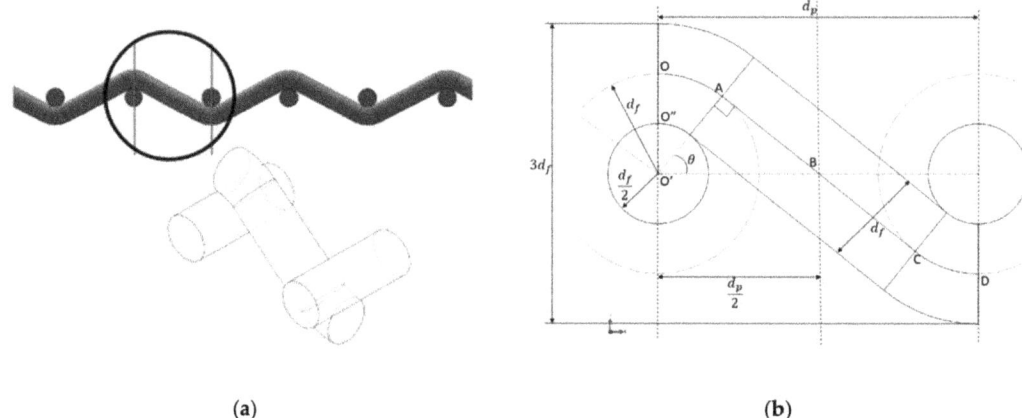

Figure 6. (a) Sketch of single cylindrical wire woven over straight cylindrical wires. (b) Description of parameters involving in expressing lengths of each woven cylindrical wire.

Therefore, angle

$$AO'O = \frac{\pi}{2} - \theta \tag{15}$$

Length of arc

$$OA = d_f\left(\frac{\pi}{2} - \theta\right) \tag{16}$$

Considering $\triangle AO'B$,

$$(AB)^2 = (O'B)^2 - (AO')^2 \tag{17}$$

Therefore,

$$AB = \sqrt{\left(\frac{d_p{}^2}{4} - d_f{}^2\right)} \tag{18}$$

However, the length $OD = OA + AB + BC + CD$, but $OA = CD$, which yields,

$$OD = 2(OA + AB) \tag{19}$$

Therefore the above equation for the length 'OD' can be rewritten, from expressions for 'OA' and 'AB' as given in Equations (16) and (18) as,

$$OD = 2\left(d_f\left(\frac{\pi}{2} - \theta\right) + \sqrt{\frac{d_p{}^2}{4} - d_f{}^2}\right) \tag{20}$$

Inserting expression for 'θ' as given in Equation (14), we get,

$$OD = \pi d_f - 2d_f\cos^{-1}\left(\frac{2d_f}{d_p}\right) + 2\sqrt{\frac{d_p{}^2}{4} - d_f{}^2} \tag{21}$$

From Equations (12) and (21) length of each woven wire can be obtained as,

$$L = \frac{\Delta X}{d_p}\left(\pi d_f - 2d_f\cos^{-1}\left(\frac{2d_f}{d_p}\right) + 2\sqrt{\left(\frac{d_p{}^2}{4} - d_f{}^2\right)}\right) \tag{22}$$

Therefore, the obtained expression for 'L' can be used in Equation (10). Now, the expression for total area of solid cylindrical wires both straight passing as well as weaving ones as expressed in Equation (6), can be rewritten by referring to Equations (4) and (8)–(10) as

$$A_{sm} = \left[\left(\pi d_f \Delta Y + \frac{\pi}{2} d_f^2\right)\frac{\Delta X}{d_p} + \left(\pi d_f \left(\frac{\Delta X}{d_p}\left(\pi d_f - 2d_f \cos^{-1}\left(\frac{2d_f}{d_p}\right) + 2\sqrt{\left(\frac{d_p^2}{4} - d_f^2\right)}\right)\right) + \frac{\pi}{2} d_f^2\right)\frac{\Delta Y}{d_p}\right]\frac{\Delta Z}{h} \quad (23)$$

Interstitial area density or specific surface area is expressed as,

$$a_{sf} = \frac{A_{sm}}{\Delta X \Delta Y \Delta Z} \quad (24)$$

Therefore, from Equation (23), the expression for 'a_{sf}' (for a woven-mesh block of spacing "h") can be deduced by simplification as

$$a_{sf} = \frac{A_{sm}}{\Delta X \Delta Y \Delta Z} = \frac{\pi d_f}{h d_p}\left[1 + \frac{d_f}{d_p}\left\{\pi - 2\cos^{-1}\left(\frac{2d_f}{d_p}\right)\right\} + \sqrt{1 - \left(\frac{2d_f}{d_p}\right)^2} + \frac{d_f}{2}\left(\frac{1}{\Delta X} + \frac{1}{\Delta Y}\right)\right] \quad (25)$$

Since term $\frac{d_f}{2}\left(\frac{1}{\Delta X} + \frac{1}{\Delta Y}\right)$ yields are of very negligible value, the above expressions can be simplified by neglecting this term as

$$\alpha_{sf} = \frac{\pi d_f}{h d_p}\left[1 + \frac{d_f}{d_p}\left\{\pi - 2\cos^{-1}\left(\frac{2d_f}{d_p}\right)\right\} + \sqrt{1 - \left(\frac{2d_f}{d_p}\right)^2}\right] \quad (26)$$

By taking $\frac{d_f}{d_p} = k$, the above equation can be rewritten as

$$\alpha_{sf} = \frac{\pi k}{h}\left[1 + k\pi - 2k\cos^{-1}(2k) + \sqrt{1 - 4k^2}\right] \quad (27)$$

Stacking of the wire-mesh layers can be made in three different ways that include staggered stacking as shown in Figure 3, inline type-a as shown in Figure 4 and inline type-b as shown in Figure 5. For the closely packed scenario, the value of 'h_{min}' varies with type of stacking. For the staggered type, it is difficult to be interpreted by mere observation and the same is expressed mathematically in the following way. The upper layer of the woven mesh screen rests on the lower layer as shown in Figure 7. It can be observed in Figure 7 that the length of 'OC' is the minimum spacing 'h_{min}' for the closely packed wire-mesh block scenario for staggered type of stacking. The expression for 'h_{min}' with reference to the Figure 7 can be obtained as follows:

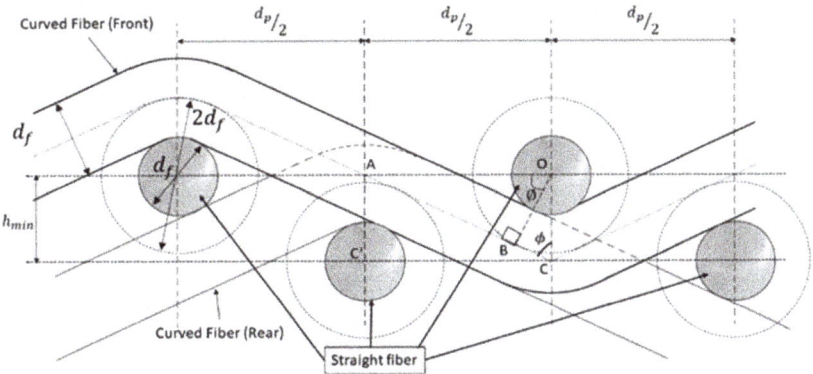

Figure 7. Description of parameters for closely packed wire-mesh layers in staggered type of stacking.

Let $OC = AC' = h_{min}$; thenb, we know that from $\triangle AOB$

$$\cos\varnothing = \frac{2d_f}{d_p} \quad (28)$$

In addition, from $\triangle OBC$,

$$\sin\varnothing = \frac{d_f}{h_{min}} \quad (29)$$

From the trigonometric relation $\cos^2\varnothing + \sin^2\varnothing = 1$, the above equations can be written as

$$\frac{4d_f^2}{d_p^2} + \frac{d_f^2}{h_{min}^2} = 1 \quad (30)$$

$$h_{min} = \frac{d_p d_f}{\sqrt{d_p^2 - 4d_f^2}} \quad (31)$$

Hence, for the closely packed woven wire-mesh scenario for the staggered stacking type, the expression for $'h'_{min}$ as given in Equation (31) can be used in Equation (25) for obtaining the corresponding value of the specific surface area. A detailed sketch of the inline type-a and type-b stacking scenarios is shown in Figures 8 and 9 following which expressions $'h'_{min} = 2d_f$ and $'h'_{min} = 3d_f$ can be used in Equation (25) for obtaining the specific surface area of the wire-mesh porous medium of inline type-a and inline type-b stacking, respectively.

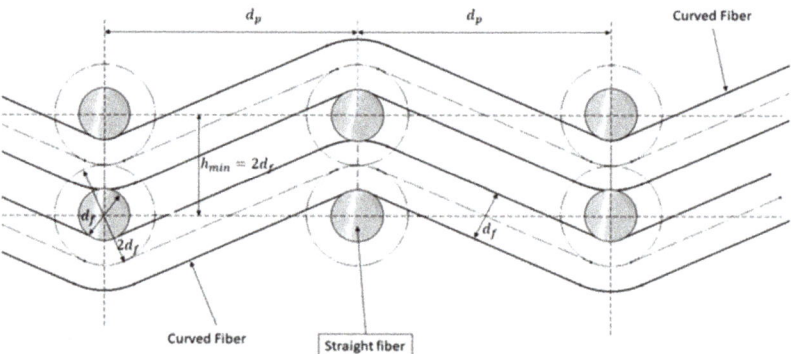

Figure 8. Description of parameters for closely packed wire-mesh layers in stacking of inline type-a.

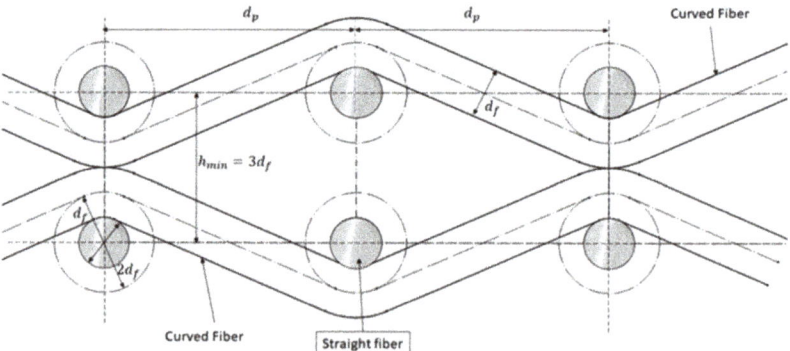

Figure 9. Description of parameters for closely packed wire-mesh layers in stacking of inline type-b.

3.2. Cross-Relations between Porosity, Fiber Diameter and Pore Width

Cross-relationships between morphological characteristic of wire-mesh layers such as fiber diameter and pore width and characteristics of stacked wire-mesh porous structures such as porosity would serve a great purpose by making it easier to avail information on unknown parameters that are difficult to evaluate (for instance, porosity and specific surface area that depend on several other parameters such as stacking manner, fiber diameter and pitch) with information on easily distinguishable parameters. The procedure for arriving at such cross-relations is mentioned below.

The volume of the solid woven mesh layer can be written analogously to Equation (5) as

$$v_1 = (v_s \times N_s) + (v_w \times N_w) \tag{32}$$

The total volume of the solid cylindrical wires (both woven as well as straight running) forming the mesh block can therefore be written as

$$V_{sm} = n((v_s \times N_s) + (v_w \times N_w)) \tag{33}$$

The volume of a single straight running cylindrical wire can be expressed as

$$v_s = \frac{\pi d_f^2 \Delta Y}{4} \tag{34}$$

Similarly, the volume of the weaving cylindrical wire can be expressed as

$$v_w = \frac{\pi d_f^2 L}{4} \tag{35}$$

Porosity is defined as [31]:

$$Porosity = \frac{Total\ volume\ of\ the\ void\ space\ (fluid\ region)}{Total\ volume\ of\ the\ whole\ block\ comprising\ of\ the\ fluid\ saturated\ porous\ medium} \tag{36}$$

which in this case can be written as

$$\varepsilon = \frac{V - V_{sm}}{V} \tag{37}$$

$$1 - \varepsilon = \frac{V_{sm}}{V} \tag{38}$$

where, 'V' is total volume of the whole fluid saturated porous mesh block expressed as $\Delta X \Delta Y \Delta Z$.

From, Equations (4), (22) and (33)–(35), V_{sm}/V can be expressed as,

$$\frac{V_{sm}}{V} = \frac{\pi d_f^2}{4hd_p}\left[1 + \frac{1}{d_p}\left(\pi d_f - 2d_f Cos^{-1}\left(\frac{2d_f}{d_p}\right) + \sqrt{d_p^2 - 4d_f^2}\right)\right] \tag{39}$$

Comparing Equations (38) and (39), the expression for porosity for a woven wire-mesh block with spacing 'h' between the mesh layers can be written as

$$\varepsilon = 1 - \frac{\pi d_f^2}{4hd_p}\left[1 + \frac{1}{d_p}\left(\pi d_f - 2d_f Cos^{-1}\left(\frac{2d_f}{d_p}\right) + \sqrt{d_p^2 - 4d_f^2}\right)\right] \tag{40}$$

For closely packed woven wire-mesh blocks of staggered type, the expression for porosity can be obtained by incorporating the expression for 'h_{min}' given in Equation (31) in Equation (40) as

$$\varepsilon_{staggered} = 1 - \frac{\pi d_f \sqrt{d_p^2 - 4d_f^2}}{4d_p^2}\left[1 + \frac{1}{d_p}\left(\pi d_f - 2d_f Cos^{-1}\left(\frac{2d_f}{d_p}\right) + \sqrt{d_p^2 - 4d_f^2}\right)\right] \tag{41}$$

Taking

$$\frac{d_f}{d_p} = k \tag{42}$$

Equation (41) can be rewritten as

$$\varepsilon_{staggered} = 1 - \frac{k\pi}{4}\sqrt{1-4k^2}\left[1 + k\pi - 2k\cos^{-1}(2k) + \sqrt{1-4k^2}\right] \tag{43}$$

valid for $0 \leq k \leq 0.5$.

For the closely packed woven wire-mesh blocks of inline type-a and inline type-b stacking scenarios, the expression for porosity can be obtained by incorporating expression for 'h_{min}' equal to $2d_f$ in Equation (40), which for inline stacking of type-a becomes

$$\varepsilon_{Inline-type-a} = 1 - \frac{\pi d_f^2}{4(2d_f)d_p}\left[1 + \frac{1}{d_p}\left(\pi d_f - 2d_f\cos^{-1}\left(\frac{2d_f}{d_p}\right) + \sqrt{d_p^2 - 4d_f^2}\right)\right] \tag{44}$$

$$\varepsilon_{Inline-type-a} = 1 - \frac{\pi d_f}{8d_p}\left[1 + \left(\pi\frac{d_f}{d_p} - 2\frac{d_f}{d_p}\cos^{-1}\left(\frac{2d_f}{d_p}\right) + \sqrt{1-\left(\frac{2d_f}{d_p}\right)^2}\right)\right] \tag{45}$$

$$\varepsilon_{Inline-type-a} = 1 - \frac{k\pi}{8}\left[1 + k\pi - 2k\cos^{-1}(2k) + \sqrt{1-4k^2}\right] \tag{46}$$

Similarly, for the stacking of inline type-b, the incorporating expression for 'h_{min}' equal to $3d_f$ generalized expression for porosity becomes

$$\varepsilon_{Inline-type-b} = 1 - \frac{\pi d_f^2}{4(3d_f)d_p}\left[1 + \frac{1}{d_p}\left(\pi d_f - 2d_f\cos^{-1}\left(\frac{2d_f}{d_p}\right) + \sqrt{d_p^2 - 4d_f^2}\right)\right] \tag{47}$$

$$\varepsilon_{Inline-type-b} = 1 - \frac{\pi d_f}{12d_p}\left[1 + \left(\pi\frac{d_f}{d_p} - 2\frac{d_f}{d_p}\cos^{-1}\left(\frac{2d_f}{d_p}\right) + \sqrt{1-\left(\frac{2d_f}{d_p}\right)^2}\right)\right] \tag{48}$$

$$\varepsilon_{Inline-type-b} = 1 - \frac{k\pi}{12}\left[1 + k\pi - 2k\cos^{-1}(2k) + \sqrt{1-4k^2}\right] \tag{49}$$

For various values of 'k' in the range of $0 \leq k \leq 0.5$, fitting a polynomial curve with $R2$ value 0.99866, 0.99963 and 0.99933 for staggered, inline type-a and inline type-b stacking scenarios and considering 501 data points obtained from the above equation provides a compact equation expressing the porosity of woven wire-mesh for the mentioned stacking types as a function of only two morphological properties (fiber diameter and pore-density/mesh-size) of the wire-mesh, which can be obtained as

$$\varepsilon_{staggered} = 1.01051 - 1.87112(k) + 1.60829(k^2) \tag{50}$$

$$\varepsilon_{Inline-type-a} = 0.99358 - 0.63809(k) - 0.68232(k^2) \tag{51}$$

$$\varepsilon_{staggered} = 0.99572 - 0.42539(k) + 0.45488(k^2) \tag{52}$$

3.3. Interstitial Heat Transfer and Flow Resistance Coefficients

In modeling any type of porous media with the LTNE approach, as can be seen in Equation (61), information on interstitial heat transfer coefficient h_{sf} is crucial in solving the energy equation. An effort to identify suitable correlations for the interstitial heat transfer coefficient can be witnessed in a study by Kotresha and Gnanasekaran [30]; the study was focused on determining the interstitial heat transfer coefficient based on correlations that were available in the literature for the packed-bed scenario [32], cross-flow through cylinders [33] and for the situation with metal foam. The study showed that the thermal phenomenon assessed (the wall heat transfer co-efficient of vertical channel) was relatively in good argument with experimental data when the interstitial heat transfer coefficient as provided by Calmidi and Mahajan [34] was used in numerical modeling. The suitability of the correlation was upheld on relative deviation of the thermal phenomenon expressed based on results obtained when the interstitial heat transfer coefficient was evaluated with respective to the correlation considered. However, in a recent study, Garg et al. [21] successfully modeled a regenerator with the wire-mesh porous modeling method by using the interstitial heat transfer coefficient obtained from relation provided by [23]. In the present study, the same relation provided by [9] as given in Equation (53) is used to the evaluate interstitial heat transfer coefficient and the thermal phenomenon expressed through numerical modeling using this relation is compared with that of experimental data in order to emphasize the suitability of this correlation. Figure 10 shows the comparison of the thermal phenomenon characterized by wall heat transfer co-efficient of the vertical channel of present numerical work to that of experimental work of Kurian et al. [28]. Good agreement in the thermal phenomenon can be witnessed here as a result of the incorporation of the interstitial heat transfer co-efficient evaluated from the relation provided by [23] to solve the energy equation. Thus, for further analysis, evaluation of the interstitial heat transfer co-efficient is carried out based on the relation provided by [23], as emphasized by [21].

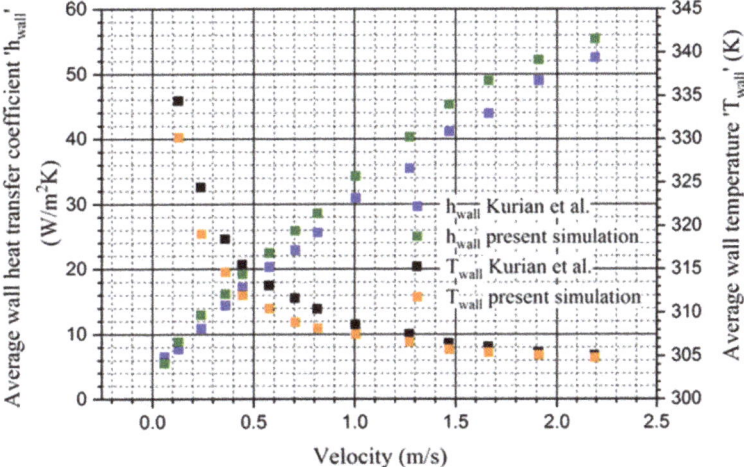

Figure 10. Comparison of wall heat transfer coefficient obtained from present numerical work to that of experimental work of [28].

The inertial and viscous resistance co-coefficients are another set of important parameters that are vital in modeling wire-mesh porous media using the Darcy-Forchheimer model. For the steady state case, Xiao et al. [9] provided the expressions for calculating these parameters by referring to friction factor expressions provided by Tanaka et al. [35] and the modified-Ergun equation provided by Gedeon and Wood [36]. The obtained relations were compared with the pressure drop as described by the Darcy-Forchheimer law and expressions for the viscous and inertial resistance coefficients were provided as given in Equations (54) and (55) respectively. It can be observed that these flow parameters are functions of both porosity as well as fiber diameter and rightly predicts flow resistance in the considered wire-mesh type of porous medium. Hence, these expressions are used for evaluating viscous and inertial resistance coefficients required for modeling the flow phenomenon in the considered wire-mesh porous media.

$$h_{sf} = \frac{\lambda_f \left(1 + 0.99(Re_{dh}Pr)^{0.66}\right)\varepsilon^{1.79}}{d_h} \quad (53)$$

$$\frac{1}{K} = \frac{134}{2\varepsilon d_h^2} \quad (54)$$

$$C = \frac{5.44}{\varepsilon^2 d_h Re_{dh}^{0.188}} \quad (55)$$

4. Numerical Simulation, Boundary Conditions, Computational Scheme and Governing Equations

For this purpose, the experimental set-up of Kurian et al. [28] to investigate stacked wire-mesh porous media is numerically modelled using commercial ANSYS FLUENT software. The sketch of the experimental set up used by Kurian et al. [28] is shown in Figure 11. However, for numerical modeling, due to the symmetric nature of the experimental domain, only one of the symmetric portions is considered for numerical modeling with the help of the symmetry boundary conditions. A sketch of the numerical domain of the present study with boundary conditions is shown in Figure 12.

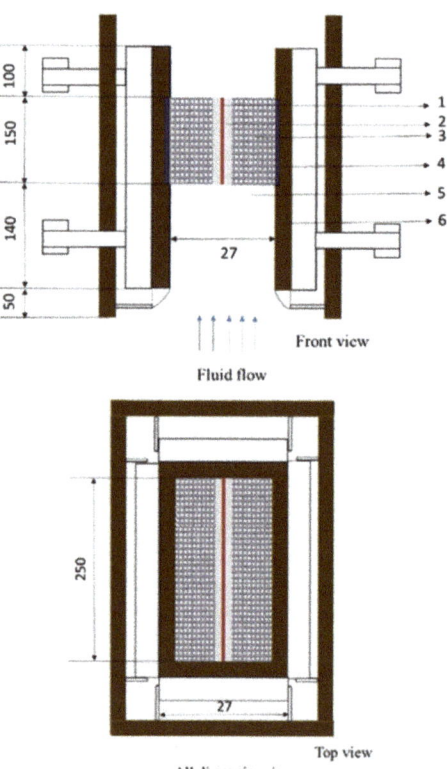

Figure 11. Sketch of experimental domain used by Kurian et al. [28]: (1) heater, (2) aluminum plate, (3) stacked wire-mesh porous media, (4) insulating material, (5) channel passage and (6) channel wall.

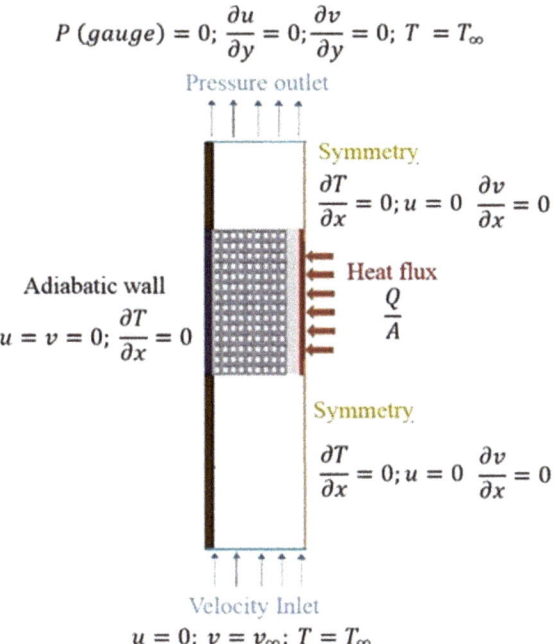

Figure 12. Sketch of numerical domain adopted in the current study.

Appropriate boundary conditions are incorporated in the present numerical domain as shown in Figure 12. Channel walls are given by adiabatic boundary conditions with no slip conditions established at the fluid-wall interface. Continuity in energy and momentum is considered at dissimilar contact regions among solid, fluid and porous zone interfaces. The velocity inlet boundary condition at inlet and pressure outlet boundary condition at outlet are specified. The heat flux boundary condition is specified in the heater region.

Momentum and continuity equations are solved in a coupled fashion by employing a pressure based coupled algorithm (COUPLED scheme) available in ANSYS FLUENT [37]. A second order upwind scheme is used for spatial discretization. As it is known, using the upwind discretization scheme, the values at upstream are used to compute the value on the faces of the cell and then used to evaluate the value at the center of the cell. The first order upwind scheme uses one point at the upstream to compute the values at cell faces and cell center whereas, the second order uses two points at the upstream for computation. Though the first order upwind scheme is most commonly used as it is easier to converge in complex domains, the second order upwind scheme is more accurate than the first order scheme; however, this could present difficulty in convergence in complex problems. Since the present study involves a less complicated 2D flow and heat transfer domain, the second order upwind scheme is used to provide more accurate results (in comparison with the first order upwind scheme) with no difficulty in convergence. Gradients are obtained through the Green-Gauss node-based method. To help ease the convergence, implicit relaxation with the pseudo-transient method is implemented. Convergence criteria are set to 10^{-5} for momentum and continuity equations and for the energy equation, it is set to 10^{-10} for higher accuracy. The outcome of the mesh independence study is shown in Table 1. A mesh with 56,700 cells resulted in the least percentage deviation in pressure drop and excess temperature values. Therefore, further numerical simulations are carried out by adopting the developed numerical domain with 56,700 cells.

Table 1. Grid independence study.

S. No.	No. of Elements	Pressure Drop ΔP, N/m²	Temperature Difference ΔT °C	Deviations ΔP, %	ΔT, %
1	26,130	20	9.75	2.30	1.10
2	56,700	19.59	9.69	0.21	0.43
3	88,400	19.55	9.65	Baseline	

The governing equations solved to obtain flow and heat transfer parameters in the wire-mesh-free region are given by Equations (56)–(58).

Continuity equation:

$$\frac{\partial(\rho_f u_i)}{\partial x_i} = 0 \tag{56}$$

Momentum equation:

$$\frac{\partial(\rho_f u_i u_j)}{\partial x_j} = -\frac{\partial p}{\partial x_j} + \frac{\partial}{\partial x_j}(\mu_f)\left(\frac{\partial u_i}{\partial x_j} + \frac{\partial u_j}{\partial x_i}\right) \tag{57}$$

Energy equation for fluid:

$$\frac{\partial(\rho_f C_{pf} u_j T)}{\partial x_j} = \frac{\partial}{\partial x_j}\left(\lambda_f \frac{\partial T_f}{\partial x_j}\right) \tag{58}$$

In the aluminum plate region, $\lambda_s(\nabla 2T) = 0$ is solved for heat transfer through conduction to take place. Furthermore, the governing equations pertaining to the wire-mesh-filled region are given in Equations (59)–(62). Flow and heat transfer influencing parameters of stacked wire-mesh porous media such as porosity, specific surface area, thermal conductivity, etc. can be observed to be well considered in the governing equations. The source term in the momentum equation given in Equation (60) can be seen to incorporate terms representing viscous and inertial effects fluid experiences while passing through the considered wire-mesh porous medium.

Continuity equation:

$$\frac{\partial(\rho_f \varepsilon u_i)}{\partial x_i} = 0 \tag{59}$$

Momentum equation:

$$\frac{\partial(\rho_f u_i u_j)}{\partial x_j} = -\varepsilon\frac{\partial p}{\partial x_j} + \frac{\partial}{\partial x_j}\left(\mu_f\left(\frac{\partial u_i}{\partial x_i} + \frac{\partial u_j}{\partial x_j}\right)\right) - \varepsilon\left(\frac{\mu_{eff}}{K}u_i + \rho_f C|u|u_i\right) \tag{60}$$

LTNE equation for fluid,

$$\varepsilon\frac{\partial(\rho_f C_{pf} u_j T)}{\partial x_j} = \lambda_{fe}\varepsilon\frac{\partial}{\partial x_j}\left(\frac{\partial T_f}{\partial x_j}\right) + h_{sf}a_{sf}(T_s - T_f) \tag{61}$$

for solid,

$$\lambda_{se}(1-\varepsilon)\frac{\partial}{\partial x_j}\left(\frac{\partial T_s}{\partial x_j}\right) = h_{sf}a_{sf}(T_s - T_f) \tag{62}$$

where

$$\lambda_{fe} = \lambda_f \cdot \varepsilon \text{ and } \lambda_{se} = \lambda_s \cdot (1-\varepsilon) \tag{63}$$

Validation of the Numerical Solution

The simulated results from the modeled numerical domain are compared with that of experimental data of Kurian et al. [28] for validation of the followed modeling methodology. From Figure 10, a good agreement between the average wall heat transfer coefficient and average wall temperature obtained through present numerical simulation and experimental results reported in the literature can be observed. An average difference of 4.2 W/m^2 K with a maximum and minimum difference of 5.13 W/m^2 K and 0.973 W/m^2 K, respectively, is observed between wall heat transfer coefficients. In terms of deviation between experimentally measured temperature and numerically simulated temperature data, an average percentage deviation of less than 0.79 percent is observed, with highest and lowest deviations of 1.65 percent and 0.09 percent, respectively. This demonstrates the correctness and suitability of the followed methodology in order to mimic heat transfer in the considered type of porous medium. In terms of flow phenomenon, variation of experimentally measured and numerically obtained pressure drop with velocity is shown in Figure 13. A good agreement between present numerically obtained data and experimentally reported data is observed with 6 percent and 8 percent average percentage deviations for the 0.71 and 0.81 porosity wire-mesh stack, respectively. Thus, the followed methodology for mimicking flow and heat transfer numerically can be seen to be appropriate for further analysis.

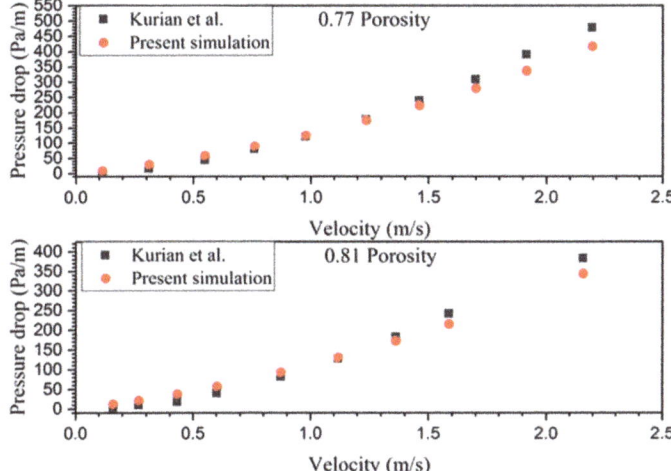

Figure 13. Comparison of pressure drop of present simulations with that of experimental work reported in literature.

5. Results and Discussion

5.1. Interpretation of Results from the Expressions Arrived

The expressions arrived at in the present study that enable the obtaining of vital morphological properties that play a key role in the thermo-hydraulic performance of stacked woven wire-mesh porous media are interpreted in this section. The present study provides cross relations between important parameters such as pore density/mesh size of the wire-mesh screen (mesh per inch/PPI) used in stacking, fiber diameter (d_f) of the wires used to weave the mesh screen, porosity and specific surface area of the stacked wire-mesh blocks corresponding to the three various kinds of stacking such as staggered, inline type-a and inline type-b. It can be noted that from the expressions provided in the present study, porosity and specific surface area, which play a prominent role in momentum and energy equations, can be evaluated with the knowledge of only fiber diameter (d_f) and pitch (d_p) of wire-mesh screens used for various stacking conditions. Thus, porous media of various characteristics that can be obtained by changing the fiber diameter and pitch of the woven screens for various stacking conditions can easily be studied using porous media formulation in a variety of applications in order to arrive at the optimum thermo-hydraulic behavior of such types of porous media.

The variation of porosity of the stacked wire-mesh porous blocks with respect to change in fiber diameter for various type of stacking and pore-density/mesh size is shown in Figure 14; for better visual observation, the variation of porosity and fiber diameter of only 15 PPI and 5 PPI randomly chosen mesh-size samples. However, the respective changes for other pore densities are also observed to follow a similar trend. It can be observed that for any type of stacking of wire-mesh screens of given mesh size (PPI/mesh per inch), porosity decreases with increase in fiber diameter as expected. The magnitude of variation in porosity with change in fiber diameter is relatively high for staggered stacking, followed by inline stacking type-a and type-b. For a given fiber diameter, a stacked block of high porosity can be achieved with inline stacking type-a, followed by inline stacking type-b and staggered stacking. In addition, it can be observed that with increase in fiber diameter, a fixed porosity can be achieved by decreasing the mesh size (MPI) or pore density (PPI). The present study enables numerical study of various stacked wire-mesh porous samples subjected to change in its morphological parameters and stacking condition by providing information on the hydro-dynamic parameters of corresponding stacked wire-mesh porous samples.

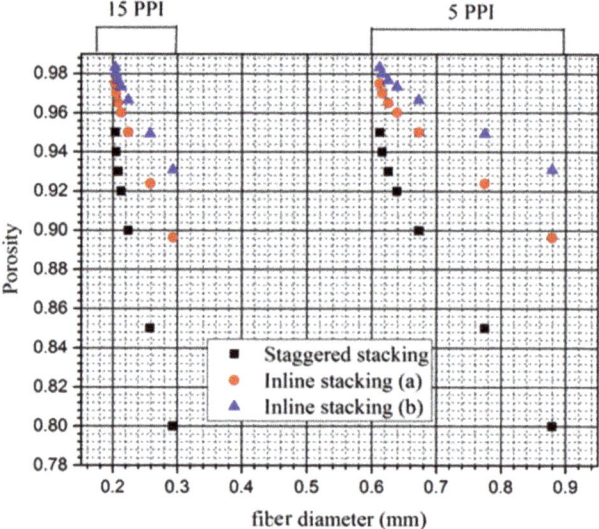

Figure 14. Variation of porosity with respect to change in fiber diameter for various stacking of wire-mesh screens of given pore density or mesh size (pores per inch/mesh per inch).

The variation of specific surface area of stacked wire-mesh porous media for various stacking conditions is shown in Figure 15; the variations are discussed by depicting randomly chosen 0.95 porosity foam samples. However, the variations show a similar trend for any other porosity case. It can be observed that for a given porosity, specific surface area increases with increase in mesh size or pore density of the stacked screens for all type of stacking; however, it is dominated by the staggered type of stacking, followed by inline type-a and type-b kinds of stacking. In addition, it has to be noted that a fixed porosity can be achieved for the considered different kinds of stacking with a compromise in the fiber diameter of weaved wires. Staggered type of stacking provides the required porosity for a given mesh size (PPI/MPI) at lesser fiber diameter compared to inline stacking of type-b and type-a scenarios for same porosity at a given mesh size or pore density condition.

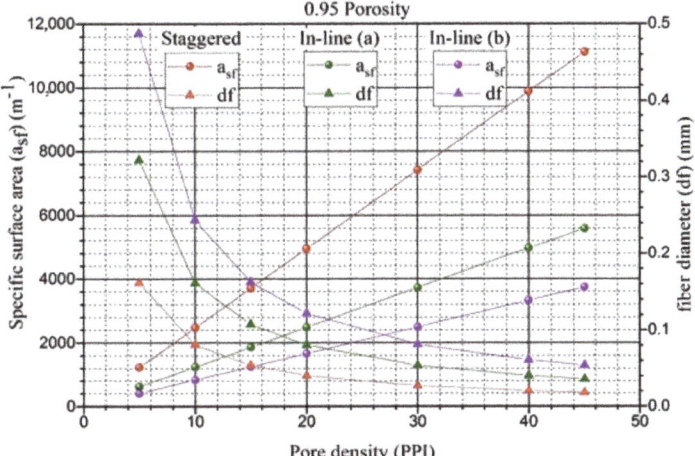

Figure 15. Variation of specific surface area of stacked wire-mesh porous media comprising of wire-mesh screens of various mesh size or pore density (pores per inch/mesh per inch).

5.2. Thermo-Hydraulic Analysis

As explained in Section 2, with the aid of the arrived expressions, stacks of woven wire-mesh samples of orderly varying porosity (0.8, 0.85, 0.9 and 0.95) for each stack of woven wire-mesh screens of 5, 10, 15, 20, and 25 PPI pore-densities/mesh-sizes were obtained. It can be noted that for a given pore density of wire-mesh screens, stacks of various porosity can be achieved by changing the fiber diameter of the wires that form the mesh screens. For the categorized stacks of woven wire mesh screens, specific area (a_{sf}) and porosity (s) are evaluated based on expressions arrived at in the present study as given by Equations (27) and (50)–(52) for considered stacking scenarios. The vertical channel numerical domain as described in Section 2 is modelled with these categorized stacked wire-mesh porous samples. With the subjected heat flux and flow velocities, the thermal performance of the categorized porous samples (based on changes in mesh-size/pore-density and porosity) is shown in Figures 16a and 17b (Figures 18 and 19) in terms of wall heat transfer coefficients for all stacking scenarios. The variation of this thermal phenomenon is demonstrated with the variation in porosity for a constant pore-density/mesh-size (PPI/MPI), as shown in Figure 16a (for 5 PPI/MPI) and Figure 16b (for 25 PPI/MPI). Similarly, variation of the same phenomenon with change in pore-density/mesh-size for a constant porosity can be seen in Figure 17a (for 0.8 porosity) and 17b (for 0.95 porosity). It can be observed that the wall heat transfer co-efficient increases with increase in both pore density as well as porosity. However, for a stack of given porosity, heat transfer greatly increases with pore density compared to the increase of the same parameter with an increase in porosity for a stack of given pore density. In LTNE modeling, the total heat transfer is accounted for by heat transfer due to fluid (this increases with porosity) and interstitial heat transfer (this increases with increase in the product $a_{sf} \times h_{sf}$) [8]. As a result, for a given pore density, the wall heat transfer coefficient increases with porosity due to enhanced availability of fluid to carry the subjected heat. Enhanced heat transfer is also witnessed with increase in pore density for a stack of given porosity, as a result of increased interstitial heat transfer (contributed by the product $a_{sf} \times h_{sf}$). The variation in the wall heat transfer coefficient is observed to be similar, corresponding to changes in porosity and mesh size for all types of stacking. However, for the same porosity and mesh size conditions, the relative deviation of the wall heat transfer of wire-mesh porous media formed by various stacking types is observed to show significant variation. This is mainly due to the increased specific surface area of staggered type for a given porosity, enabling higher heat transfer, followed by inline type-a and inline type-b, which have relatively lesser specific surface area for a given mesh-size/pore-density and porosity condition. As a result, interstitial heat transfer is increased for the staggered type compared to wire-mesh porous samples formed as a result of inline stacking of type-a and type-b conditions. In Figure 16a,b, the thermal performance is depicted for varying porosity condition for 5 PPI/MPI and 25 PPI/MPI mesh-size/pore-density situation only. However, the relative trend is observed to be the same for other 10 PPI/MPI and 15 PPI/MPI mesh-size/pore- densities of the considered wire mesh porous samples of various stacking scenarios. Similarly, in Figure 17a,b, though the variation of the thermal phenomenon is shown with reference to only 0.8 and 0.95 porosity samples, the relative deviation for other porosity conditions such as 0.85 and 0.9 porosity is observed to follow a similar trend. This implies that domination of heat transfer follows a similar trend for any other porosity and pore-density/mesh-size combinations for the considered different stacking scenarios, agreeing well with the explained reasons.

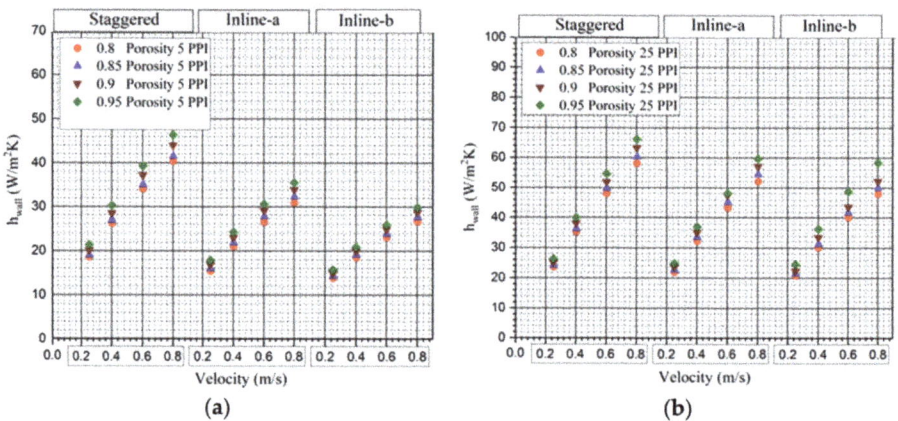

Figure 16. (**a**) Variation of average wall heat transfer coefficient for 5 PPI (pore- density/mesh-size) wire mesh samples of different porosity for all three stacking scenarios. (**b**) Variation of average wall heat transfer coefficient for 25 PPI (pore-density/mesh-size) wire mesh samples of different porosity for all three stacking scenarios.

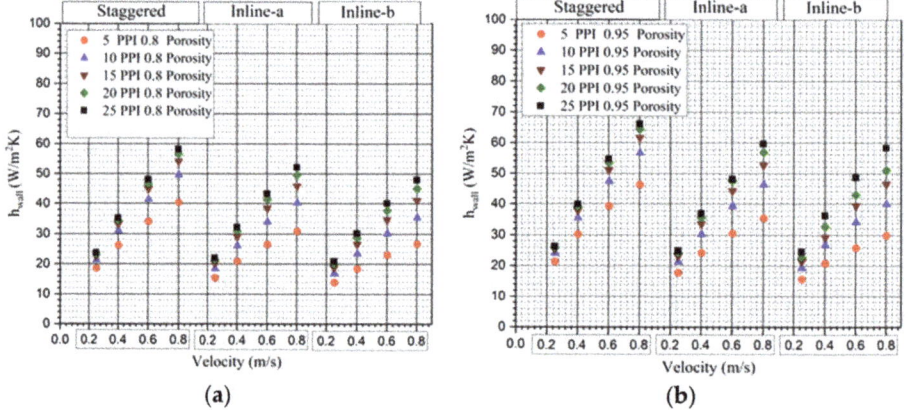

Figure 17. (**a**) Variation of average wall heat transfer coefficient for 0.8 porosity wire mesh samples of different pore-density/mesh-size for all three stacking scenarios. (**b**) Variation of average wall heat transfer coefficient for 0.95 porosity wire mesh samples of different pore-density/mesh-size for all three stacking scenarios.

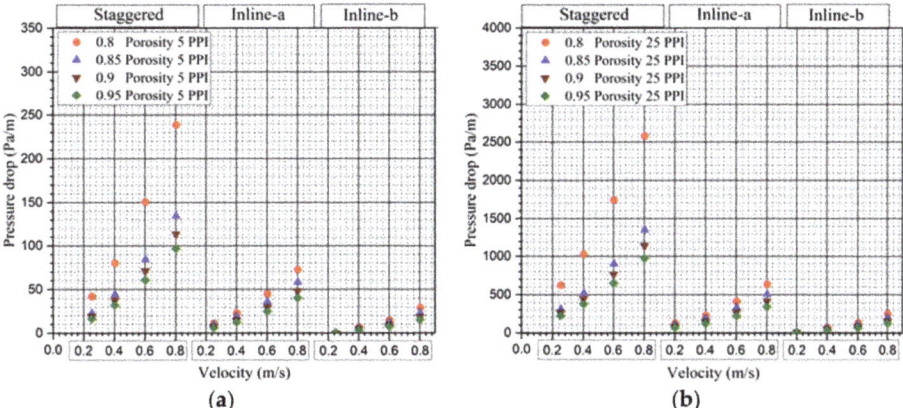

Figure 18. (**a**) Variation of pressure drop for 5 PPI (pore-density/mesh-size) wire mesh samples of different porosity for all three stacking scenarios. (**b**) Variation of pressure drop for 25 PPI (pore-density/mesh-size) wire mesh samples of different porosity for all three stacking scenarios.

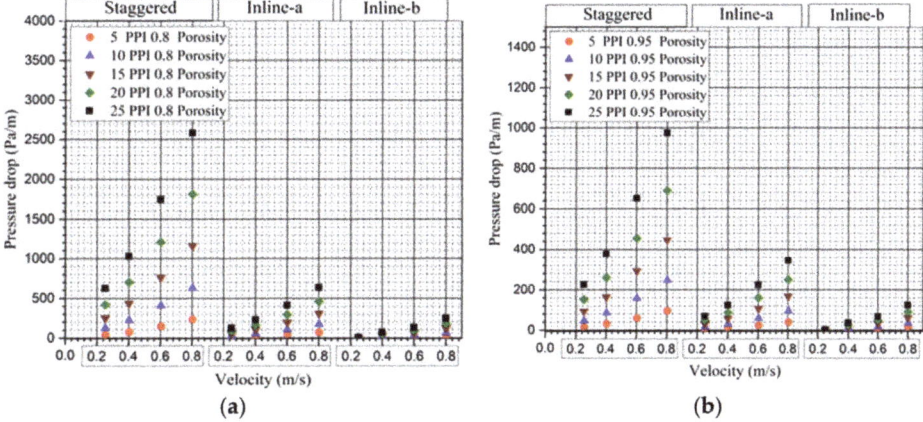

Figure 19. (**a**) Variation of pressure drop for 0.8 porosity wire mesh samples of different pore-density/mesh-size for all three stacking scenarios. (**b**) Variation of pressure drop for 0.95 porosity wire mesh samples of different pore-density/mesh-size for all three stacking scenarios.

Flow resistance, characterized by the pressure drop phenomenon in the considered wire-mesh type of porous medium corresponding to mesh size and porosity of samples under various stacking scenarios, is shown in Figures 18a and 19b. Variation of this phenomenon for wire mesh samples of various stacking conditions under varying porosities at constant mesh-size/pore-density is shown in Figure 18a,b. Similarly, variation of pressure drop for wire mesh samples of various stacking types under varying pore-densities/mesh-sizes at a constant porosity is depicted in Figure 19a,b. It can be observed that, for a given stacking scenario, the pressure drop decreases with increase in porosity for a given mesh size, as shown in Figure 18a,b. This is due to increased permeability of the fluid through the wire mesh structures with an increase in porosity. However, for a given porosity with an increase in mesh size, the pressure drop can be observed to be increasing due to increased obstruction to the flow with an increase in mesh-size/pore-density, as shown in Figure 19a,b. In terms of the stacking scenario, staggered stacking that offers low permeability to the fluid flow shows an increased pressure drop for a given porosity and mesh size, followed by inline stacking type-a and type-b. In addition, an increase in pressure drop is observed to be relatively more significant at higher mesh size than at lower mesh sizes. Though variation of flow resistance, characterized by the pressure drop for various stacking scenarios with changes in porosity is shown with reference to only 5 PPI/MPI pore-densities/mesh-sizes, the relative deviation is also observed to follow similar trends for 10 and 15 PPI/MPI wire-mesh samples. The argument is the same with reference to deviations of pressure drop with respect to change in pore-density/mesh-size. That is, though these deviations are shown with reference to only 0.8 and 0.85 porosity samples, the relative deviation is also observed to follow a similar trend for other 0.85 and 0.9 porosity wire-mesh samples, agreeing well with the given reasons. Unlike heat transfer at higher mesh sizes, a relative increase in pressure drop is observed to be higher for the staggered stacking type compared to the other two types of stacking. In the present study, it is also observed that with variation in stacking scenario, the variation in flow phenomena is larger compared to the variation in heat transfer. It is quite evident from this study that heat transfer enhancement with change in porosity and mesh size for various stacking types of wire-mesh porous media is accompanied by individually varying flow resistance phenomena; hence, optimum selection of such types of porous media with respect to mesh size, porosity and stacking types has to be the prime aspect for achieving their desired thermo-hydraulic performance in any heat exchange applications.

6. Conclusions

In the present work, a set of cross-relations that can describe/classify a given stacked wire-mesh porous medium is arrived at based on the mesh-size, fiber (wire) diameter and type of stacking involved. Expressions presented from this study enable evaluation of heat transfer and flow influencing morphological properties such as porosity and specific surface area that are unique to wire-mesh porous types of given mesh-size/pore-density, fiber diameter and stacking condition. Such thermo-hydraulic properties of stacked woven wire-mesh samples of orderly varying mesh-size/pore density and porosity are categorized corresponding to each type of various stacking scenario, namely staggered, inline type-a and type-b. The mentioned key parameters that highly influence thermo-hydraulic performance of the categorized porous medium are evaluated using the expressions provided in the present study and are incorporated in momentum equations (Darcy-Forchheimer model) and energy equations (LTNE model) for evaluation of the thermal and hydraulic performance of this type of porous media in a vertical channel subjected to constant heat flux and flow conditions. Heat transfer, characterized by a wall heat transfer coefficient, is observed to increase with porosity as well as mesh-size/pore-density for stacked wire mesh of all stacking types. However, due to the relative increase in specific surface area for a given porosity and mesh size, the wall heat transfer coefficient is observed to be dominant in the staggered type followed by inline stacking of type-a and type-b. In terms of flow resistance characterized by pressure drop, the staggered type of stacking showed the highest pressure drop for a given porosity and mesh size condition due to highly restricted flow path as a result of high compactness in stacking of the wire-mesh screens in this type of stacking, followed by inline type-a and type-b stacking scenarios. In addition, for any given type of stacking, it is observed that the pressure drop increases with increase in mesh-size/pore-density for a given porosity scenario due to increased flow obstruction and contrarily, with increase in porosity for a given mesh size, the pressure drop is observed to be decreasing as a result of reduced flow obstruction. Required changes in porosity and specific surface area that highly influence flow and heat transfer through porous media can be achieved with most ease in stacked wire-mesh type of porous media compared to other types of porous media such as metal foams. The expressions arrived at in the present study are intended to be useful in numerical modeling of flow and heat transfer through such types of porous media of various morphological characteristics such as porosity and mesh size in a variety of applications.

Author Contributions: Conceptualization, T.G. and G.N.; methodology, T.G., K.E.S.S., G.N. and M.M.; software, T.G., D.H. and G.N.; validation, T.G., D.H. and G.N.; formal analysis, T.G., K.E.S.S. and G.N.; investigation, T.G., G.N. and M.M.; resources, T.G. and G.N.; data curation, T.G. and D.H.; writing—original draft preparation, T.G.; writing—review and editing, T.G., K.E.S.S., D.H., G.N. and M.M.; visualization, T.G.; supervision, G.N. and M.M.; project administration, T.G. and G.N.; funding acquisition, G.N. All authors have read and agreed to the published version of the manuscript.

Funding: This research was funded by SERB, grant number EEQ/2018/000322.

Acknowledgments: Science and Engineering Research Board (SERB), DST No: EEQ/2018/000322, India, supported and funded this work.

Conflicts of Interest: The authors declare no conflict of interest. The funders had no role in the design of the study; in the collection, analyses, or interpretation of data; in the writing of the manuscript, or in the decision to publish the results.

Nomenclature

A	Surface area of Aluminum plate (m^2)
A_{sm}	Total area of the solid mesh in the whole porous block
a_1	Area of total solid fibers constituting an individual mesh layer
a_s	Area of straight running wire fiber (m^2)
a_w	Area of weaving wire fiber (m^2)
a_{sf}	Area density (m^{-1})
C	Inertial resistance coefficient (m^{-1})
C_p	Specific heat (J/kgK)
d_p	Pore diameter (m)
d_f	Fiber diameter (m)
d_h	Hydraulic diameter
h_{wall}	Wall heat transfer coefficient (W/m^2 K)
h_{sf}	Interstitial heat transfer coefficient (W/m^2 K)
h	Spacing (distance) between individual mesh layers.
H_{min}	Minimum possible distance measured from centre to centre of straight running fibres of consecutive mesh layers for closely packed condition
K	Permeability (m^2)
$1/K$	Viscous resistance coefficient (inverse of permeability)
k	Dimensionless ratio (d_f/d_p)
Nu	Nusselt number
MPI	Mesh per inch
N_s	Number of straight running wire fibers
N_w	Number of weaving wire fibers
P	Pressure (N/m^2)
Pr	Prandtl number of fluid
Q	Heat input (W)
Re_{d_h}	Reynolds number based on hydraulic diameter
T	Temperature (K)
T_∞	Ambient temperature (K)
w	Pore width
Greek symbols	
ε	Porosity
ω	pore density
λ	Thermal conductivity (W/mK)
μ	Dynamic viscosity (N-s/m^2)
v	Kinematic viscosity (m^2/s)
ρ	Density (kg/m^3)
Subscript	
f	Fluid
fe	Fluid effective
s	Solid
se	Solid effective

References

1. Donmus, S.; Mobedi, M.; Kuwahara, F. Double-layer metal foams for further heat transfer enhancement in a channel: An analytical study. *Energies* **2021**, *14*, 672. [CrossRef]
2. Hossain, M.S.; Shabani, B. Experimental study on confined metal foam flow passage as compact heat exchanger surface. *Int. Commun. Heat Mass Transf.* **2018**, *98*, 286–296. [CrossRef]
3. Boyd, B.; Hooman, K. Air-cooled micro-porous heat exchangers for thermal management of fuel cells. *Int. Commun. Heat Mass Transf.* **2012**, *39*, 363–367. [CrossRef]
4. Qureshi, Z.A.; Al-Omari, S.A.B.; Elnajjar, E.; Al-Ketan, O.; Al-Rub, R.A. Using triply periodic minimal surfaces (TPMS)-based metal foams structures as skeleton for metal-foam-PCM composites for thermal energy storage and energy management applications. *Int. Commun. Heat Mass Transf.* **2021**, *124*, 105265. [CrossRef]
5. Odabaee, M.; Hooman, K. Application of metal foams in air-cooled condensers for geothermal power plants: An optimization study. *Int. Commun. Heat Mass Transf.* **2011**, *38*, 838–843. [CrossRef]
6. Wang, F.; Guan, Z.; Tan, J.; Ma, L.; Yan, Z.; Tan, H. Transient thermal performance response characteristics of porous-medium receiver heated by multi-dish concentrator. *Int. Commun. Heat Mass Transf.* **2016**, *75*, 36–41. [CrossRef]
7. Lotfizadeh, H.; Mehrizi, A.A.; Motlagh, M.S.; Rezazadeh, S. Thermal performance of an innovative heat sink using metallic foams and aluminum nanoparticles-Experimental study. *Int. Commun. Heat Mass Transf.* **2015**, *66*, 226–232. [CrossRef]
8. Trilok, G.; Gnanasekaran, N. Numerical study on maximizing heat transfer and minimizing flow resistance behavior of metal foams owing to their structural properties. *Int. J. Therm. Sci.* **2021**, *159*, 106617. [CrossRef]
9. Xiao, G.; Peng, H.; Fan, H.; Sultan, U.; Ni, M. Characteristics of steady and oscillating flows through regenerator. *Int. J. Heat Mass Transf.* **2017**, *108*, 309–321. [CrossRef]
10. Zhao, J.; Sun, M.; Zhang, L.; Hu, C.; Tang, D.; Ynag, L.; Song, Y. Forced Convection Heat Transfer in Porous Structure: Effect of Morphology on Pressure Drop and Heat Transfer Coefficient. *J. Therm. Sci.* **2021**, *30*, 363–393.
11. Kang, K.J. Wire-woven cellular metals: The present and future. *Prog. Mater. Sci.* **2015**, *69*, 213–307. [CrossRef]
12. Sypeck, D.J.; Introduction, I. Multifunctional microtruss laminates: Textile synthesis and properties. *J. Mater. Res.* **2001**, *16*, 890–897.
13. Tian, J.; Lu, T.J.; Hodson, H.P.; Queheillalt, D.T.; Wadley, H.N.G. Cross flow heat exchange of textile cellular metal core sandwich panels. *Int. J. Heat Mass Transf.* **2007**, *50*, 2521–2536. [CrossRef]
14. Costa, S.C.; Barrutia, H.; Esnaola, J.A.; Tutar, M. Numerical study of the pressure drop phenomena in wound woven wire matrix of a Stirling regenerator. *Energy Convers. Manag.* **2013**, *67*, 57–65. [CrossRef]
15. Costa, S.C.; Barrutia, H.; Esnaola, J.A.; Tutar, M. Numerical study of the heat transfer in wound woven wire matrix of a Stirling regenerator. *Energy Convers. Manag.* **2014**, *79*, 255–264. [CrossRef]
16. Bussière, W.; Rochette, D.; Clain, S.; André, P.; Renard, J.B. Pressure drop measurements for woven metal mesh screens used in electrical safety switchgears. *Int. J. Heat Fluid Flow* **2017**, *65*, 60–72. [CrossRef]
17. Jadhav, P.H.; Trilok, G.; Gnanasekaran, N.; Mobedi, M. Performance score based multi-objective optimization for thermal design of partially filled high porosity metal foam pipes under forced convection. *Int. J. Heat Mass Transf.* **2022**, *182*, 121911. [CrossRef]
18. Trilok, G.; Kumar, K.K.; Gnanasekaran, N.; Mobedi, M. Numerical assessment of thermal characteristics of metal foams of orderly varied pore density and porosity under different convection regimes. *Int. J. Therm. Sci.* **2022**, *172*, 107288. [CrossRef]
19. Trilok, G.; Gnanasekaran, N.; Mobedi, M. Various Trade-Off Scenarios in Thermo-Hydrodynamic Performance of Metal Foams Due to Variations in Their Thickness and Structural Conditions. *Energies* **2021**, *14*, 8343. [CrossRef]
20. Armour, J.C. Fluid Through Woven Screens. *AIChE J.* **1968**, *14*, 415–420.
21. Garg, S.K.; Premachandran, B.; Singh, M. Numerical study of the regenerator for a miniature Stirling cryocooler using the local thermal equilibrium (LTE) and the local thermal nonequilibrium (LTNE) models. *Therm. Sci. Eng. Prog.* **2019**, *11*, 150–161. [CrossRef]
22. Nam, K.; Jeong, S. Novel flow analysis of regenerator under oscillating flow with pulsating pressure. *Cryogenics (Guildf)*. **2005**, *45*, 368–379. [CrossRef]
23. Geodeon, D. *Baseline Stirling Modeling*; Gedeon Associates: Athens, OH, USA, 1999.
24. Iwaniszyn, M.; Sindera, K.; Gancarczyk, A.; Korpy, M.; Roman, J.J.; Ko, A.; Jod, J. Experimental and CFD investigation of heat transfer and flow resistance in woven wire gauzes. *Chem. Eng. Process.—Process. Intensif.* **2021**, *163*, 108364. [CrossRef]
25. Wang, Y.; Yang, G.; Huang, Y.; Huang, Y.; Zhuan, R.; Wu, J. Analytical model of flow-through-screen pressure drop for metal wire screens considering the effects of pore structures. *Chem. Eng. Sci.* **2021**, *229*, 116037. [CrossRef]
26. Xu, J.; Tian, J.; Lu, T.J.; Hodson, H.P. On the thermal performance of wire-screen meshes as heat exchanger material. *Int. J. Heat Mass Transf.* **2007**, *50*, 1141–1154. [CrossRef]
27. Zhao, Z.; Peles, Y.; Jensen, M.K. Properties of plain weave metallic wire mesh screens. *Int. J. Heat Mass Transf.* **2013**, *57*, 690–697. [CrossRef]
28. Kurian, R.; Balaji, C.; Venkateshan, S.P. Experimental investigation of convective heat transfer in a vertical channel with brass wire mesh blocks. *Int. J. Therm. Sci.* **2016**, *99*, 170–179. [CrossRef]
29. Kurian, R.; Balaji, C.; Venkateshan, S.P. An experimental study on hydrodynamic and thermal performance of stainless steel wire mesh blocks in a vertical channel. *Exp. Therm. Fluid Sci.* **2017**, *86*, 248–256. [CrossRef]

30. Kotresha, B.; Gnanasekaran, N. Determination of interfacial heat transfer coefficient for the flow assisted mixed convection through brass wire mesh. *Int. J. Therm. Sci.* **2019**, *138*, 98–108. [CrossRef]
31. Kaviany, M. *Principles of Heat Transfer in Porous Media*; Springer: New York, NY, USA, 1995.
32. Wakao, N.; Kaguei, S.; Funazkri, T. Effect of fluid dispersion coefficients on particle-to-fluid heat transfer coefficients in packed beds: Correlation of nusselt numbers. *Chem. Eng. Sci.* **1979**, *34*, 325–336.
33. Zukauskas, A. *Convective Heat Transfer in Cross Flow. Handbook of Single-Phase Convective Heat Transfer*; John Wiley &Sons: Hoboken, NJ, USA, 1987.
34. Calmidi, V.V.; Mahajan, R.L. Forced convection in high porosity metal foams. *J. Heat Transfer* **2000**, *122*, 557–565. [CrossRef]
35. Tanaka, M.; Yamashita, I.; Chisaka, F. Flow and heat transfer characteristics of the stirling engine regenerator in an oscillating flow. *JSME Int. J.* **1990**, *33*, 283–289.
36. Gedeon, D.; Wood, J.G. *Oscillating-Flow Regenerator Test Rig: Hardware and Theory with Derived Correlations for Screens and Felts*; National Aeronautics and Space Administration: Washington, DC, USA, 1996.
37. *A F.U. Guide*; Version 12; Ansys Inc.: Canonsburg, PA, USA, 2009.

Article

Heat and Mass Transfer in Structural Ceramic Blocks: An Analytical and Phenomenological Approach

Stephane K. B. M. Silva [1], Carlos J. Araújo [1], João M. P. Q. Delgado [2,*], Ricardo S. Gomez [1], Hortência L. F. Magalhães [3], Maria J. Figueredo [4], Juliana A. Figueirôa [5], Mirenia K. T. Brito [1], José N. O. Neto [6], Adriana B. C. Pereira [3], Leonardo P. L. Silva [7] and Antonio G. B. Lima [1]

1. Department of Mechanical Engineering, Federal University of Campina Grande, Campina Grande 58429-900, Brazil
2. CONSTRUCT-LFC, Department of Civil Engineering, Faculty of Engineering, University of Porto, 4200-465 Porto, Portugal
3. Department of Chemical Engineering, Federal University of Campina Grande, Campina Grande 58429-900, Brazil
4. Department of Agro-Industrial Management and Technology, Federal University of Paraíba, Bananeiras 58220-000, Brazil
5. Federal Institute of Education, Science and Technology of the Sertão Pernambucano, Petrolina 56316-686, Brazil
6. Integrated Colleges of Ceará, Iguatú 63508-025, Brazil
7. Department of Industry, Federal Institute Education, Science and Technology of Paraiba, Cajazeiras 58900-000, Brazil
* Correspondence: jdelgado@fe.up.pt; Tel.: +351-225081404

Abstract: The ceramic industry is one of the pillars of the Brazilian economy, characterized by making low-cost products and an obsolete manufacturing process from a technological point of view. Among the various stages of production of ceramic materials, drying is one of the most energy-consuming and, in general, causes structural damage to the product, compromising its mechanical performance and final quality. Despite the relevance, studies on the drying of ceramic materials are mostly conducted at the experimental level and limited to some specific operational conditions. In this scenario, this research aims to theoretically study the heat and mass transfers in industrial ceramic blocks during drying. Based on the lumped analysis method, and considering the dimensional variations of the material, new phenomenological mathematical models and their respective analytical solutions are proposed to describe the kinetics of mass loss and heating of the material. The predicted results referring to the thermal and gravimetric behavior of the block during the oven drying process under different conditions are compared with the experimental data, obtaining excellent agreement between the results. Furthermore, the transport coefficients were estimated, proving the dependence of these parameters on the drying air conditions. The convective mass transfer coefficient ranged from 6.69×10^{-7} to 15.97×10^{-7} m/s on the outer surface of the block and from 0.70×10^{-7} to 1.03×10^{-7} m/s on the inner surface of the material when the drying air temperature ranged from 50 to 100 °C. The convective heat transfer coefficient ranged from 4.79 to 2.04 W/(m$^2 \cdot$°C) on the outer surface of the block and from 1.00 to 0.94 W/(m$^2 \cdot$°C) on the inner surface of the material when air temperature ranged from 50 to 100 °C.

Keywords: ceramic block; drying; lumped analysis; analytical; simulation

1. Introduction

Having been used for millennia, ceramics are characterized by their ability to conform when hydrated, which allows them to be molded for different uses, thus making them the main components for both artistic and industrial activities [1]. Considering its use in several activities, the ceramic sector is quite diversified, being divided into numerous segments and with emphasis on the red ceramic segment, which is linked to the civil construction industry, thus occupying a prominent place in the Brazilian economy [2,3] and worldwide.

The red ceramic segment is composed of numerous industries, characterized by different levels of technological development and production potential whose most representative elements are materials for civil construction (roof tiles, blocks, bricks, etc.). It is a sector that encompasses a series of different processes and raw materials, giving rise to several products which can be classified according to their physicochemical properties, chemical compositions, and applications [4–6].

Ceramic blocks are essential components for structural masonry, since they are responsible for providing mechanical strength, durability, and thermal comfort to buildings. These materials are characterized by having a low manufacturing cost, encompassing the stages of exploration of deposits, beneficiation, homogenization, conformation, drying, and firing [7–9].

As one of the steps in the manufacturing of ceramic materials that presents the most hydric and thermal problems, drying is defined as a thermodynamic phenomenon in which there is an exchange of energy either by conduction, convection, or radiation, as well as a change in the state of water, where it changes from the liquid state to the vapor state and is then dragged by the fluid that surrounds the material, phenomena that occur simultaneously. The purpose of this step is the hygroscopic removal of the moisture inside the product. However, when done incorrectly, it can cause structural damage to the part, thus promoting the loss of the product. Therefore, an in-deep understanding in the moisture and heat transfer mechanisms inside these materials becomes of paramount importance for the ceramic industry [10–13] and academy.

Despite its importance, studies on the drying of ceramic materials are conducted mostly at an experimental level with specific operating conditions, and there are few studies that involve phenomenological mathematical modeling and numerical simulation. The use of mathematical simulations associated with drying kinetics enables the optimization of the drying process, allowing it to be carried out economically and in a shorter time. In this way, the use of computer codes can promote greater control of the drying process, thus allowing optimization in production so that raw material losses and process costs are reduced, and the productivity of companies in the ceramic sector is increased. In this context, the following research can be cited: (1) at the industrial dryer level [14–16], (2) at the ceramic block level using CFD [17–19], and (3) at the ceramic block level using a lumped model [20–22]. In these works, details about the drying period, drying rate, moisture migration mechanisms, and many others interesting topics in drying process theory can be found. Furthermore, we can be cite the following works: Brown [23], Gualtieri et al. [24], Dogru et al. [25], Ndukwu et al. [26], Karagiannis et al. [27], and Pujari et al. [28].

Although the use of numerical simulations to describe phenomena such as the drying process of a material is of great value, there is a scarcity of mathematical models that describe the simultaneous heat and mass transfer in hollow solids with three-dimensional geometries, mainly for models that consider the external effects of heating and evaporation of the vapor produced on the surface of the solid, including dimensional variations in the material, a frequent phenomenon in this process. Furthermore, we can mention the high computational time spent in simulations with commercial CFD software. Thus, it is necessary to deepen the study with the use of more complete models that incorporate parameters which explain the internal and external phenomena intrinsic to the process, especially when applied to solids with complex geometries, and that provide results in a short computational time, which is of great importance for decision making in the industry.

Given the above, the present study aims to analytically describe the convective drying process of structural ceramic blocks based on lumped analysis. The use of mathematical models to describe the phenomena of energy and mass transfer in these blocks aims to control and optimize the drying process, preventing this process from being carried out incorrectly and thus affecting the performance and quality of the final part. The central idea is to reduce the cost of the process, achieved by controlling the drying time, increasing the quality of the final product, and minimizing the waste of raw materials, which directly contributes to the reduction in environmental impact.

As innovative aspects of this research, the following can be highlighted: (a) the use of more complete phenomenological models, which consider several simultaneous physical phenomena such as heat and mass convection, evaporation, and dimensional variations in the product throughout the process, (b) the reduced computational time to obtain the results, allowing the effects of several process variables to be analyzed in a short time, (c) the versatility of the proposed model being applied to different materials and product shapes, (d) the application of the study to a product with industrial dimensions, supplied by the Brazilian industry's ceramic sector, and (e) the theme of worldwide interest, which will certainly reach the interest of readers. The choice of the structural brick occurred due to its large application in different countries, including Brazil.

2. Methodology

2.1. The Geometry of the Ceramic Block and the Physical Problem

This investigation presents a numerical analysis of the convective drying process (heat and mass transfer) of structural ceramic blocks, as shown in Figure 1. The analysis was carried out from the mathematical modeling and computational simulation of the process where, to describe it, a lumped analysis method was used, considering the energy and mass balances for the ceramic block and drying air.

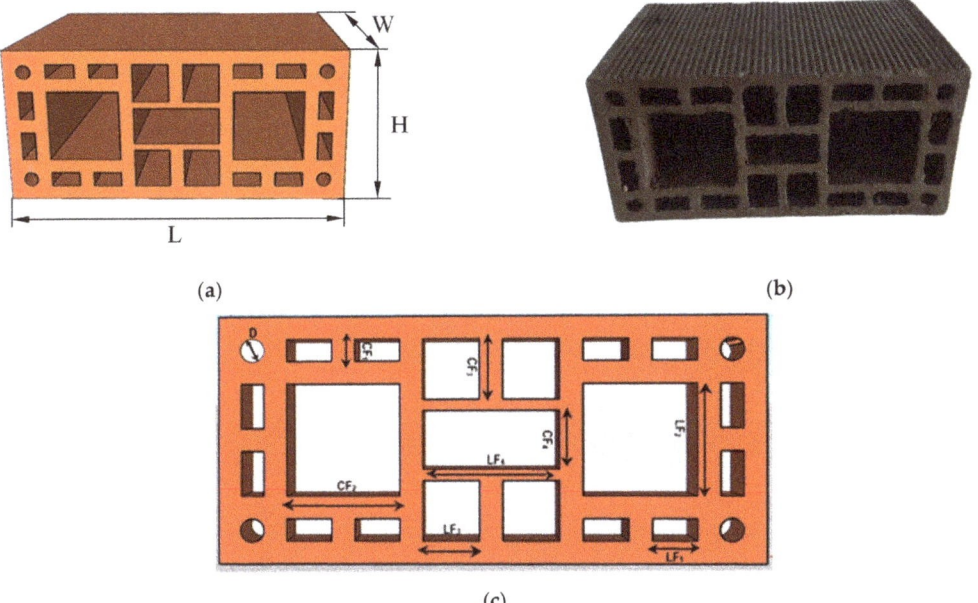

Figure 1. Structural ceramic block. (a) 3D drawing. (b) Photograph of the wet material. (c) Front view indicating the main dimensions of the material.

Table 1 summarizes the values of the characteristic dimensions of the ceramic block under study at the beginning of the drying process.

Table 1. Dimensions of the ceramic block before drying at different temperatures [14].

T (°C)	H (mm)	W (mm)	L (mm)	D (mm)	CF_1 (mm)	CF_2 (mm)	LF_1 (mm)	LF_2 (mm)	CF_3 (mm)	LF_3 (mm)	CF_4 (mm)	LF_4 (mm)
50	143	199	303	14	15	73	33	75	38	24	31	62
60	140	200	302	15	15	72	33	78	37	28	30	64
70	142	200	300	15	33	74	15	75	36	26	31	62
80	144	200	300	15	15	74	33	74	35	26	31	63
90	143	199	300	15	15	73	34	75	36	26	30	63
100	142	201	301	14	15	73	31	75	36	26	30	63

2.2. Mathematical Modeling

As is well known, the type of analysis chosen to describe a process limits the level of understanding that can be gained from this process [29]. In this research, the method of lumped analysis was chosen. In this method, the rates of heat and mass transfer between the product and air during the drying process neglect the intrinsic resistance of the product to these transfers.

Thus, to mathematically describe the process of energy and mass transfer in ceramic blocks, some hypotheses were considered: (a) The structural ceramic block is isotropic and consists of solid matter and water. (b) The mechanical and thermophysical properties of the material are constant throughout the process. (c) The water contained in the ceramic block, in its liquid state, migrates from the interior of the material, evaporating on its surface. (d) Drying occurs in the falling drying rate period (based on the experimental data of the moisture content throughout the process). (e) Drying occurs by the transport of heat and mass inside the solid, the convection of heat and mass, and evaporation at the surface. (f) Generations of energy and mass inside the material are negligible. Finally, (g) the drying process occurs with dimensional variation of the ceramic block.

2.2.1. Dimensional Analysis

In addition to the previously presented hypotheses, to numerically describe the process of heat and mass transfer in a ceramic block, as described in Figure 1, the dimensional variation models of the block during drying, proposed by Silva et al. [29], were used. The equations governing this model are as follows:

$$S_1(t) = S_{10} \left[a_1 + b_1 \times \text{Exp}\left(-k_1^2 \times t\right) \right], \qquad (1)$$

$$S_2(t) = S_{20} \left[a_2 + b_2 \times \text{Exp}\left(-k_2^2\, t\right) \right], \qquad (2)$$

$$V(t) = V_0 \left[a_3 + b_3 \times \text{Exp}\left(-k_3^2 \times t\right) \right], \qquad (3)$$

where t is the time, V_0 and S_{i0} correspond to the volume and surface areas (external, i = 1 and internal, i = 2) at t = 0, respectively, and the variables a_1, b_1, k_1, a_2, b_2, k_2, a_3, b_3, and k_3 are parameters estimated from the non-linear adjustment to the experimental data in each drying condition. These parameters are described in Table 2.

Thus, based on the hypotheses established in the dimensional variation models, and considering the lumped analysis method, a phenomenological mathematical model was developed in order to describe the energy and mass transfers in the material. It is a complex but innovative model, since in the involved parameters, it considers the dimensional variations of the material during the drying process, as well as the phenomena of convective heat and mass transport and evaporation on the surface of the material. Details about the experimental procedure for measuring the dimensions of the brick along the drying process can be found in [14,20]. Details about the non-linear regression of Equations (1)–(3) to the experimental data are reported in [29]. However, we state that the determination coefficient ranged from 0.988 to 0.995, and the explained variance ranged from 97.6% to 99.1%.

Table 2. Statistical parameters of Equations (1–3), obtained after fitting to experimental data of the volume and surface area of the ceramic block throughout drying.

Parameter	T (°C)					
	50	60	70	80	90	100
a_1	0.901906	0.916317	0.913417	0.906832	0.918261	0.924380
b_1	0.094942	0.089238	0.090522	0.095390	0.081259	0.078334
k_1	−0.118153	−0.107296	−0.136887	−0.142611	−0.145465	−0.144425
a_2	0.934331	0.916679	0.936533	0.933251	0.930446	0.95068
b_2	0.078248	0.092013	0.076191	0.085057	0.077706	0.065325
k_2	−0.099675	−0.117138	−0.092472	−0.108381	−0.116544	−0.106260
a_3	0.827963	0.883788	0.845889	0.836882	0.868853	0.877361
b_3	0.156506	0.118260	0.168594	0.156642	0.127073	0.125252
k_3	−0.153099	−0.103415	−0.193381	−0.185467	−0.195847	−0.202705

2.2.2. Mass Transfer Model

Assuming a convective condition and performing a mass balance on the material, we can write

$$V\frac{d\overline{M}}{dt} = -h_{m1}S_1(\overline{M} - \overline{M}_e) - h_{m2}S_2(\overline{M} - \overline{M}_e) + \dot{M}V, \quad (4)$$

where V represents the volume of the block, h_{m1} and h_{m2} refer to the convective mass transfer coefficients on the outer and inner surfaces of the block, respectively, S_1 and S_2 represent the external and internal surface areas of the material, respectively, M and M_e are the moisture contents at time t and in the hygroscopic equilibrium condition, respectively, t is time, and $\dot{M}$ refers to mass generation inside the block.

Considering that there are no reactions that can generate moisture inside the block ($\dot{M} = 0$), and that M = M_e at t = 0, when substituting in the mass balance model equations proposed by Silva et al. [29] for S_1, S_2, and V and solving the resulting ordinary differential equation, the following equation is obtained:

$$\frac{\overline{M} - \overline{M}_e}{\overline{M}_o - \overline{M}_e} = \text{Exp}(A_M + B_M + C_M + D_M + E_M), \quad (5)$$

or

$$\overline{M} = (\overline{M}_o - \overline{M}_e)\text{Exp}(A_M + B_M + C_M + D_M + E_M) + \overline{M}_e, \quad (6)$$

where
$A_M = \left(\frac{1}{a_{33}k_{11}k_{22}k_{33}}\right)\left[(-b_{11}k_{22}k_{33})_2F_1\left(1, \frac{k_{11}}{k_{33}}, \frac{k_{11}+k_{33}}{k_{33}}, -\frac{b_{33}}{a_{33}}\right)\right]$

$B_M = b_{11}\text{Exp}(k_{11}t)k_{22}k_{33\,2}F_1\left(1, \frac{k_{11}}{k_{33}}, \frac{k_{11}+k_{33}}{k_{33}}, -\frac{b_{33}\text{Exp}(k_{33}t)}{a_{33}}\right)$

$C_M = k_{11}\left[-b_{22}k_{33\,2}F_1\left(1, \frac{k_{22}}{k_{33}}, \frac{k_{22}+k_{33}}{k_{33}}, -\frac{b_{33}}{a_{33}}\right)\right]$

$D_M = b_{22}\text{Exp}(k_{22}t)k_{33\,2}F_1\left(1, \frac{k_{22}}{k_{33}}, \frac{k_{22}+k_{33}}{k_{33}}, -\frac{b_{33}\text{Exp}(k_{33}t)}{a_{33}}\right)$

$E_M = (a_{11} + a_{22})k_{22}\{[k_{33}t + \ln(a_{33} + b_{33})] - \ln[a_{33} + b_{33}\text{Exp}(k_{33}t)]\}$

where $a_{11} = h_{m_1}S_{10}a_1$, $b_{11} = h_{m_1}S_{10}b_1$, $k_{11} = -k_1^2$, $a_{22} = h_{m_2}S_{20}a_2$, $b_{22} = h_{m_2}S_{20}b_2$, $k_{22} = -k_2^2$, $a_{33} = V_0a_3$, $b_{33} = V_0b_3$, and $k_{33} = -k_3^2$. The term $_2F_1(a, b, c, z)$ is the ordinary hypergeometric function or hypergeometric Gauss series. For a better understanding, in the interval of $|z| < 1$, the hypergeometric function is defined by the following infinite series:

$$_2F_1(a,b,c,z) = \sum_{n=0}^{\infty}\left[\frac{(a)_n \times (b)_n}{(c)_n} \times \frac{z^n}{n!}\right] = 1 + \frac{a \times b}{c} \times \frac{z}{1!} + \frac{a \times (a+1) \times b \times (b+1)}{c \times (c+1)} \times \frac{z^2}{2!} + \ldots, \quad (7)$$

where the term $(m)_n = m(m+1)(m+2)\ldots$ is the Pochhammer symbol. By definition, the value $(m)_0 = 1$ [30,31].

From Equation (6), it is possible to predict the behavior of mass transfer in a structural ceramic block by considering all the dimensional variations of the block during the drying process.

2.2.3. Heat Transfer Model

Analogous to the mass transfer, in the analysis of the heat transfer, it is assumed that the phenomena of thermal convection, evaporation, and heating of the produced vapor occur simultaneously on the surface of the block. In this way, the energy balance equation will be given by

$$\rho_u V c_p \frac{d\overline{\theta}}{dt} = h_{c1}S_1(\overline{\theta}_\infty - \overline{\theta}) + h_{c2}S_2(\overline{\theta}_\infty - \overline{\theta}) + \rho_s V \frac{d\overline{M}}{dt}\left[h_{fg} + c_v(\overline{\theta}_\infty - \overline{\theta})\right] + \dot{q}V, \quad (8)$$

where ρ_u is the density of the wet ceramic block, c_p is the specific heat of the block, h_{c1} and h_{c2} correspond to the convective heat transfer coefficients (1 (external) and 2 (internal)), ρ_s refers to the density of the dry block, h_{fg} is the latent heat of vaporization of the free water, c_v is the specific heat of the water vapor, $\overline{\theta}_\infty$ and $\overline{\theta}$ are the temperatures of the external medium and the block at any time t of the process, respectively, and $\dot{q}$ corresponds to the internal heat generation. The term $\rho_s V \frac{d\overline{M}}{dt}$ represents the amount of water evaporated from the surface of the material per unit of time.

By neglecting the effects of mass transfer on heat transfer and internal energy generation (i.e., ($\frac{d\overline{M}}{dt} = 0$ and $\dot{q} = 0$)), substituting Equations (1–3) into Equation (8), considering the initial condition $\overline{\theta} = \overline{\theta}_o$ at t = 0, and performing the solution of the resulting ordinary differential equation, the following mathematical equation was obtained:

$$\frac{\overline{\theta}_\infty - \overline{\theta}}{\overline{\theta}_\infty - \overline{\theta}_o} = \mathrm{Exp}(A_T + B_T + C_T + D_T + E_T), \quad (9)$$

or

$$\overline{\theta} = (\overline{\theta}_\infty - \overline{\theta}_o)\mathrm{Exp}(A_T + B_T + C_T + D_T + E_T) + \overline{\theta}_\infty, \quad (10)$$

where

$A_T = \left(\frac{1}{\hat{a}_{33}\hat{k}_{11}\hat{k}_{22}\hat{k}_{33}}\right)\left[\left(-\hat{b}_{11}\hat{k}_{22}\hat{k}_{33}\right){}_2F_1\left(1, \frac{\hat{k}_{11}}{\hat{k}_{33}}, \frac{\hat{k}_{11}+\hat{k}_{33}}{\hat{k}_{33}}, -\frac{\hat{b}_{33}}{\hat{a}_{33}}\right)\right]$

$B_T = \hat{b}_{11}\mathrm{Exp}(\hat{k}_{11}t)\hat{k}_{22}\hat{k}_{33}{}_2F_1\left(1, \frac{\hat{k}_{11}}{\hat{k}_{33}}, \frac{\hat{k}_{11}+\hat{k}_{33}}{\hat{k}_{33}}, -\frac{\hat{b}_{33}\mathrm{Exp}(\hat{k}_{33}t)}{\hat{a}_{33}}\right)$

$C_T = \hat{k}_{11}\left[-\hat{b}_{22}\hat{k}_{33}{}_2F_1\left(1, \frac{\hat{k}_{22}}{\hat{k}_{33}}, \frac{\hat{k}_{22}+\hat{k}_{33}}{\hat{k}_{33}}, -\frac{\hat{b}_{33}}{\hat{a}_{33}}\right)\right]$

$D_T = \hat{b}_{22}\mathrm{Exp}(\hat{k}_{22}t)\hat{k}_{33}{}_2F_1\left(1, \frac{\hat{k}_{22}}{\hat{k}_{33}}, \frac{\hat{k}_{22}+\hat{k}_{33}}{\hat{k}_{33}}, -\frac{\hat{b}_{33}\mathrm{Exp}(\hat{k}_{33}t)}{\hat{a}_{33}}\right)$

$E_T = (\hat{a}_{11} + \hat{a}_{22})\hat{k}_{22}\{[\hat{k}_{33}t + \ln(\hat{a}_{33} + \hat{b}_{33})] - \ln[\hat{a}_{33} + \hat{b}_{33}\mathrm{Exp}(\hat{k}_{33}t)]\}$

In Equation (10), the terms $\hat{a}_{11} = h_{c_1}S_{10}a_1$, $\hat{b}_{11} = h_{c_1}S_{10}b_1$, $\hat{k}_{11} = -k_1^2$, $\hat{a}_{22} = h_{c_2}S_{20}a_2$, $\hat{b}_{22} = h_{c_2}S_{20}b_2$, $\hat{k}_{22} = -k_2^2$, $\hat{a}_{33} = \rho_u C_p V_0 a_3$, $\hat{b}_{33} = \rho_u C_p V_0 b_3$, and $\hat{k}_{33} = -k_3^2$.

From Equation (10), it is possible to predict the behavior of heat transfer in a structural ceramic block while considering all the dimensional variation occurring in the block during the drying process.

2.3. Application to Structural Ceramic Blocks

In this research, emphasis is given to the drying of structural ceramic blocks. In order to determine the convective heat and mass transfer coefficients on the internal and external surfaces of the block, the results of the moisture content and temperature in the block, predicted by the proposed mathematical models, were compared with the experimental data of oven drying, as reported in [14,20].

According to Silva [14] and Silva et al. [20], in the beginning, during (at predetermined time intervals), and at the end of the drying process of the ceramic blocks, measurements of the dimensions, mass, and temperature (upper right vertex of the front face) of the ceramic block and the relative humidity, temperature, and velocity of the drying air inside the oven were made. Six (6) experimental tests were performed with different temperatures. At the end of each experiment, the dimensions, equilibrium mass, equilibrium temperature of the block, and total process time were obtained. Some data obtained by the authors can be found in Tables 3 and 4. These data were used in the computer simulation stage. Details about the data accuracy can be found in [14,20].

Table 3. Experimental parameters for the air and the ceramic block at the beginning of the drying process [14,20].

Test	Air			Ceramic Block			
	T (°C)	RH (%)	v (m/s)	M_0 (kg/kg, d.b.)	θ_0 (°C)	ρ (kg/m^3)	c_p (J/kgK)
1	50	18.39	1.0	0.172319	31.5	1920	1673.51
2	60	12.27	1.0	0.173163	32.0	1920	1673.51
3	70	7.72	1.0	0.170186	29.8	1920	1673.51
4	80	4.99	1.0	0.172723	30.5	1920	1673.51
5	90	3.56	1.0	0.167900	27.6	1920	1673.51
6	100	2.34	1.0	0.169366	27.5	1920	1673.51

Table 4. Experimental parameters obtained after drying the ceramic block [14,20].

Test	M_e (kg/kg, d.b.)	θ_e (°C)	t (min)
1	0.002685	50.5	1170
2	0.001834	58.0	1050
3	0.001189	64.7	990
4	0.000826	68.6	930
5	0.000511	76.2	930
6	0.000054	95.1	750

To obtain the simulated results, the computer codes were developed in Mathematica® software. The comparison between the results predicted by the proposed mathematical model with the experimental data of the moisture content and temperature of the ceramic block was performed until a minimum error was reached. The squared deviations between the experimental and calculated values and the variance for the moisture content and temperature were obtained as follows:

$$\text{MSE}_M = \sum_{i=1}^{n} \left(\overline{M}_{i,\text{Num}} - \overline{M}_{i,\text{Exp}} \right)^2, \tag{11}$$

$$\overline{S}_M^2 = \frac{\text{ERMQ}_M}{(n - \hat{n})}, \tag{12}$$

$$\text{MSE}_\theta = \sum_{i=1}^{n} \left(\frac{\overline{\theta}_{i,\text{Num}} - \overline{\theta}_{i,\text{Exp}}}{\overline{\theta}_\infty - \overline{\theta}_0} \right)^2, \quad (13)$$

$$\overline{S}_\theta^2 = \frac{\text{ERMQ}_\theta}{(n - \hat{n})}, \quad (14)$$

where n is the number of experimental points and n̂ is the number of fitted parameters (number of degrees of freedom) [32].

3. Results

3.1. Mass Transfer Analysis

As mentioned before, the estimation of the convective mass transfer coefficients on the external and internal surfaces of the ceramic block was made through a comparison between the moisture content data obtained by the proposed mathematical model and the data obtained experimentally by Silva [14] and Silva et al. [20]. The values obtained from these parameters as well as the mean squared error (MSE) are reported in Table 5.

Table 5. External (h_{m1}) and internal (h_{m2}) convective mass transfer coefficients for each drying temperature.

T (°C)	Convective Mass Transfer Coefficient		MSE_M (kg/kg)^2
	hm_1 (m/s)	hm_2 (m/s)	
50	6.69×10^{-7}	0.70×10^{-7}	0.00174
60	8.80×10^{-7}	0.80×10^{-7}	0.00053
70	9.28×10^{-7}	1.00×10^{-7}	0.00030
80	11.84×10^{-7}	1.30×10^{-7}	0.00119
90	11.69×10^{-7}	1.03×10^{-7}	0.00027
100	15.97×10^{-7}	1.03×10^{-7}	0.00036

When evaluating the physical behavior of the convective mass transfer coefficient (Table 5), an increase in this parameter was observed with the increase in the drying temperature. It is well known that the increase in this thermo-physical parameter increases the drying rates of the ceramic block, causing hygroscopic equilibrium conditions to be reached more quickly. In addition, it is also possible to observe that the convective mass transfer coefficient obtained on the external surface of the ceramic block was greater than the coefficient of the internal surface of this block. This was due to the fact that the external surface area of the block in contact with the drying air was greater than the internal one, favoring the mass exchange between the product and the heated fluid surrounding the block. Additionally, it was noticed that the position of the block inside the oven contributed to the behavior of these parameters in each drying condition.

Regarding the values obtained for the mean squared error, low values of this statistical parameter can be observed, an indication of an excellent fit between the results predicted by the model and the experimental data and that the proposed mathematical model can adequately describe the drying process of ceramic blocks.

Confirming the data presented in Table 5, Figure 2 illustrates the predicted and experimental curves [14,20] of the moisture content of the ceramic block as a function of the drying time for different drying temperatures. When analyzing the behavior of the moisture content as a function of the processing time, it was possible to identify excellent agreement between the values predicted by the mathematical model and the experimental data, with a small deviation in the final process times mainly for low temperatures of drying. This can be attributed to the consideration of a constant mass transfer coefficient throughout the process being imposed on the model.

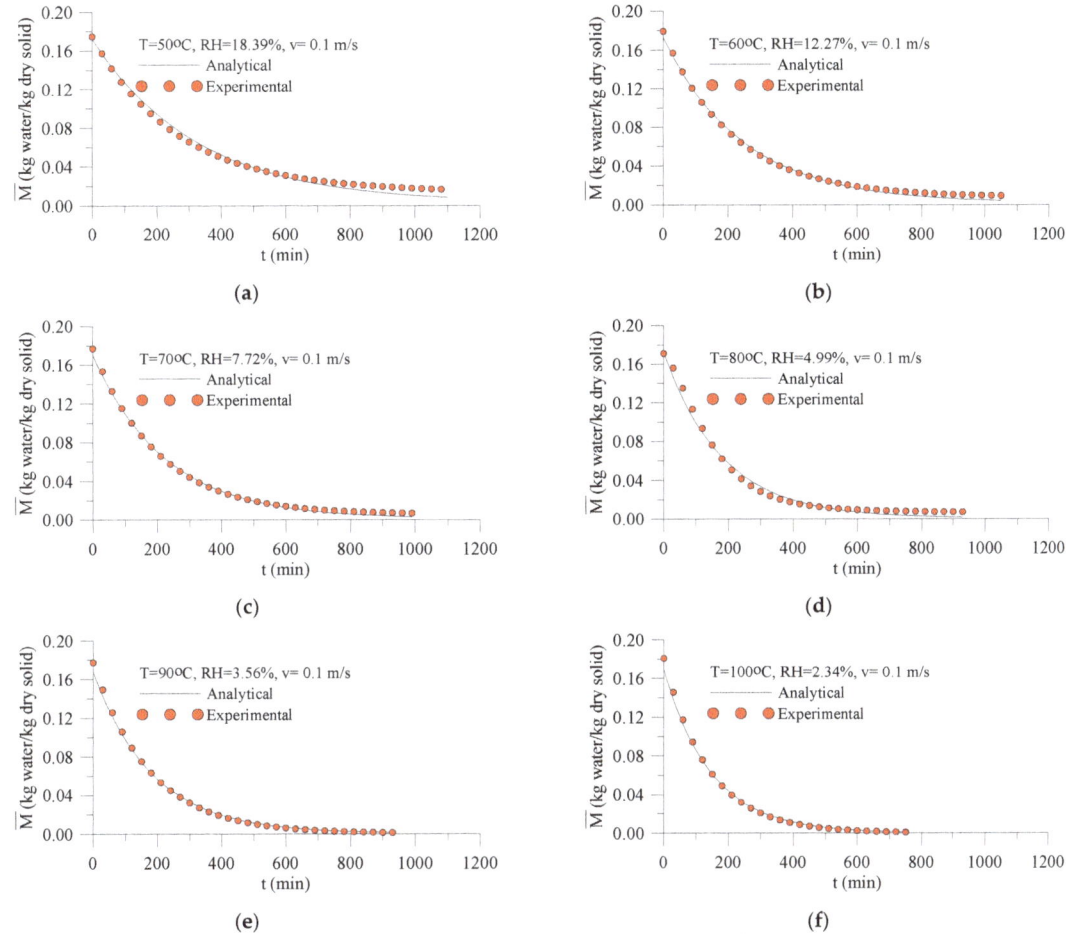

Figure 2. Transient behavior of predicted and experimental [14] moisture content of the ceramic block for different drying temperatures: (**a**) T = 50 °C, (**b**) T = 60 °C, (**c**) T = 70 °C, (**d**) T = 80 °C, (**e**) T = 90 °C, and (**f**) T = 100 °C.

When analyzing the moisture content curves from a physical point of view, there was a significant reduction in the moisture content in the first minutes of drying (high falling drying rate), as reported in [11,12,14,22].

For the lowest drying temperatures (from 50 °C to 70 °C), it was noted that more intense reduction in the moisture content occurred until the first 500 min (approximately 8.3 h) of the process. For higher temperatures (above 70 °C), this intense reduction occurred in a shorter process time, occurring up to the first 300 min (approximately 5 h). Thus, higher drying temperatures contributed to the material reaching its hygroscopic equilibrium in a shorter process time.

The variation in the moisture removal rate from the ceramic block was influenced by both the drying temperature and geometry of the material. It is well known that hollow ceramic blocks have better air circulation inside, which leads to a faster reduction in moisture. However, although its geometric shape provides a more accentuated reduction in the processing time, abrupt reductions in the moisture content can cause structural defects in the material, such as cracks and deformations, reducing the quality of the product at the end of drying. It is also important to highlight that in the first minutes of the drying stage, there is a large reduction in the dimensions of the solids (higher shrinkage speed), tending to remain constant after a long drying period, which can affect the quality of the product at the end of drying [13,15,17,18,21,24].

3.2. Heat Transfer Analysis

In a similar way to the procedure performed in the analysis of the mass transfer, the obtaining of the convective heat transfer coefficients on the external and internal surfaces of the ceramic block was given through a comparison between the predicted data and the experimental data [14,20] of the temperature at the vertex of the ceramic block until a smaller least square error was obtained. Table 6 summarizes the values of the convective heat transfer coefficients and the least square error for each drying temperature.

Table 6. Convective heat transfer coefficients for the external (h_{c1}) and internal (h_{c2}) areas for each drying temperature.

T (°C)	Convective Heat Transfer Coefficient		MSE_θ (-)
	h_{c1} (W/m^2·°C)	h_{c2} (W/m^2·°C)	
50	4.79	1.00	1.18412
60	2.19	1.06	4.08616
70	2.72	1.00	18.1275
80	2.41	0.37	4.99504
90	2.03	0.60	2.33426
100	2.04	0.94	1.12158

When analyzing Table 6, some observations can be made from the values obtained for the temperature and convective heat transfer coefficients. Differences between the predicted and experimental data of the vertex temperature can be attributed to the possible temperature measurement errors which occurred during the collection of experimental data, the fact that the model considered that the temperature of the material was the same throughout its volume at each time, and the neglected effects of the mass transfer on the heat transfer. It was observed that the convective coefficients obtained for the external surface of the ceramic block were higher than the convective coefficients obtained for the internal surface, proving that the block heated up more slowly inside. This is because the external surface of the block is more directly exposed to the drying air, allowing the convective heat flux to act with greater intensity in this region. However, even though the values obtained for the convective coefficients were low, typical of natural convection, it is noted that these coefficients did not assume an increasing behavior with the increase in the drying temperature. The divergence in the behavior of these thermo-physical parameters can be associated with possible temperature measurement errors which occurred during the collection of experimental data and the fact that the model considers that the temperature of the material is the same throughout its volume at each time, which clearly does not occur in practice, particularly for drying at high temperatures and a low relative humidity. Other factors can be listed: drying time, air relative humidity, shrinkage, and the different dimensions and moisture contents of the ceramic block at the beginning of drying.

It is known that higher values of the convective heat transfer coefficients provoke an increase in the heating rates, causing the material to reach the thermal equilibrium condition more quickly. On the other hand, high heating rates cause high temperature variation rates inside the material and, with this, high drying rates and thermal and hydric stresses, which in turn can cause severe drying problems such as cracks, warping, deformations, and ruptures, which considerably reduce the quality of the product in the firing stage [13,15,17,18,21].

Complementing the heat transfer analysis, Figure 3 illustrates the heating curves of the ceramic block for different drying temperatures (from 50 °C to 100 °C).

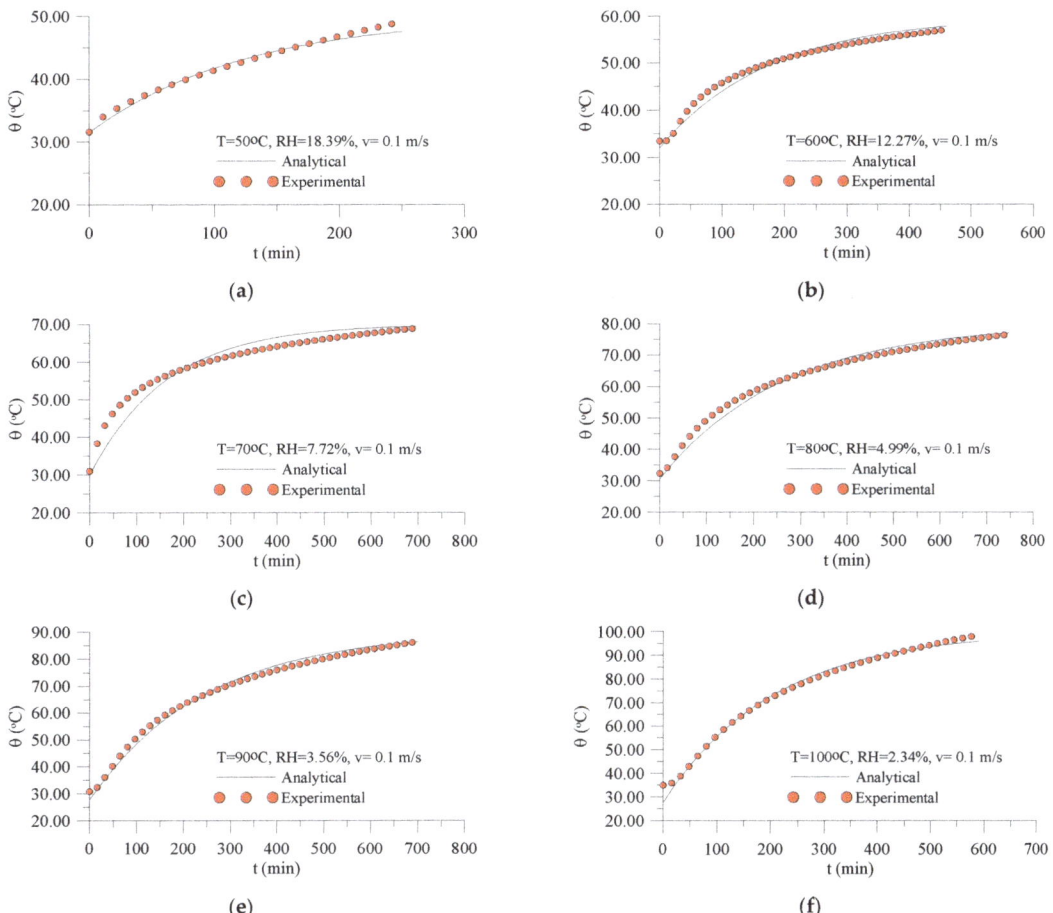

Figure 3. Transient behavior of experimental [14] and predicted temperatures of the ceramic block for different temperatures: (**a**) T = 50 °C, (**b**) T = 60 °C, (**c**) T = 70 °C, (**d**) T = 80 °C, (**e**) T = 90 °C, and (**f**) T = 100 °C.

Although the heat transfer coefficients showed a different behavior than expected, when analyzing the heating curves, a concordance between the values predicted by the model and the experimental data was observed, confirming once again that the proposed modeling was efficient for describing the drying process.

When analyzing the graphs, it was observed that the heating curves presented a behavior similar to that presented by the moisture content curves (Figure 2) since, for the lower drying air temperatures, a significant increase occurred for the temperature of the block until the first 300 min (5 h) of the process, and it then assumed an almost constant behavior once the block had almost completely reached its thermal equilibrium.

Although high drying temperatures imply a shorter process time, it is necessary to have strict control regarding this drying condition. As mentioned earlier, performing convective drying with high heating rates in an atmosphere with low relative humidity can lead to greater removal of the existing moisture in the solid, causing damage to the ceramic product and thus affecting its mechanical performance and quality when in operation.

In general, a comparison between the transport coefficients reported in the literature is very difficult due to the different parameter estimation methods used, the variation in the chemical composition of the ceramic material, and its physical and chemical structures.

The precision and accuracy of the measurements are important factors to consider when a method is good enough to be used in the estimation of some process parameters. In general, systematic errors around 5% and standard errors around 3% are expected due to, for example, limited instrument precision. Therefore, these errors are directly passed on to the estimation of transport coefficients. Other errors can be attributed to the different nature of the materials and the uncertainty in the experimental measurement, methodology, and geometric consideration of the body.

Despite this, the convective heat and mass transfer coefficients obtained in this study, compared with those reported in the literature and which are listed in Table 7, corroborated the effectiveness of the methodology used.

Table 7. Heat and mass transfer coefficients obtained in the drying of ceramic bricks reported in the literature.

Geometry	T (°C)	RH (%)	Convective Mass Transfer Coefficient (m/s)		Convective Heat Transfer Coefficient (W/m²·°C)		Source
			hm_1	hm_2	hc_1	hc_1	
	50	20.8	2.87×10^{-7}		0.50		Silva et al. [22]
		20	1.67×10^{-7}	1.59×10^{-7}	5.34	5.11	Lima et al. [13]
	60	13.5	2.92×10^{-7}		0.58		Silva et al. [22]
		20	2.05×10^{-7}	1.96×10^{-7}	5.51	5.26	Lima et al. [13]
	70	7.6	4.95×10^{-7}		0.98		Silva et al. [22]
		20	2.46×10^{-7}	2.35×10^{-7}	5.64	5.39	Lima et al. [13]
	80	4.6	6.01×10^{-7}		0.96		Silva et al. [22]
		20	2.91×10^{-7}	2.78×10^{-7}	5.75	5.49	Lima et al. [13]
	90	3.3	6.03×10^{-7}		1.07		Silva et al. [22]
		20	3.43×10^{-7}	3.28×10^{-7}	5.84	5.58	Lima et al. [13]
	100	1.8	8.17×10^{-7}		1.74		Silva et al. [22]
		20	4.10×10^{-7}	3.92×10^{-7}	5.91	5.66	Lima et al. [13]
		1.8	6.47×10^{-7}	6.13×10^{-7}	6.89	6.54	Lima et al. [21]

Silva et al. [22]: lumped model considering a single heat and mass transfer coefficient estimated by non-linear regression and without dimensional variations. Lima et al. [13] and Lima et al. [21]: lumped model considering the existence of two heat and mass transfer coefficients determined by a proposed formulation and without dimensional variations.

In all the cases listed in Table 7, lumped models were used. Thus, the highest differences in the convective heat and mass transfer coefficient could mainly be attributed to the following factors:

(a) The type of lumped model;
(b) Considerations of the dimensional variations adopted in the model;
(c) The chemical composition of the product;
(d) The product's geometry;
(e) The equilibrium moisture content (hysteresis phenomenon in the sorption isotherm);
(f) The drying conditions (relative humidity and air velocity);
(g) Variations in the physical structure of the product (porosity, tortuosity, and permeability);
(h) The probable formation of pores by evaporation of water.

The convective heat transfer coefficient is not really a property of a material. It is used to quantify the rate of heat transfer at the surface of the body. It is dependent on the fluid velocity, fluid properties, surface roughness, and body shape, as well as the temperature difference between the surface and the fluid surrounding the body. In practice, the heat flux and surface temperature are very difficult to measure without disturbing the heat transfer and the flow of the heated fluid in the thermal and hydrodynamic boundary layers. In some materials with high moisture contents, heat transfer is accompanied by mass transfer, further complicating the measurement of the convective heat transfer coefficient.

Finally, as already mentioned, the convective heat and mass transfer coefficients are dependent on the velocity of the free air stream, but this effect is not explicitly stated in the model since the air velocity, in all experiments, remained practically constant (natural convection inside the oven). However, the low results of these parameters reflect this experimental condition, mainly inside the holes, where the air is confined, in contrast to what occurs on the external surface, where the air can flow by buoyant forces with no boundaries to impede its movement, which resulted in higher results for this parameter.

4. Conclusions

This study presents a numerical analysis of the drying process of structural ceramic blocks, where through phenomenological mathematical models based on lumped analysis, it was possible to analyze the transient heat and mass transfers that occurred in the ceramic block during the process. Thus, based on the analysis of the obtained results, it was verified that the phenomenological mathematical modeling used was effective for describing the heat and mass transfers of the ceramic block during the drying process. The effectiveness of the proposed models for predicting the heat and mass transfer phenomena, including dimensional variations of the ceramic block, was proven when it was observed that the discrepancies between the predicted and experimental results (reported in the literature) of the moisture content and temperature at the vertex of the ceramic block were small. Thus, it easily becomes possible to estimate the drying times of ceramic products with different shapes in different drying conditions in order to assist academy and industry in the decision making related to this complex physical problem.

From the results obtained, it was possible to estimate the convective mass transfer coefficients involved in the process. It was found that the convective mass transfer coefficient ranged from 6.69×10^{-7} to 15.97×10^{-7} m/s on the outer surface of the block and from 0.70×10^{-7} to 1.03×10^{-7} m/s on the inner surface of the material when the drying air temperature ranged from 50 to 100 °C. The convective heat transfer coefficient ranged from 4.79 to 2.04 W/(m^2.°C) on the outer surface of the block and from 1.00 to 0.94 W/(m^2.°C) on the inner surface of the material while the temperature of the drying air ranged from 50 to 100 °C. The low values for the convective heat and mass transfer coefficients are a strong indication that the mass removal and heating processes at the block surface occurred by natural convection.

Author Contributions: Conceptualization, S.K.B.M.S. and J.A.F.; methodology, S.K.B.M.S. and M.K.T.B.; software, S.K.B.M.S., J.N.O.N. and L.P.L.S.; validation, S.K.B.M.S. and M.J.F.; formal analysis, S.K.B.M.S.; investigation, S.K.B.M.S.; writing—original draft preparation, S.K.B.M.S. and H.L.F.M.; writing—review and editing, R.S.G. and J.M.P.Q.D.; visualization, A.B.C.P.; supervision, C.J.A. and A.G.B.L. All authors have read and agreed to the published version of the manuscript.

Funding: This research was funded by CAPES, CNPq, and FAPESQ-PB (Brazilian research agencies). In addition, this work was financially supported by base funding (UIDB/04708/2020) and programmatic funding (UIDP/04708/2020) of the Instituto de I&D em Estruturas e Construções (CONSTRUCT), funded by national funds through the FCT/MCTES (PIDDAC), and by the Fundação para a Ciência e a Tecnologia (FCT) through the individual Scientific Employment Stimulus 2020.00828.CEECIND.

Data Availability Statement: The data that support the findings of this study are available upon request from the authors.

Acknowledgments: The authors are grateful to CNPq, CAPES, FINEP, and FAPESQ-PB for the financial support and the researchers mentioned in the manuscript who, with their research, helped in the development of this work.

Conflicts of Interest: The authors declare no conflict of interest.

Nomenclature

c_p	Specific heat of the block	(J/kgK)
c_v	Specific heat of the water vapor	(J/kgK)
h_{c1}	Convective heat transfer coefficients (1, external)	(W/m²K)
h_{c2}	Convective heat transfer coefficients (2, internal)	(W/m²K)
h_{fg}	Latent heat of vaporization of free water	(J/kgK)
h_{m1}	Convective mass transfer coefficients (1, external)	(m/s)
h_{m2}	Convective mass transfer coefficients (2, internal)	(m/s)
H	Height	(mm)
L	Length	(mm)
M	Moisture content	(kg/kg, d.b.)
$\dot{M}$	Mass generation inside the block	(kg/kg/s/m³)
M_e	Hygroscopic equilibrium condition	(kg/kg, d.b.)
M_0	Initial moisture content	(kg/kg, d.b.)
MSE_M	Squared deviations for the moisture content	(-)
MSE_θ	Squared deviations for the temperature	(-)
n	Number of experimental points	(-)
$\hat{n}$	Number of fitted parameters	(-)
$\dot{q}$	Internal heat generation	(W/m³)
RH	Air relative humidity	(%)
ρ	Density of the wet block	(kg/m³)
ρ_s	Density of the dry block	(kg/m³)
ρ_u	Density of the wet ceramic block	(kg/m³)
S	Total surface area of the material	(m²)
S_1	External surface area of the material	(m²)
S_2	Internal surface area of the material	(m²)
S_{i0}	Surface areas (external, i = 1 and internal, i = 2) at t = 0	(m²)
$\bar{S}_M^2$	Variance for the moisture content	(-)
$\bar{S}_\theta^2$	Variance for the temperature	(-)
$\bar{\theta}$	Temperatures at any time t of the process	(°C)
$\bar{\theta}_\infty$	Temperatures of the external medium	(°C)
θ_e	Thermal equilibrium temperature	(°C)
θ_o	Initial temperature	(°C)
t	Time	(min)
T	Air temperature	(°C)
v	Air velocity	(m/s)
V	Volume	(m³)
V_0	Initial volume	(m³)
W	Width	(mm)

References

1. Cavalcanti, M.S.L. Development of Ceramic Masses for Sanitary Stoneware Using Flat Glass Residue as a Flux in Partial Replacement of Feldspar. Ph.D. Thesis, Process Engineering, Federal University of Campina Grande, Campina Grande, Brazil, 2010. Available online: http://dspace.sti.ufcg.edu.br:8080/jspui/handle/riufcg/1778 (accessed on 10 November 2021). (In Portuguese)
2. Motta, J.F.M.; Zanardo, A.; Cabral, M., Jr. Ceramic raw materials. Part I: The profile of the main ceramic industries and their products. *Cerâm. Ind.* **2021**, *6*, 28–39. Available online: https://ceramicaindustrial.org.br/article/5876570b7f8c9d6e028b4643/pdf/ci-6-2-5876570b7f8c9d6e028b4643.pdf (accessed on 6 June 2022). (In Portuguese)
3. Bellingieri, J.C. The origins of the ceramic industry in São Paulo. *Cerâm. Ind.* **2005**, *10*, 19–23. Available online: http://host-article-assets.s3.amazonaws.com/ci/587657237f8c9d6e028b46d1/fulltext.pdf (accessed on 7 June 2022). (In Portuguese)

4. Almeida, G.S. Simulation and Experimentation of the Drying of Red Ceramics in Industrial Thermal Systems. Ph.D. Thesis, Process Engineering, Federal University of Campina Grande, Campina Grande, Brazil, 2009. Available online: http://dspace.sti.ufcg.edu.br:8080/xmlui/handle/riufcg/11369 (accessed on 12 November 2021). (In Portuguese)
5. Prado, U.S.; Bressiani, J.C. Overview of the Brazilian ceramic industry in the last decade. *Cerâm. Ind.* **2013**, *18*, 7–11. Available online: http://repositorio.ipen.br/bitstream/handle/123456789/17430/18423.pdf?sequence=1 (accessed on 7 June 2022). (In Portuguese) [CrossRef]
6. Cabral, M., Jr.; Azevedo, P.B.M.; Cuchierato, G.; Motta, J.F.M. Strategic study of the ceramic industry production chain in the state of São Paulo: Part I—Introduction and the red ceramic industry. *Cerâm. Ind.* **2019**, *24*, 20–34. [CrossRef]
7. Martinez, P.H. Influence of Ceramic Block Geometry on the Behavior of Structural Masonry Wall Beams. Master's Thesis, Civil Engineering, Paulista State University, Ilha Solteira, Brazil, 2017. Available online: https://repositorio.unesp.br/handle/11449/148961 (accessed on 15 November 2021). (In Portuguese)
8. Santos, R.S. Study of the Drying Process of Structural Ceramic Blocks: Modeling and Simulation. Ph.D. Thesis, Process Engineering, Federal University of Campina Grande, Campina Grande, Brazil, 2019. Available online: http://dspace.sti.ufcg.edu.br:8080/xmlui/handle/riufcg/10549 (accessed on 18 November 2021). (In Portuguese)
9. Santos, I.A.P. *Structural Masonry and a Comparative Analysis between Ceramic and Concrete Blocks*; Course Final Report (Undergraduate Course in Civil Engineering); Federal University of Campina Grande: Pombal, Brazil, 2021; Available online: http://dspace.sti.ufcg.edu.br:8080/xmlui/handle/riufcg/21732 (accessed on 21 November 2021). (In Portuguese)
10. Ibrahim, M.H.; Daud, W.R.W.; Talib, M.Z.M. Drying Characteristics of Oil Palm Kernels. *Dry. Technol.* **1997**, *15*, 1103–1117. [CrossRef]
11. de Brito, M.T.; de Almeida, D.T.; de Lima, A.B.; Rocha, L.A.; de Lima, E.S.; de Oliveira, V.B. Heat and Mass Transfer during Drying of Clay Ceramic Materials: A Three-Dimensional Analytical Study. *Diffus. Found.* **2017**, *10*, 93–106. [CrossRef]
12. Nascimento, J.J.D.S.; Luna, C.B.B.; Costa, R.F.; Barbieri, L.F.P.; Bezerra, E.D.O.T. Evaluation of red clay and ball clay drying using transient three-dimensional mathematical modeling: Volumetric shrinkage and moisture content. *Mater. Res. Express* **2019**, *6*, 95206. [CrossRef]
13. Lima, E.S.; Lima, W.M.P.B.; Silva, S.K.B.M.; Magalhães, H.L.F.; Nascimento, L.P.C.; Gomez, R.S.; Lima, A.G.B. Drying of industrial ceramic bricks and process parameter estimation: An advanced concentrated approach. *Res. Soc. Dev.* **2020**, *9*, e48391211391. [CrossRef]
14. Silva, A.M.V. Drying of Industrial Ceramic Blocks: Modeling, Simulation and Experimentation. Ph.D. Thsis, Process Engineering, Federal University of Campina Grande, Campina Grande, Brazil, 2018. Available online: http://dspace.sti.ufcg.edu.br:8080/jspui/handle/riufcg/7173 (accessed on 23 November 2021). (In Portuguese)
15. Tavares, F.V.D.S.; Cavalcante, A.M.D.M.; de Figueiredo, M.J.; Vilela, A.F.; de Lima, A.R.C.; Azerêdo, L.P.M.; de Lima, A.G.B. On the Modeling of Drying Process in an Industrial Tunnel Dryer. *Defect Diffus. Forum* **2020**, *400*, 51–56. [CrossRef]
16. Gomez, R.S.; Porto, T.R.N.; Magalhães, H.L.F.; Guedes, C.A.L.; Lima, E.S.; Wanderley, D.M.A.; Lima, A.G.B. Transient Thermal Analysis in an Intermittent Ceramic Kiln with Thermal Insulation: A Theoretical Approach. *Adv. Mater. Sci. Eng.* **2020**, *2020*, 6476723. [CrossRef]
17. Santos, R.S.; de Farias Neto, S.R.; de Lima, A.G.B.; Silva Júnior, J.B.; da Silva, A.M.V. Drying of Ceramic Bricks: Thermal and Mass Analysis via CFD. *Diffus. Found.* **2020**, *25*, 133–153. [CrossRef]
18. Araújo, M.D.V.; Correia, B.R.D.B.; Brandão, V.A.A.; De Oliveira, I.R.; Santos, R.S.; Neto, G.L.D.O.; Silva, L.P.D.L.; De Lima, A.G.B. Convective Drying of Ceramic Bricks by CFD: Transport Phenomena and Process Parameters Analysis. *Energies* **2020**, *13*, 2073. [CrossRef]
19. Santos, R.S.; Delgado, J.M.P.Q.; Silva, F.A.N.; Azevedo, A.C.; Neto, S.R.F.; Farias, F.P.M.; de Lima, A.G.B.; de Lima, W.M.P.B.; Lima, E.S. Behind the Manufacturing of Industrial Clay Bricks: Drying Stage Predictions Using CFD. *Adv. Mater. Sci. Eng.* **2022**, *2022*, 5530362. [CrossRef]
20. Da Silva, A.V.; Delgado, J.; Guimarães, A.; De Lima, W.B.; Gomez, R.S.; De Farias, R.P.; De Lima, E.S.; De Lima, A.B. Industrial Ceramic Blocks for Buildings: Clay Characterization and Drying Experimental Study. *Energies* **2020**, *13*, 2834. [CrossRef]
21. Lima, E.S.; Delgado, J.M.P.Q.; Guimarães, A.S.; Lima, W.M.P.B.; Santos, I.B.; Gomes, J.P.; Santos, R.S.; Vilela, A.F.; Viana, A.D.; Almeida, G.S.; et al. Drying and Heating Processes in Arbitrarily Shaped Clay Materials Using Lumped Phenomenological Modeling. *Energies* **2021**, *14*, 4294. [CrossRef]
22. Silva, S.K.B.M.; Araújo, C.J.; Lima, A.G.B. Parameters estimation of the drying process of industrial ceramic bricks via concentrated analysis model. *Res. Soc. Dev.* **2021**, *10*, e57210716913. [CrossRef]
23. Brown, A.E. *Brick Drying: A Practical Treatise on the Drying of Bricks and Similar Clay Products*; Franklin Classics: Minneapolis, MN, USA, 2018; p. 236.
24. Gualtieri, A.F.; Ricchi, A.; Gualtieri, M.L.; Maretti, S.; Tamburini, M. Kinetic study of the drying process of clay bricks. *J. Therm. Anal. Calorim.* **2015**, *123*, 153–167. [CrossRef]
25. Dogru, A.K.; Camcioglu, E.; Ozgener, O.; Ozgener, L. Thermodynamics analysis of biomass fired brick drying process. *Int. J. Exergy* **2021**, *35*, 421. [CrossRef]
26. Ndukwu, M.C.; Bennamoun, L.; Simo-Tagne, M.; Ibeh, M.I.; Abada, U.C.; Ekop, I.E. Influence of drying applications on wood, brick and concrete used as building materials: A review. *J. Build. Pathol. Rehabil.* **2021**, *6*, 24. [CrossRef]

27. Karagiannis, N.; Karoglou, M.; Bakolas, A.; Krokida, M.; Moropoulou, A. Drying kinetics of building materials capillary moisture. *Constr. Build. Mater.* **2017**, *137*, 441–449. [CrossRef]
28. Pujari, A.S.; Bhosale, C.H.; Wagh, M.M.; Shinde, N.N. Effect of Temperature on Drying Rate of Various Types of Bricks. *Int. Res. J. Eng. Technol. IRJET* **2016**, *3*, 2793.
29. Silva, S.K.B.M.; Araújo, C.J.; Lima, A.G.B. Numerical analysis of the dimensional variation in structural ceramic blocks during the drying process. *Res. Soc. Dev.* **2021**, *10*, e33310615680. [CrossRef]
30. Robalo, T.C.G. Generalized Hypergeometric Series in the Context of the Theory of Hypercomplex Functions. Master´s Thesis, Mathematic, University of Aveiro, Aveiro, Portugal, 2006. Available online: https://ria.ua.pt/handle/10773/2878?locale=pt_PT (accessed on 6 June 2022).
31. Seaborn, J.B. *Hypergeometric Functions and Their Applications*; Springer Science & Business Media: New York, NY, USA, 2013. [CrossRef]
32. Figliola, R.S.; Beasley, D.E. *Theory and Design for Mechanical Measurements*; John Wiley and Sons: New York, NY, USA, 1995.

Article

Numerical Study for Enhancement of Heat Transfer Using Discrete Metal Foam with Varying Thickness and Porosity in Solar Air Heater by LTNE Method

Rawal Diganjit [1], N. Gnanasekaran [1,*] and Moghtada Mobedi [2,*]

[1] Department of Mechanical Engineering, National Institute of Technology, Karnataka, Surathkal, Mangalore 575025, India
[2] Mechanical Engineering Department, Faculty of Engineering, Shizuoka University, 3-5-1 Johoku, Naka-ku, Hamamatsu-shi 432-8561, Japan
* Correspondence: gnanasekaran@nitk.edu.in (N.G.); moghtada.mobedi@shizuoka.ac.jp (M.M.)

Abstract: A two-dimensional rectangular domain is considered with a discrete arrangement at equal distances from copper metal foam in a solar air heater (SAH). The local thermal non-equilibrium model is used for the analysis of heat transfer in a single-pass rectangular channel of SAH for different mass flow rates ranging from 0.03 to 0.05 kg/s at 850 W/m^2 heat flux. Three different pores per inch (PPI) and porosities of copper metal foam with three different discrete thicknesses at equal distances are studied numerically. This paper evaluates the performance of SAH with 10 PPI 0.8769 porosity, 20 PPI 0.8567 porosity, and 30 PPI 0.92 porosity at 22 mm, 44 mm, and 88 mm thicknesses. The Nusselt number for 22 mm, 44 mm, and 88 mm thicknesses is 157.64%, 183.31%, and 218.60%, respectively, higher than the empty channel. The performance factor for 22 mm thick metal foam is 5.02% and 16.61% higher than for 44 mm and 88 mm thick metal foam, respectively. Hence, it is found that metal foam can be an excellent option for heat transfer enhancement in SAH, if it is designed properly.

Keywords: metal foam; local thermal nonequilibrium model (LTNE); forced convection; performance factor; solar air heater; single pass

Citation: Diganjit, R.; Gnanasekaran, N.; Mobedi, M. Numerical Study for Enhancement of Heat Transfer Using Discrete Metal Foam with Varying Thickness and Porosity in Solar Air Heater by LTNE Method. *Energies* **2022**, *15*, 8952. https://doi.org/10.3390/en15238952

Academic Editor: Gianpiero Colangelo

Received: 9 November 2022
Accepted: 22 November 2022
Published: 26 November 2022

Publisher's Note: MDPI stays neutral with regard to jurisdictional claims in published maps and institutional affiliations.

Copyright: © 2022 by the authors. Licensee MDPI, Basel, Switzerland. This article is an open access article distributed under the terms and conditions of the Creative Commons Attribution (CC BY) license (https://creativecommons.org/licenses/by/4.0/).

1. Introduction

Solar energy is readily available in the environment, free of cost. Using fossil fuels for energy production affects our environment severely, and fossil fuels are non-renewable energy sources. We have to use clean energy sources to avoid these harmful effects. By using solar air collectors, solar energy can be converted into thermal energy. The solar air heater is simple in design and requires little maintenance [1–3]. Corrosion and leakage problems do not occur in solar air heaters. Because of their simplicity and low cost, they are widely used worldwide. The solar air heater involves low heat-capacity air and its efficiency is lower than that of the water heater [2,3]. Metal foam has a lower density, high structural strength, and high superficial area, and can increase convection due to which the heat transfer increases. Recently, the application of partially filled porous media, graded metal foam [4,5], or triangular porous media [6,7] have been required to increase the heat transfer in the system and to lower the cost. Nowadays, in solar applications, different porous media like metal foam [1,8–11], wire mesh [12–18], and spherical pebbles [15] are used to increase the temperature of the working fluid. Porosity and pore density are responsible for the enhancement of heat transfer, in addition to the thermal conductivity of the metal foam. These structural properties not only enhance heat transfer, but also increase the pressure drop [18]. In recent times, nano wires with carbon have enhanced the thermal conductivity due to large aspect ratio, and there has been an increase in the heat transfer rate [19]. The geometry of nanowires and their location affect absorption intensities in solar applications [20]. The silver nanowires are also a good option to increase the heat transfer in the solar power system. Ref. [21] Metal foam is one type of porous media used to enhance the temperature in the system. Metal foam has two types: open and closed cells. These types depend on whether the pores are sealed or not sealed. Open cell foam is a very homogenous structure that has almost constant properties. Copper and Aluminum metal foam are widely used in the system to enhance the heat transfer due to high thermal conductivity [8,22,23]. Dukhan and Quinones [24] observed that the effective conductivity and heat transfer of porous aluminum metal foam are more than a solid fin by 4% and 1.5%, respectively. They studied 10 PPI, 20 PPI, and 30 PPI with the porosity of 95% aluminum metal foam for heat transfer enhancement in SAH. Further, it has been observed that the number of pores per inch increases when heat transfer is more for the same porosity. Mancin et al. [25] studied the experimental heat transfer coefficient and pressure drop for five different copper metal foam samples. They noticed that heat transfer coefficient was more for higher mass flow rate and that pressure drop reduced with reduced porosity. The copper metal foam with 10 PPI and 0.905 porosity was found to be the best option for electronic cooling applications compared to 5, 20, and 40 PPIs and different porosities. Chen and Huang [26] reported a computational study of the heat transfer rate for solar water collector with the application of metal foam. They studied copper metal foam of different PPI with the same porosity at different heights of metal foam blocks. As the height of the metal foam increased, the Nusselt number also increased. Because of its higher thermal conductivity, copper has a higher heat transfer rate than aluminum or nickel. Kamath et al. [22] studied the heat transfer enhancement of aluminum and copper metal foams in the application of vertical channels. They conducted an experimental study for metal foam thicknesses of 10 mm, 20 mm and 30 mm and porosity ranging from 0.95 to 0.87. The 0.87 porosity of copper metal foam and 0.95 porosity of aluminum metal foam provides similar results for the same velocity and heat flux value. Bayrak and Oztop [23] studied the thermal performance of a solar air heater with aluminum metal at different thicknesses experimentally. They concluded that the 6 mm thickness aluminum metal foam has higher efficiency than 10 mm thickness metal foam for 0.025 kg/s mass flow rate. It is noted that aluminum metal foam gives better results than empty channel solar air heater for the same velocity and heat flux conditions.

Jouybari et al. [27] experimentally investigated the use of metal foam with the addition of nanofluid to improve thermal performance. The performance evaluation criteria are to reduce the pressure drop and increase the heat transfer. With the help of metal foam and nanofluid, the performance evaluation criteria increased more than 1% for lower flow rate. Further, the increase in nanofluid concentration increases the performance evaluation criteria. Saedodin et al. [28] reported experimental and numerical analysis of porous media in a Flat Plate Solar Collector (FPSC). The thermal efficiency increases by 18.5% with metal foam as porous media in a FPSC. Hussien and Farhan [29] investigated the thermohydraulic performance of SAH with three types of metal foam configurations. The corrugated metal foam gives higher thermal and effectiveness efficiencies, rather than longitudinal and staggered. A high heat transfer rate obtained for a higher PPI. Baig and Ali [30] proposed an experimental study on thermal storage in solar air heaters with the help of paraffin wax combined with aluminum metal foam. The analysis included four different configurations: flat plate, two copper ducts, four copper ducts, and the fourth configuration as a flat plate with pre-heat. Using two and four copper ducts gives more heat transfer than the other two configurations. A maximum efficiency of 97% was achieved with the help of a flat plate pre-heat configuration without a fan and with the help of aluminum foam and paraffin wax. Anirudh and Dhinakran [31] numerically studied metal foam blocks in the solar water heater (SWH). Different heights of metal foam blocks and 0.2 H, 0.6 H, and H of metal foam in the channel were considered where H is the channel height. It is observed that as the height of the metal foam increased, the performance of the SWH reduced, because pressure drop increases with height. Farhan et al. [32] performed a comparative study on the solid fin and metal foam. These types were further arranged in longitudinal, staggered, and corrugated configurations to check the heat transfer enhancement rate. It was noticed that the corrugated arrangement gives more heat transfer than does the other arrangements. Also, it was observed that the exergy loss and efficiency depend on the solar intensity and the velocity of air flowing through the channel.

Anirudh and Dhinakara [33] investigated the optimum performance of FPSC using different heights at inlet, test, and outlet sections of metal foam. The height at the inlet should be lower than the height at the outlet of the metal foam. Due to this arrangement, it gives less pressure drop, and improved efficiency. Kansara et al. [1] performed experiments in FPSC with internal fins and porous media. The authors showed that porous media has the highest heat transfer compared to fins and conventional SAH. Jadhav et al. [11] conducted numerical analysis of forced convection in the horizontal pipe in the presence of metal foam. They emphasized that the computational modelling of forced convection heat dissipation in the presence of high porosity and high thermal conductivity metallic foam. Rajarajeswari et al. [13] investigated numerical and experimental studies using single-pass flow. The diagonal arrangement of two wire mesh having different porosity was considered. The increase in thermal efficiency for 92.5% and 84.5% porosity is about 5–17% and 5–20%, respectively, with the mass flow rate ranging from 0.01 kg/s to 0.055 kg/s. A diagonal arrangement gives higher heat transfer compared to a parallel one. Jadhav [34] et al. studied the performance of the copper, aluminum, and nickel metal foams in a horizontal pipe. The performance factor increased with an increase in PPI. Table 1 gives the summary of previous arrangements of metal foams used in solar air heater.

The present study assumes that the test is conducted for open loop in clear sky days. The specified limit of the solar radiation, ambient temperature, air flow rate, air inlet temperature and temperature rise across solar air heater are ± 50 W/m^2, ± 1 °C, $\pm 1\%$, ± 0.1 °C, and ± 0.1 °C, respectively, for a 15 min duration. For example, RT-PT100 (manufactured by Heatron Indl. Heaters, Mangaluru, India) with tolerance of ± 0.1 °C and 16 Channel universal data logger (manufactured by Sunsui-DL-35, Pune, India) with accuracy of ± 0.08 °C are generally used to measure temperature at different points of the experimental apparatus in the study of Rajarajeswari et al. [9]. Hence the solar air heater is operated in steady state condition for the present study. Also, the average flux falling on the absorber plate for the month of April is 850 W/m^2. Here, 15 April is the mean of the value of solar intensity (IT) for the month of April. Similar assumptions are mentioned for test in [3]. The present study is selected for 0.3779, 0.3401, 0.3023, 0.2646, 0.2268, m/s, as mentioned in Rajarajeswari et al. [9]. The range of air velocity is less than 30% of the Mach number. Hence, the density variation is very much less, due to a velocity which is below 5%. So the flow is assumed to be a steady-state, incompressible turbulent flow. Similar assumptions are mentioned in [35].

The above literature shows that the metal foam arrangement in SAH improves the heat transfer rate, while at the same time the pressure drop increases when the inlet velocity increases. Instead of fully filled metal foams in the SAH, a discrete arrangement of metal foam reduces the pressure drop with reasonable heat transfer. There always exists a trade-off between the heat transfer and pressure drop as the inlet velocity of the fluid increases. Hence, to underline this situation, discrete metal foams with different thermal conductivity have been considered. Since the thickness of the metal foam in the discrete arrangement plays a significant role in heat transfer, the same has been varied while the distance between the discrete metal foams was kept constant. Moreover, the PPI of the metal foam is changed to see its effect in heat transfer and pressure drop. Hence in this paper, the following objectives are accomplished numerically: (i) to numerically design the SAH in ANSYS and use copper metal foam with discrete arrangement and different thicknesses, (ii) to compare different porosity of copper metal foam with different PPI and (iii) to analyze the best suitable metal foam amongst copper, aluminum, and nickel metal foams according to performance evaluation factor and pressure drop.

Table 1. Literature of metal foam as a porous media in solar air heater.

Ref.	LTE/LTNE	Methodology	Metal Foam Material	Pore Density	Pore Dia.	Porosity	Type of Arrangement of Metal Foam in SAH
[1]	LTNE	Expt. and Num-3D	Al	NM	2	0.92	Horizontal
[23]	NM	Expt.	Al	NM	NM	NM	Vertical–Staggered
[24]	NM	Theoretical–1D	Al	10 PPI, 20 PPI, 30 PPI	NM	0.95	Horizontal
[27]	LTE	Expt.	Cu	20	NM	0.93	Horizontal
[28]	LTE	Expt. and Num-2D	Cu	20	NM	0.93	Horizontal
[29]	NM	Expt.	Cu	15 PPI, 20 PPI	NM	NM	Fin configuration (i) longitudinal (ii) corrugated (iii) staggered
[30]	NM	Expt.	Al	NM	NM	NM	Horizontal
[32]	NM	Expt.	Cu	NM	NM	NM	(i) longitudinal, (ii) corrugated, (iii) staggered
[36]	LTE	Expt. and Num-2D	Al	NM	NM	0.90	Horizontal

Ref—Reference, LTE—local thermal equilibrium, LTNE—local thermal non equilibrium, NM—Not mentioned, Num—Numerical, Expt.—Experimental.

2. Theoretical (Analytical) Design of Solar Air Heater (SAH)

The theoretical design of the SAH was developed at the location of Mechanical Engineering Department, National Institute of Technology Surathkal Karnataka, India. For the conventional SAH, the material and properties are considered as mentioned in [13,37]. The dimensions mentioned in [13] are considered additional design parameters. As given in [2], based on Klein's recommendation, the mean value for the month in April is 15. Hence the analytical solution for the empty channel is done on 15 April at 13:00 PM, because, at this time, the solar radiation is maximum. The latitude and longitude of further study are 12°54′ N, 74°51′ E for the National Institute of Technology Karnataka, Surathkal. The analytical readings are considered during clear sky days in April 2022. Analytical studies are calculated under the climatic conditions of Surathkal, Karnataka, India (12.99° N, 74.81° E). The tilt angle of 13° with the ground surface facing south is taken for testing the SAH to achieve maximum solar radiation. The angle of tilt is equal to the latitude of that location, as mentioned in [13]. The constant a and b for monthly average daily global radiation are obtained for Mangalore city at 0.27 and 0.43, respectively, as mentioned in [2]. For the Surathkal location, wind speed, V_∞ is assumed as 1 m/s. The mean plate temperature is assumed as 323 K.

In this study, the absorber plate is considered as aluminum plate with 0.5 mm thickness. Aluminum is light in weight compared to copper, and its cost is also less than the copper plate. The insulation and frame are considered to be polyurethane foam and wood, respectively, for the present study. The toughened glass with 4 mm thickness is attached above the aluminum plate. The space between the glass and aluminum plate is 120 mm. The air flows through the space between glass and absorber plate. The detailed schematic diagram of the SAH is shown in Figure 1. Table 2 shows the material properties used during the simulation.

The present study is evaluated with similar velocities to [13]. The Reynolds number varies from 3287 to 5479. The material properties are considered to be isotropic. The detailed procedure followed for analytical calculation as explained in [2,3].

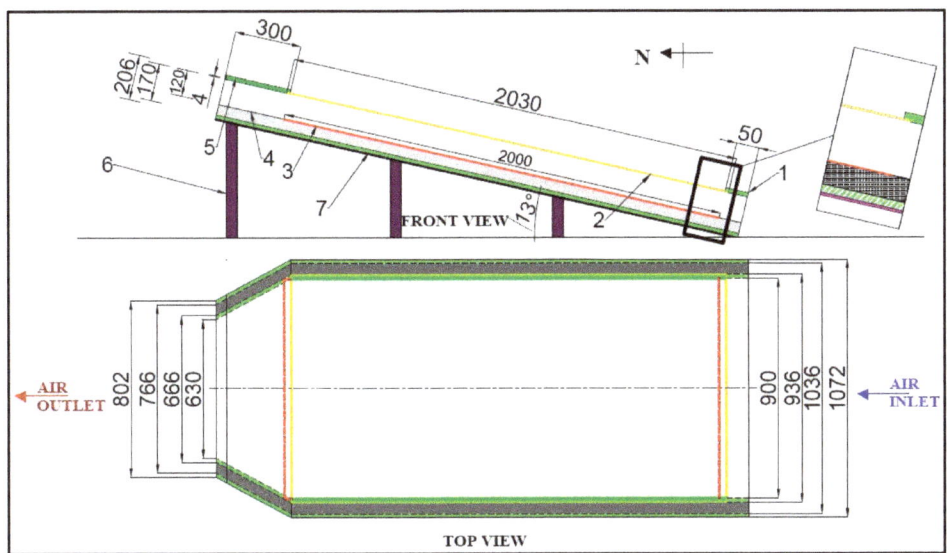

Figure 1. Detailed schematic layout of SAH: (1) wooden material for entrance section (in Green line), (2) toughened glass (in Yellow line), (3) aluminum absorber plate (Red line), (4) polyurethane foam (Grey hatch line), (5) wooden material for exit section (Green line), (6) M S steel stand for support (Purple line), and (7) wooden material (Green line) for the frame as an outer box of the solar air heater (All dimensions are in mm).

Table 2. Material properties considered for simulation [37,38].

Material	Density kg/m^3	Specific Heat (J/kg K)	Thermal Conductivity (W/m K)	Kinematic Viscosity (m^2/s)	Prandtl Number	Emissivity	Absorptivity
Air	1.225	1006.43	0.0242	1.79×10^{-5}	0.702	-	-
Alumium	2719	871	202.4	-	-	0.8	0.95
Glass	2500	670	0.7443	-	-	0.9	-
Wood	700	2310	0.173	-	-	-	-
Copper	8978	381	387.6	-	-	-	-
Nickel	8900	460.6	91.74	-	-	-	-

The following assumptions [13,37,39] are considered for analytical and numerical analysis of SAH:

1. The flow is considered steady state, two-dimensional and incompressible.
2. The thermo-physical properties of air are considered to be constant.
3. Inlet fluid temperature = 300 K.
4. Outlet pressure = P_{atm}.
5. I = 850 W/m^2.
6. Side walls are considered to be adiabatic. Negligible heat loss from the bottom plate and the periphery envelope to the surroundings. Negligible heat loss from the inlet and outlet surfaces.
7. The metal foam is an isotropic and homogeneous porous medium.

The analytical calculations of conventional solar air heater are done by the procedure mentioned in [2,3] as follows-

The monthly average daily inclined irradiance is calculated by following Equation (1) as

$$\delta \text{ (in degree)} = 23.45 \sin [0.9863(284 + n)] \quad (1)$$

where n is the day of the year, the present study for analytical is 15 April hence, n = 105. δ is the declination.

Equation (2) below calculates the value of the angle between an incident solar beam flux and the normal to a plane surface. Considering surface of solar air heater is facing south ($\gamma = 0^0$)

$$\cos \theta = \sin \delta \sin (\phi - \beta) + \cos \delta \cos \omega \cos(\phi - \beta) \tag{2}$$

where θ is the angle between an incident solar beam flux and the normal to a plane surface. Φ is latitude of a location. β is the slope of the solar air heater with the horizontal surface.

The magnitude of ω_{st} for an inclined surface facing south is calculated by Equation (3)

$$|\omega_{st}| = \min[|\cos^{-1}(-\tan\varnothing \tan\delta)|, |\cos^{-1}\{-\tan(\varnothing - \beta)\tan\delta\}|] \tag{3}$$

The daily sunlight or sunshine hours per day is calculated from Equation (4) as

$$S_{max} = \frac{2}{15}\omega_{st} \tag{4}$$

The daily radiation fall on a horizontal surface at the location is calculated by Equation (5) as

$$H_0 = \frac{24}{\pi}I_{SC}(1 + 0.033\cos(\frac{360\,n}{365}))(\sin\omega_s \sin\phi \sin\delta + \cos\phi \cos\delta \sin\omega_s) \tag{5}$$

From Sukhatme et al. [2] constant a and b for Mangalore city in India are 0.27 and 0.43, respectively. Assuming the average sunshine hours per day are 9.5 h for April month. The monthly average of the daily global radiation a horizontal surface is calculated by Equation (6) as

$$\frac{\overline{H_g}}{\overline{H_o}} = a + b\left(\frac{\overline{S}}{\overline{S_{max}}}\right) \tag{6}$$

The monthly average daily diffuse radiation is calculated by Equation (7)

$$\frac{\overline{H_d}}{\overline{H_g}} = 0.8677 - 0.7365\left[\frac{\overline{H_g}}{\overline{H_o}}\right] \tag{7}$$

The hourly radiation on an inclined surface on nth day between 1 h is calculated by Equation (8) as

$$I_o = 1.367\left(1 + 0.033\cos\left(\frac{360\,n}{365}\right)\right)\sin\delta\sin(\phi - \beta) \\ + \cos\delta\cos\omega\cos(\phi - \beta)\frac{kW}{m^2} \tag{8}$$

Normalizing factor f_c is mentioned in Equations (9) and (10)

$$\frac{\overline{I_g}}{\overline{H_g}} = \frac{\overline{I_o}}{\overline{H_o}}\frac{(a + b\cos\omega)}{f_c} \; kJ/m^2\text{-h} \tag{9}$$

where

$$f_c = a + 0.5b\left[\frac{\frac{\pi \omega_s}{180} - \sin\omega_s \cos\omega_s}{\sin\omega_s - \frac{\pi \omega_s}{180}\cos\omega_s}\right] \tag{10}$$

The monthly average hourly diffuse radiation is calculated by Equation (11) as

$$\frac{\overline{I_d}}{\overline{H_d}} = \frac{\overline{I_o}}{\overline{H_o}} \tag{11}$$

The diffuse radiation is calculated by Equation (12) as

$$\frac{\overline{I_{dg}}}{\overline{H_d}} = \frac{\overline{I_o}}{\overline{H_o}} \qquad (12)$$

Choose the maximum value of diffuse radiation (I_d) between Equations (11) and (12) for further calculations.

The beam radiation is calculated by Equation (13) as

$$I_b = I_g - I_d \qquad (13)$$

The tilt factor for beam radiation (r_b) is calculated by Equation (14)

$$r_b = \frac{\cos \theta}{\cos \theta_z} = \frac{\sin \delta \, \sin(\varnothing - \beta) + \cos \delta \cos \omega \cos(\varnothing - \beta)}{\sin \varnothing \sin \delta + \cos \varnothing \cos \delta \cos \omega} \qquad (14)$$

The tilt factor for diffuse radiation (r_d) is calculated by Equation (15)

$$r_d = \frac{(1 + \cos \beta)}{2} \qquad (15)$$

The tilt factor for reflector radiation (r_r) is calculated by Equation (16)

$$r_r = \frac{\rho(1 - \cos \beta)}{2} \qquad (16)$$

Assume ground reflectivity be 0.2. [2] The total flux (I_T) falling on tilted surface at any instant is calculated by Equation (17) as

$$I_T = I_b \, r_b + I_d r_d + (I_b + I_d) \, r_r \; (W/m^2) \qquad (17)$$

The total flux (I_T) falling on tilted surface at any instant is calculated by flux coming on the surface of absorber plate i.e., flux incident on the transparent glass is passing through glass towards the black painted absorber plate. This flux is the addition of beam and diffuse radiation coming directly on the absorber plate and the radiation reflected onto the surface from surroundings. Here, all the solar radiation coming from the sun is absorbed by the absorber plate. The heated absorber plate transfers heat as heat flux to moving air from inlet to outlet with help of conduction, a convection mechanism neglecting radiation heat transfer. As mentioned in Sukhatme and Nayak [2], it is assumed that the heat flux i.e., solar intensity falling on the absorber plate is not more than ±50 W/m² for a 15 min duration. Hence the solar air heater is working under a steady state condition.

The number of covers is considered for this solar air heater to be 1. The spacing between the plate is 120 mm. The top loss coefficient of solar air heater (U_t) is calculated by Equation (18)

$$U_t = \left[\frac{M}{\left(\frac{C}{T_{pm}}\right)\left(\frac{T_{pm} - T_a}{M+f}\right)^{0.252}} + \frac{1}{h_w}\right]^{-1} + \left[\frac{\sigma \left(T_{pm}^2 + T_a^2\right)(T_{pm} + T_a)}{\frac{1}{\varepsilon_p + 0.0425 \, M \, (1 - \varepsilon_p)} + \frac{2M + f - 1}{\varepsilon_c} - M}\right] \qquad (18)$$

where

$$f_t = \left(\frac{9}{h_w} - \frac{30}{h_w^2}\right)\left(\frac{T_a}{316.9}\right)(1 + 0.091M) \qquad (19)$$

$$C_t = 204.429 \, (\cos \beta)^{0.252} / d^{\,0.24} \qquad (20)$$

d is spacing (in m) between cover plate and absorber plate, h_w is the convective heat transfer coefficient at the top cover. The convective heat transfer coefficient at transparent cover is calculated by Equation (21)

$$h_W = 5.7 + 3.8 V\infty \qquad (21)$$

where σ is the Stefan Boltzmann constant, ε_p and ε_c is the emissivity of the absorber plate surface and bottom surface respectively.

The bottom loss coefficient of solar air heater (U_b) is calculated by Equation (22)

$$U_b = \frac{k_i}{\delta_b} \qquad (22)$$

where k_i is the thermal conductivity of the insulation material and δ_b is the thickness of the insulation material.

The side loss coefficient is assumed as zero.

The overall loss coefficient (U_L) is calculated by Equation (23)

$$U_L = U_t + U_b + U_S \qquad (23)$$

The transmissivity of the cover system of a solar air heater is calculated by Equation (24)

$$\tau = \tau_r \, \tau_a \qquad (24)$$

where τ_r is the transmissivity obtained by considering only reflection and refraction, τ_a is the transmissivity obtained by considering only absorption.

The value of the convective heat transfer coefficient h_{fp} is calculated by using Equation (25)

$$h_{fp} = Nu \left(\frac{k_{air}}{\text{Hydraulic diameter } (d_h)} \right) \qquad (25)$$

where Nu is Nusselt number, and k_{air} is the thermal conductivity of air

The Hydraulic diameter is calculated by Equation (26)

$$\text{Hydraulic diameter}(d_h) = \frac{4 \, (W \times d)}{2 \, (W + d)} \qquad (26)$$

where, W is the width of the absorber plate and d is the spacing between the glass and absorber plate.

The average air velocity is calculated by Equation (27)

$$\text{Average air velocity} = \frac{\dot{m}}{\rho \, (W \times L)} \qquad (27)$$

The Reynold number (Re) is calculated by Equation (28)

$$Re = \frac{\rho \, V \, d_h}{\mu} \qquad (28)$$

The radiative heat transfer coefficient (h_r) is calculated as Equation (29)

$$h_r = \frac{\sigma}{\left(\frac{1}{\varepsilon_p} + \frac{1}{\varepsilon_b} - 1 \right)} (T_{pm} + T_{bm})(T_{pm}^2 + T_{bm}^2) \qquad (29)$$

where h_r is the radiative heat transfer coefficient, T_{pm} and T_{bm} is the mean temperature of the absorber plate and the bottom plate. It can be taken to be equal to the mean fluid temperature T_{fm}.

The effective heat transfer coefficient (h_e) between the absorber plate and the air stream is calculated by Equation (30)

$$h_e = h_{fp} + \frac{h_r\, h_{fb}}{h_r + h_{fb}} \qquad (30)$$

The solar air heater efficiency factor is calculated Equation (31)

$$\acute{F} = \left(1 + \frac{U_L}{h_e}\right)^{-1} \qquad (31)$$

The useful heat gain rate for the solar air heater is calculated by Equation (32)

$$q_u = F_R\, A_P\, [S - U_l(T_{fi} - T_a)] \qquad (32)$$

where F_R is the solar air heater heat removal factor, S is the flux absorbed in the absorber plate.

$$F_R = \frac{\dot{m}\, C_P}{U_L\, A_P}\left[1 - \exp\left\{-\frac{F'\, U_l\, A_P}{\dot{m}\, C_P}\right\}\right] \qquad (33)$$

$$S = I_T(\tau\alpha)_{avg} \qquad (34)$$

The instantaneous efficiency of the solar air heater is calculated by Equation (35)

$$n_i = \frac{q_u}{I_T A_C} \qquad (35)$$

The outlet temperature of the solar air heater is obtained by Equation (36)

$$q_u = \dot{m} C_P (T_{fo} - T_{fi}) \qquad (36)$$

The pressure drop across the collector is calculated by Equation (37)

$$\text{Pressure drop } (\Delta P) = \frac{4\, f\, \rho L V^2}{2\, d_h} \qquad (37)$$

where f is the friction factor, L is the length of SAH.

The detailed information of analytical calculation is mentioned in [2,3]. All the calculations are done with the help of Microsoft Excel.

3. Numerical Modelling and Meshing

All the design and analysis are performed in ANSYS Fluent 2022 R2 software. The empty channel and porous bed analysis are done for the same heat flux, i.e., the same solar intensity falling on the SAH. The dimensions and material properties of SAH, governing equations, methodology, and assumptions are considered as mentioned in [9,10,37,40]. Figure 2 shows the metal foam arrangement adopted for numerical study. Figure 3 is meshing done for 88 mm metal foam thicknesses. The detailed boundary conditions used during simulation are mentioned in Table 3. k-ε viscous model is used in ANSYS Fluent for this study. The planar-space steady-state pressure-based solver with double precision is considered for 2D analysis. A Green Gauss node-based method is used for the gradient to discretize the convection and diffusion terms. A second-order upwind scheme is applied to discretize pressure, momentum, Turbulent kinetic energy, turbulent dissipation

The under-relaxation factors for pressure, momentum, turbulent kinetic energy, turbulent dissipation rate, turbulent viscosity, and energy are taken as 0.3, 0.7, 0.8, and 1, respectively. The relaxation factors for other terms are kept in unity by default. In solution initialization, standard initialization method is selected with computing from the inlet. The convergence criteria set for energy is 10^{-6}, while for other terms it is set as 10^{-5}.

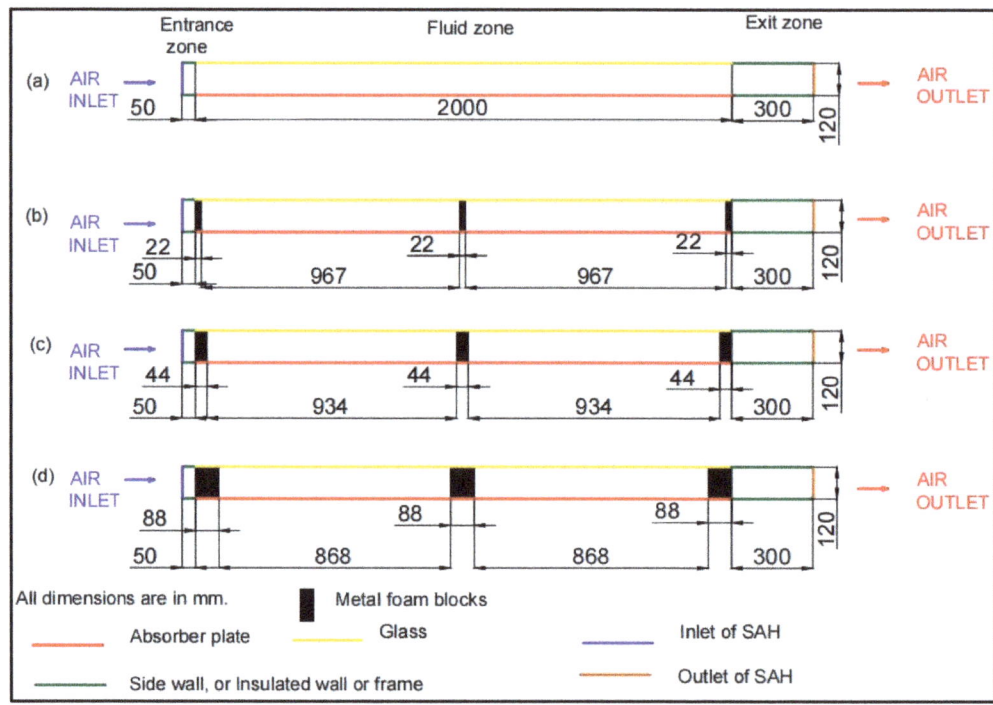

Figure 2. Schematic of SAH: (**a**) empty channel SAH, (**b**) 22 mm filled metal foam SAH, (**c**) 44 mm filled metal foam SAH (**d**) 88 mm filled metal foam SAH.

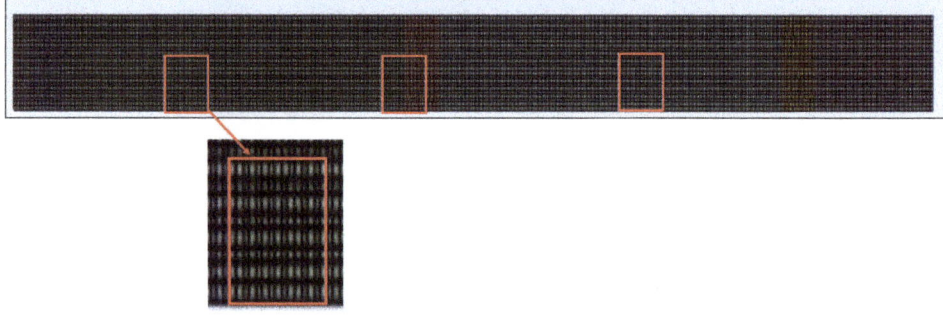

Figure 3. Quadrilateral mesh of 88 mm metal, foam block, solar air heater.

Grid Sensitivity Analysis

The minimum size of the mesh is achieved by grid sensitivity analysis. Table 4 shows the details of the number of elements and its skewness. The simulations are performed for four different mesh sizes. The temperature variation and change in pressure are shown in Table 4. The maximum number of elements is set as baseline and other elements are compared with it. From the results, 125,280 elements are preferred for further computational investigation because it has less deviation than other mesh sizes.

Table 3. Boundary conditions used during simulation in SAH [2,13,14,17,37].

	Momentum	Thermal	
Absorber plate	Stationary wall No slip shear condition	Heat flux = 850 W/m² Material = Aluminium Wall Thickness = 0.0005 m Bottom of the wall	
Glass	Stationary wall No slip shear condition	Mixed Heat transfer coefficient (HTC) = 9.5 W/m² as wind speed assumed as 1 m/s Free stream temperature = 300 K External emissivity = 0.88 External radiation temperature = 300 K Wall thickness = 0.004 m	Equation considered as $h_w = 5.7 + (3.8\ V_\infty)$
Side wall and other wall	Stationary wall No slip shear condition	Heat flux = zero W/m² i.e., adiabatic wallMaterial = wood Wall thickness = 0.018 m	
Inlet	Velocity magnitude as 0.3779, 0.3401, 0.3023, 0.2646, 0.2268, m/s	Inlet temperature = 300 K	
Outlet	Pressure outlet as zero	Back flow temperature = 300 K	

Table 4. Mesh generation.

Number of ELEMENTS	Max Skewness	Outlet Temperature, T_{out}, K	Pressure Drop ΔP, Pa	T_{out} Deviation (%)	ΔP Deviation (%)
70,499	0.273	334.41	0.053	0.2	0
92,652	0.278	334.45	0.053	0.009	0
125,280	0.004	334.47	0.053	0.002	0
180,480	0.0036	334.48	0.053	Baseline	

4. Governing Equations and Turbulence Modelling

For fluid flow in solar air heater, continuity and Reynolds-Averaged-Navier-Stocks (RANS) equations are used. In this study, the Renormalization group (RNG) k-ε turbulence model with enhanced wall treatment [13,14,17,38] is used, as it improves the performance for rotation and streamline curvature.

Continuity equation for empty channel is mentioned in Equation (38a)

$$\frac{\partial(\rho u_j)}{\partial x_i} = 0 \qquad (38a)$$

Continuity equation for metal foam is mentioned in Equation (38b)

$$\frac{\partial(\rho \varepsilon u_j)}{\partial x_i} = 0 \qquad (38b)$$

Momentum equation for empty channel is mentioned in Equation (39a)

$$\frac{\partial}{\partial x_j}(\rho u_i u_j) + \frac{\partial p}{\partial x_i} = \frac{\partial}{\partial x_j}\left[\mu\left(\frac{\partial u_i}{\partial x_j} + \frac{\partial u_j}{\partial x_i}\right)\right] \qquad (39a)$$

Momentum equation for metal foam channel is mentioned in Equation (39b)

$$\frac{\partial}{\partial x_j}(\rho u_i u_j) + \varepsilon\frac{\partial p}{\partial x_j} = \frac{\partial}{\partial x_j}\left[\mu\left(\frac{\partial u_i}{\partial x_j} + \frac{\partial u_j}{\partial x_i}\right) - \varepsilon\left(\frac{\mu_{eff}}{K}\mu_i + \rho C|u|u_i\right)\right] \qquad (39b)$$

Here, K is the permeability and C is the inertia coefficient.

Energy equation for fluid in empty channel,

$$\frac{\partial}{\partial x_i}(\rho u_j T) - \frac{\partial}{\partial x_j}\left[\lambda_f \frac{\partial T}{\partial x_j}\right] = 0 \qquad (40)$$

To model flow through porous media in non-equilibrium thermal model, for simulations solid porous zone and fluid zone are not in thermal equilibrium. Hence, these two zones are interacted with heat transfer only.

For fluid zone equation as:

$$\varepsilon \frac{\partial(\rho C_p u_j T)}{\partial x_j} = \lambda_{fe}\varepsilon \frac{\partial}{\partial x_j}\left(\frac{\partial T_f}{\partial x_j}\right) + h_{sf} a_{sf}(T_s - T_f) \qquad (41)$$

For solid zone equation as:

$$\lambda_{se}(1-\varepsilon)\frac{\partial}{\partial x_j}\left(\frac{\partial T_s}{\partial x_j}\right) = h_{sf} a_{sf}(T_s - T_f) \qquad (42)$$

where,

$$\lambda_{fe} = \lambda_f \cdot \varepsilon \text{ and } \lambda_{se} = \lambda_s \cdot (1-\varepsilon)$$

In this study, to obtain the characteristics of porous media for solar air heater, a Darcy Extended Forchheimer (DEF) flow model is considered. The source term is added with the help of a viscous loss term and an inertial loss term. The DEF model is further joined with momentum equation as a source term. The inertial and viscous loss terms are calculated with the help of permeability and form drag coefficient of porous media. Calmidi and Mahajan [40] have proposed metal foam properties as superficial area density and interfacial heat transfer coefficient, which are given by Equations (43) and (44).

Superficial area density

$$a_{sf} = \frac{3\pi d_f (1 - \exp^{-(\frac{1-\varepsilon}{0.04})})}{(0.59 d_p)^2} \qquad (43)$$

Interfacial heat transfer coefficient

$$\frac{h_{sf} d_f (1 - \exp^{-(\frac{1-\varepsilon}{0.04})})}{\lambda_f} = \begin{cases} 0.76 Re_{d_f}^{0.4} Pr^{0.37}, & (1 \leq Re_{d_f} \leq 40) \\ 0.52 Re_{d_f}^{0.5} Pr^{0.37}, & (40 \leq Re_{d_f} \leq 10^3) \\ 0.26 Re_{d_f}^{0.6} Pr^{0.37}, & (10^3 \leq Re_{d_f} \leq 2 \times 10^5) \end{cases} \qquad (44)$$

where λ_f is the thermal conductivity of working fluid, Pr is the Prandtl number, Re_{d_f} is known as Reynolds number calculated by the fiber diameter of the metal foam. It is calculated from following Equation (45).

$$Re_{d_f} = \left\{ u d_f \left(\frac{1 - \exp^{-(\frac{1-\varepsilon}{0.04})}}{\varepsilon v}\right) \right\} \qquad (45)$$

where d_f is the fiber diameter in m, and d_P is the pore diameter in m.

The properties of metal foam, for example fiber diameter, permeability, pore size and inertial coefficient are determined by Table 5. The detailed information on porous media metal foam is described in [11,22,34]. Table 6 gives the copper metal foam properties considered for present study. The volume of the present porous metal foam block is considered a continuum with homogenous properties with respect to porosity and pore size. The similar homogeneous properties are considered in previous literature. The solid metal foam assumed here is gray and optically thick considering its absorption, isotropic scattering and emission properties throughout the length is same. The representative elementary volume (REV) analysis is important to get more information about heat and/or fluid flow in the porous medium or to determine volume average transport parameters (such as permeability, inertia coefficient, interfacial heat transfer coefficient etc.) or do a pore-scale study including voids and struts in the computational domain requiring extremely long computational time. To reduce the computational time and complexities in smaller size of pores in present porous media, it has uniform mixed medium of air as fluid and metal foam. As per REV scale simulation, it is not necessary to detailed accurate dimensions of porous block. Hence, the flow of air in metal foam is laminar and incompressible. The volume difference between metal foam before heating and after heating due to solar intensities are ignored [41,42].

Table 5. Properties and its correlations of metal foam [11,40].

Sr. No	Properties	Correlations
1	Pore size (d_p)	$d_p = \frac{0.0254}{PPI}$
2	Fiber diameter (d_f)	$\frac{d_f}{d_p} = 1.18\sqrt{\frac{(1-\varepsilon)}{3\pi}}\left(\frac{1}{1-e^{\left(\frac{(1-\varepsilon)}{0.04}\right)}}\right)$
3	Permeability (K)	$K = 0.00073(1-\varepsilon)^{-0.224}\left(\frac{d_f}{d_p}\right)^{-1.11} d_p^2$
4	Inertial/form coefficient (CI)	$CI = 0.00212(1-\varepsilon)^{-0.132}\left(\frac{d_f}{d_p}\right)^{-1.63}$

Table 6. Properties of metal foam [11,40].

PPI	Fiber Diameter	Pore Diameter	Porosity	Viscous Resistance	Inertial Resistance	Interfacial Area Density	Heat Transfer Coefficient
10	0.687	4.644	0.8769	1.742×10^{-7}	176.75	824.2496	85.8858
20	0.619	3.837	0.8567	2.490×10^{-7}	217.04	1106.8362	91.2402
30	0.703	4.732	0.92	1.644×10^{-7}	148.97	936.38	178.908

For all the cases, the inlet temperature is kept constant as the ambient temperature. The outlet is modelled as a pressure outlet with gauge pressure as zero Pascal. The turbulent intensity is specified as Equation (46)

$$I = 0.16 \, (Re)^{-1/8} \text{ in percentage} \quad (46)$$

The bulk mean fluid temperature is calculated as mentioned in Equation (47)

$$T_{bulk\,mean} = \frac{T_i + T_o}{2} \quad (47)$$

where T_i is the inlet temperature of the air in K, T_o is the outlet temperature of the air in K. The convective heat transfer coefficient (h) in W/m^2 K is calculated by Equation (48) as

$$h = \frac{q_W}{T_{abs} - T_{bulk\,mean}} \quad (48)$$

where q_W is the useful heat gain for solar air heater in W/m^2, T_{abs} is the absorber plate temperature in K,

The average heat transfer coefficient ($\overline{h}$) is calculated by Equation (49),

$$\overline{h} = \frac{\sum_1^N h}{N} \quad (49)$$

where the N is the total number of samples or heat transfer coefficient obtained at the particular velocity.

The Nusselt number is calculated by the Equation (50) as

$$Nu = \frac{hD_h}{k_{air}} \qquad (50)$$

where Nu is the Nusselt number, h is the heat transfer coefficient in W/(m² K), D_h is the hydraulic diameter in m, and k_{air} is the thermal conductivity of air in W/(m K).

The average Nusselt number is calculated by the Equation (51) as

$$\overline{Nu} = \frac{\overline{h}D_h}{k_{air}} \qquad (51)$$

where $\overline{Nu}$ is the average Nusselt number, and $\overline{h}$ is the average heat transfer coefficient in W/(m²K).

D_h is the hydraulic diameter in m, k_{air} is the thermal conductivity of air.

The friction factor (f) across the SAH is calculated by Equation (52) with the help of pressure drop across the rectangular channel i.e., inlet pressure and outlet pressure.

$$f = \frac{2\rho_f \Delta P D_h}{u^2 L} \qquad (52)$$

where the ρ_f is the density of fluid in kg/m³, ΔP is the difference of pressure between inlet pressure and outlet pressure in Pa, D_h is the hydraulic diameter in m, u is velocity of air in m/s, and L is the length of the SAH in m.

The heat transfer enhancement ratio for a solar air heater is calculated based on Equation (53)

$$\text{Heat transfer enhancement ratio} = \frac{Nu_P}{Nu_E} \qquad (53)$$

where Nu_P is the Nusselt number of porous media and Nu_E is the Nusselt number of empty channels.

The performance factor is calculated by Equation (54) as

$$\eta_P = \frac{j}{f^{1/3}} \qquad (54)$$

where η_P is the performance factor, j is the Colburn j factor, and f is the friction factor.

5. Results and Discussion

5.1. Verification and Validation of Empty Channel Solar Air Heater

For accurate analysis, the flow of working fluid in the empty channel within the glass and the absorber plate of the test section is essential. The solar radiation first falls on the glass then is transmitted through the glass. Further, this solar intensity is absorbed by the black-painted absorber plate. The air is flowing through the space available between the glass plate and absorber plate, which is 120 mm in the present study. Consequently, air gets heated from the glass as well as the absorber plate. The effect of it shows that the outlet temperature increases. For the Nusselt number relations, when air as a fluid is passing through the two parallel smooth plates, i.e., glass and absorber plate for lower Reynolds number (3000 to 7500) are calculated by relation of the Gnielinski equation as mentioned in [43]. Hence, the correlation of Gnielinski in terms of Nusselt number (Nu) and the correlation of Blasius and Petukhov in terms of friction factor (f) is applied to validate the flow characteristic of turbulent flow in the test section. The validated results of heat transfer and friction factor are shown in Figure 4a,b, respectively. A comparison between Nu and f obtained from the CFD results with the correlation given in Table 7. In Figure 4a,b, the CFD results are in good agreement with the correlations, and the results also showed that the Nu number is directly proportional to the Reynolds number and the friction factor is inversely proportional to the Reynolds number. The correlation and numerical results have similar trends for Nusselt number and friction factor in Figure 4a,b, respectively.

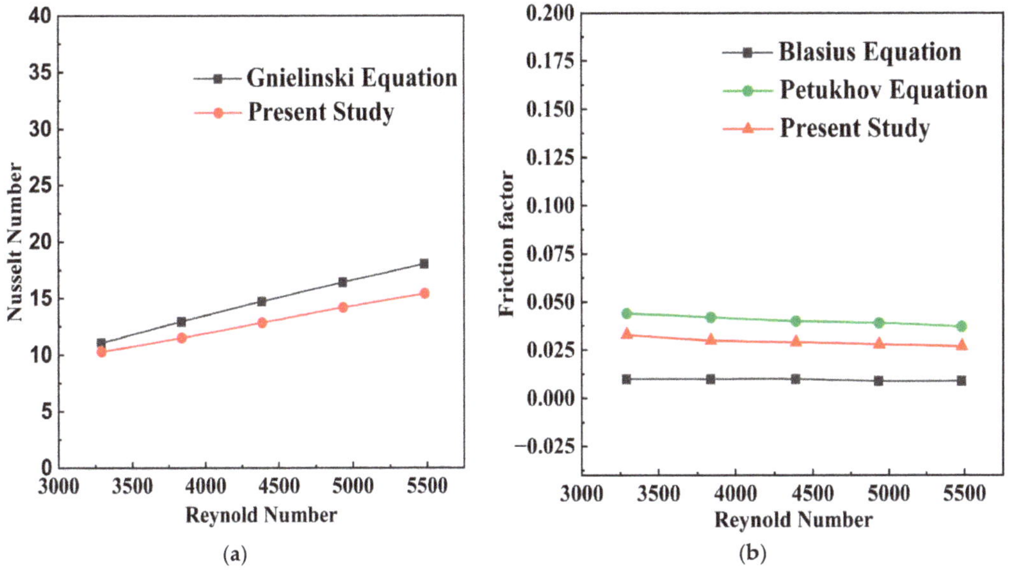

Figure 4. (**a**) Verification of Nusselt number for empty channel of solar air heater. (**b**) Verification of friction factor for empty channel of solar air heater.

Table 7. Correlation equations for the verification of the empty channel SAH.

Name	Correlation Equation	Reference
Gnielinski	$Nu = \frac{(f/8)(Re-1000)\,Pr}{1+12.7(f/8)^{0.5}\,(Pr^{2/3}-1)}$ for $3000 < Re < 7500$	
Petukhov	$f = (0.790 \ln Re - 1.64)^{-2}$ for $3000 < Re < 5\times10^6$	[2,3,14,17,37,43]
Blasius	$f = 0.079 Re^{0.25}$	

5.2. Validation Part

The analytical and numerical results of the present study are similar to the conventional SAH [13]. The average deviation between the analytical and CFD results with K Rajarajeswari et al. [13] is 9.66%. Figure 5 shows that as the mass flow rate increases, the temperature difference between outlet and inlet gives less deviation. The analytical and numerical studies show a similar trend. The average deviation between the analytical and CFD results is 2.78%. The detailed procedure followed for analytical calculation is as explained by Equations (1) to (37) and mentioned in [2,3].

5.3. Effect of Velocity Distribution along the Length of the Channel

The velocity distribution for 0.3779, 0.3401, and 0.3023 m/s of 30 PPI 0.92 porosity is presented in Figure 6. The figure shows that the velocity in the middle of the channel is maximum. The line path for all the velocities shows a parabolic curve for 88 mm thick copper metal foam.

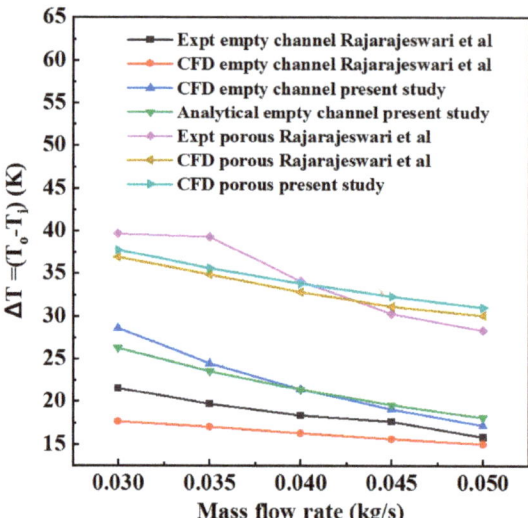

Figure 5. Validation of analytical and CFD results of present empty channel SAH with Rajarajeswari et al. [13].

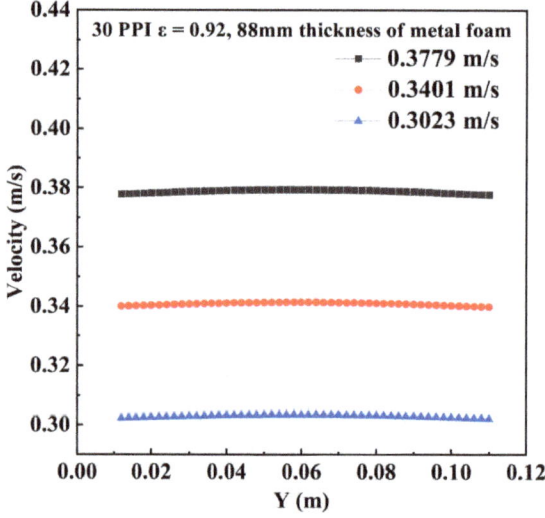

Figure 6. Velocity profile for 30 PPI 0.92 porosity with 88 mm thickness metal foam at three different velocities.

5.4. Temperature Distribution and Velocity Distribution for Different Thickness

Figure 7a–c presents the temperature contour relative to 0.92 porosity 30 PPI Copper metal foam for 0.3779 m/s velocity for the thicknesses of 22 mm, 44 mm, and 88 mm. In the case of porous media such as metal foam, the maximum temperature represents the temperature near the absorber plate. In the case of the lower thickness of the metal foam, the absorber plate temperature is higher, as shown in Figure 7. The temperature is uniform throughout the channel except near the absorber plate. For the same PPI and porosity, as the thickness of the metal foam increases, the absorber plate temperature decreases due to more heat transfer area.

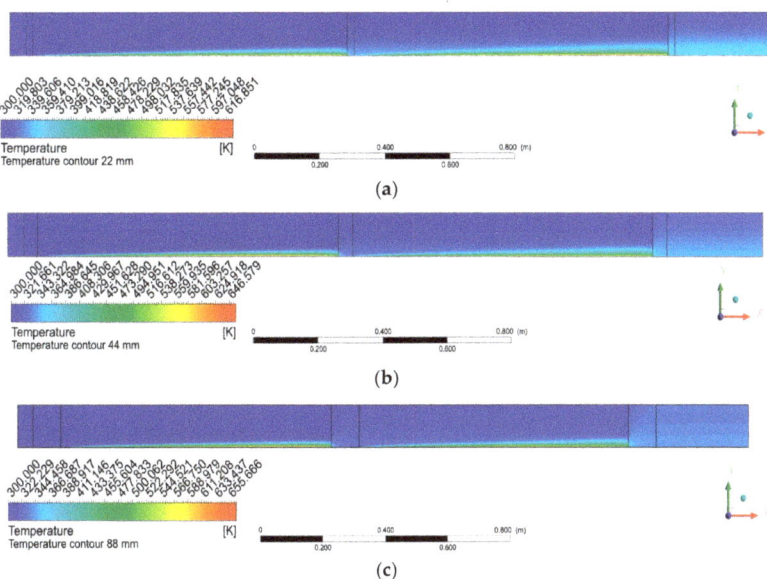

Figure 7. Contour of temperature for 30 PPI 0.92 porosity copper metal foam at 0.3779 velocity of (**a**) 22 mm, (**b**) 44 mm and (**c**) 88 mm thick metal foam.

5.5. Velocity Distribution for Different Thickness

Figure 8 represents the velocity distribution for 0.3779 m/s velocity for 30 PPI 0.92 porosity copper metal foam at (a) 22 mm, (b) 44 mm and (c) 88 mm thick metal foam. Figure 8a–c shows the maximum velocity in the middle of the channel. The velocity near the wall is close to zero because of the shear resistance effect.

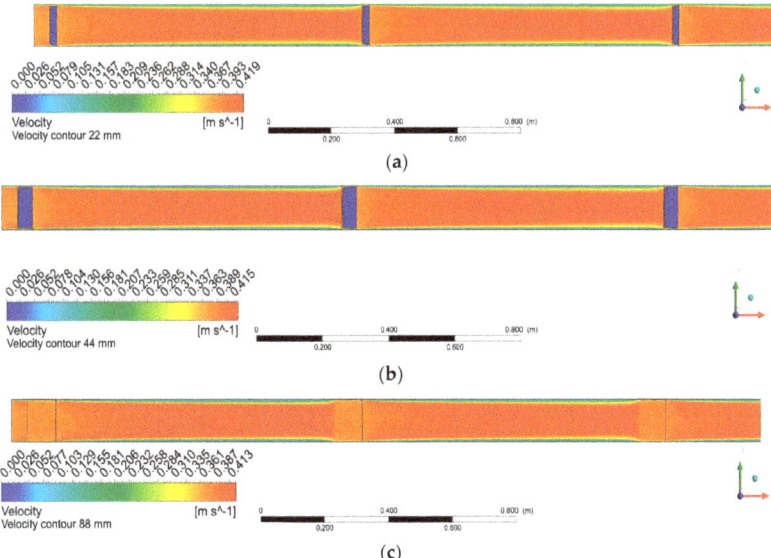

Figure 8. Contour of velocity distribution for 30 PPI 0.92 porosity copper metal foam at 0.3779 m/s velocity for (**a**) 22 mm, (**b**) 44 mm and (**c**) 88 mm thick metal foam.

5.6. Effect of Outlet Temperature and Absorber Plate Temperature

Figure 9a,b shows the variation of temperature with varying mass flow rate from 0.03 to 0.05 kg/s. With 10 PPI with porosity of 0.8769, 20 PPI with a porosity of 0.8567, and 30 PPI with porosity of 0.92 for 22 mm, 44 mm, and 88 mm metal foam thicknesses are considered. It shows that as the mass flow rate and thickness of the metal foam increases, the difference between the absorber plate temperature and bulk fluid temperature reduces. A similar trend is observed for the difference in the absorber plate and outlet temperatures with an increase in mass flow rate. Due to turbulent flow and velocity of air, the difference in the absorber plate and bulk fluid temperature changes. The lower velocity takes more time to travel in the channel, so that the temperature difference increases. Figure 9a shows that the 22 mm 10 PPI copper metal foam has an 8.79% and 11.45% higher average temperature difference of absorber plate temperature and bulk fluid temperature compared to the metal foam of 20 and 30 PPI, respectively. The 44 mm 10 PPI copper metal foam has the same percentage of increase in an average temperature difference of absorber plate temperature and bulk fluid temperature, which is about 3.30% and 3.04% increase for 20 PPI and 30 PPI, respectively. The same trend is observed in 88 mm thickness for 10 PPI compared to 20 PPI and 30 PPI, which is 2.03% and 2.46% higher than 20 PPI and 30 PPI, respectively.

Figure 9b shows that the average temperature difference for 22 mm thickness 10 PPI is higher than all other PPI and all other thicknesses. As the thickness of metal foam increases, the temperature difference between the absorber plate and outlet temperature decreases. As the mass flow rate increase, the temperature difference also decreases. The 22 mm thickness metal foam is having 10.86% and 14.32% more average temperature difference than 20 PPI and 30 PPI, respectively, for the same thickness. The 44 mm, 10 PPI copper metal foam has 4.22% and 3.90% increase in average temperature difference than 20 PPI and 30 PPI, respectively, for the same thickness. The average temperature difference between absorber temperature and outlet temperature of 10 PPI is 2.26% and 3.26% higher than the 20 PPI and 30 PPI of 88 mm metal foam thicknesses.

The above discussion concludes that 10 PPI has a higher temperature difference than 20.

PPI and 30 PPI because of more interfacial surface area of the metal foam. As the thickness of metal foam increases, more conduction occurs near the absorber plate, and hence more heat is transferred to metal foam. So, it is noticed that the average temperature reduces as the thickness of metal foam increases. The empty channel has high average absorber plate temperature than the porous media channel.

5.7. Effect of Heat Transfer Coefficient

Figure 10 shows the heat transfer coefficient variation with respect to different mass flow rate for 10 PPI of 0.8769 porosity, 20 PPI of 0.8567 porosity, and 30 PPI of 0.92 porosity with 22 mm, 44 mm, and 88 mm thicknesses of the metal foam. It is observed that the heat transfer coefficient increases as the mass flow rate increases. As the thickness of the metal foam increases, the heat transfer coefficient also increases. The heat transfer coefficient for 20 PPI 0.8567 porosity and 30 PPI 0.92 porosity is almost in the same range as compared to 10 PPI 0.8769 porosity. The heat transfer coefficient for 10 PPI 0.8769 porosity is less than 20 PPI 0.8567 porosity and 30 PPI 0.92 porosity for all the thicknesses of 22 mm, 44 mm, and 88 mm. The 30 PPI 0.92 porosity has a higher heat transfer coefficient than 10 PPI 0.8769 porosity which is 11.70% for 22 mm thickness, 2.86% for 44 mm thickness, and 2.32% for 88 mm thickness. It is also observed that placing the discrete metal foam and with an increase in thickness of the metal foam the heat dissipation in SAH increases.

5.8. Effect of Nusselt Number

Figure 11 shows that the Nusselt number is directly proportional to the mass flow rate. As the mass flow rate increases, the Nusselt number also increases. With an increase in thickness, the Nusselt number also increases. Figure 11 observes that the 20 PPI of 0.8567 and 30 PPI of 0.92 porosity has a higher Nusselt number compared to 10 PPI of

0.8769 porosity metal foam for 22 mm, 44 mm, and 88 mm thickness of the metal foam. The Nusselt number for 30 PPI of 0.92 porosity is 11.70%, 2.86%, and 2.32% more compared to 10 PPI of 0.8769 porosity for 22 mm, 44 mm, and 88 mm, respectively. The Nusselt number for 20 PPI 0.8567 porosity and 30 PPI 0.92 porosity is almost in the same range for 22 mm, 44 mm, and 88 mm thickness of the metal foam. The results show that with an increase in porosity, the Nusselt number increases because more fluid is flowing through the metal foams.

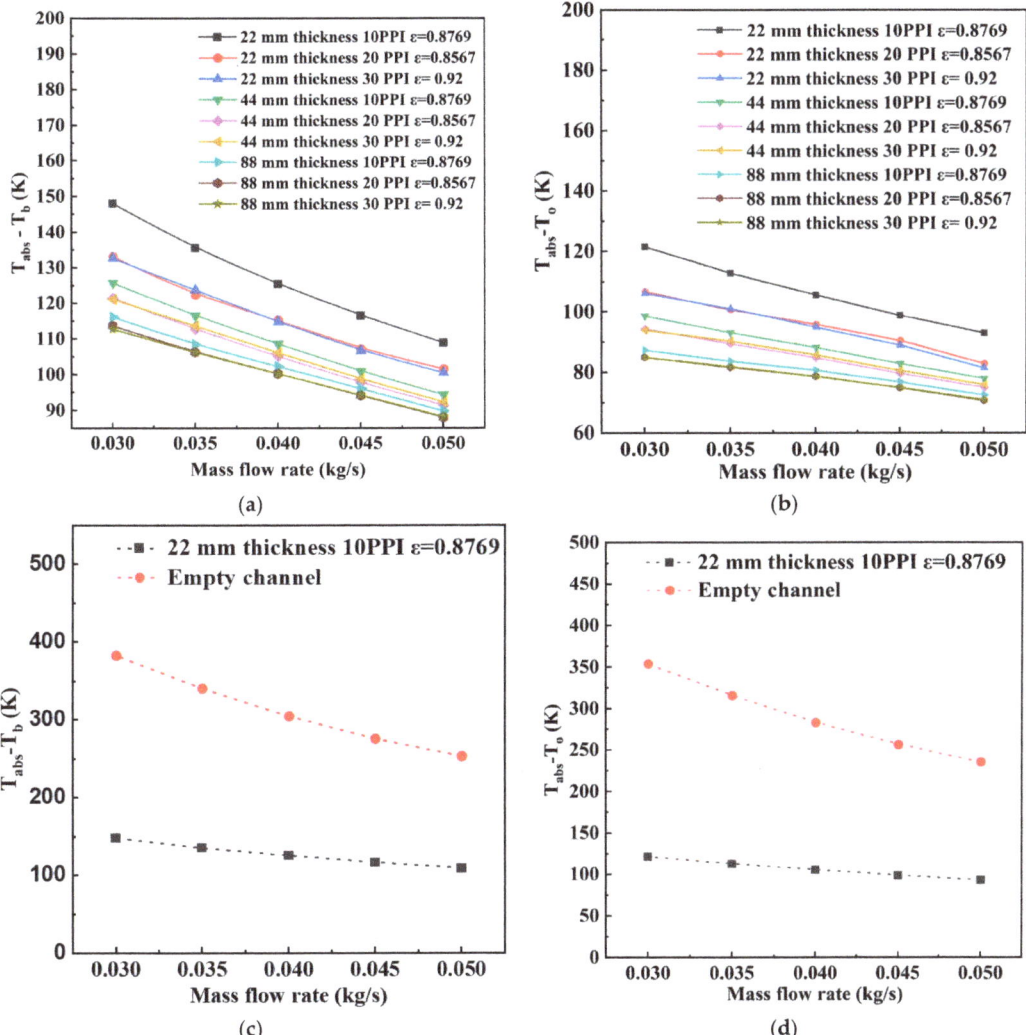

Figure 9. (**a**) Variation of temperature between absorber plate and bulk mean fluid temperature vs. mass flow rate. (**b**) Variation of temperature between absorber plate and outlet temperature vs. mass flow rate. (**c**) Variation of a temperature difference between absorber plate and bulk fluid temperature for empty channel and 22 mm thickness 10 PPI 0.8769 porosity. (**d**) Variation of the temperature difference between absorber plate and outlet temperature for empty channel and 22 mm thickness 10 PPI 0.8769 porosity.

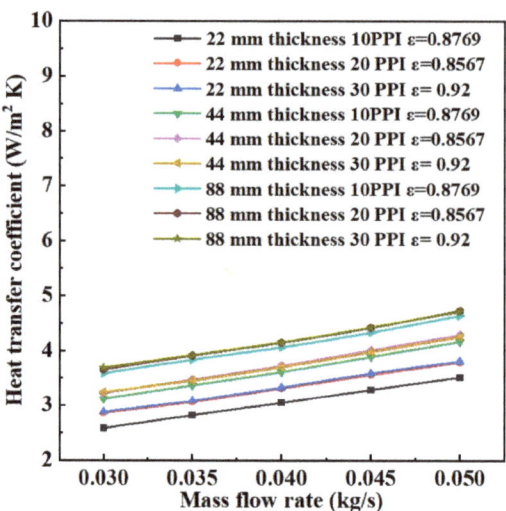

Figure 10. Variation of heat transfer coefficient for different mass flow rates for different PPI and different thickness.

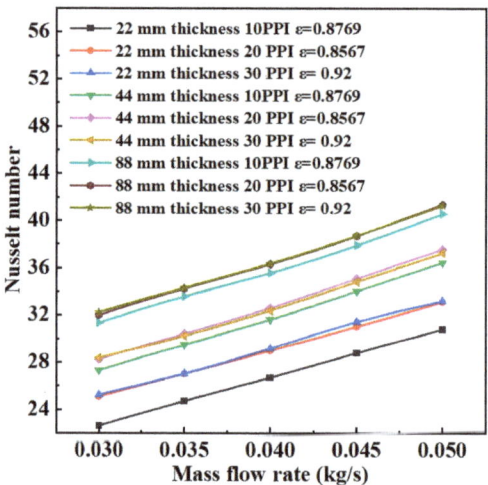

Figure 11. Variation of Nusselt number for different mass flow rates for different PPI and different thicknesses.

5.9. Effect of Pressure Drop

Figure 12 represents that the average pressure drops increase with an increase in the mass flow rate and the thickness of the metal foam 10 PPI, 20 PPI, and 30 PPI of copper metal foam. The average pressure drop is the same for 10 PPI 0.8769 porosity and 30 PPI of 0.92 porosity for 22 mm, 44 mm, and 88 mm thick metal foam. The 30 PPI of 0.92 porosity has 28% and 2% more average pressure drop than 20 PPI 0.8567 porosity and 10 PPI 0.8769 porosity, respectively, for 22 mm, 44 mm, and 88 mm thickness. Hence it is concluded that with increase in heat transfer rate, the pressure drop also increases.

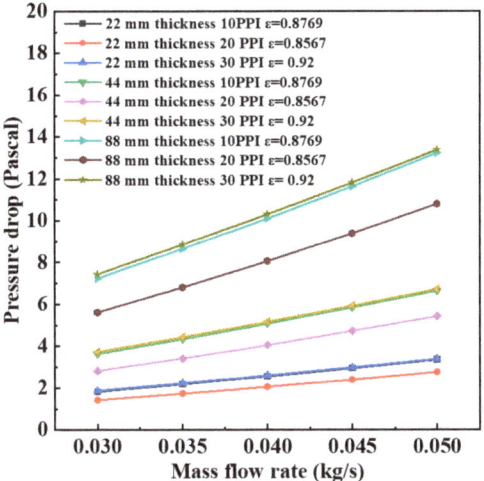

Figure 12. Variation of pressure drop for different mass flow rates for different PPI and different thicknesses.

5.10. Effect of Friction Factor

Figure 13 shows that the friction factor decreases with an increase in mass flow rate. As the thickness of metal foam increases, the friction factor also increases. The 30 PPI 0.92 porosity metal foam has 31% and 2.62% higher friction factor than 20 PPI 0.8567 porosity and 10 PPI 0.8769 porosity copper metal foam for 22 mm, 44 mm, and 88 mm thickness of the metal foam. As the thickness of metal foam increases with twice the value of the previous thickness, the friction factor increases with the same percentage. Hence, it shows that more the PPI, the higher the disturbance to flow, which gives a higher friction factor. The more the thickness of the metal foam, the greater the disturbance of the fluid flow, hence an increase in friction factor.

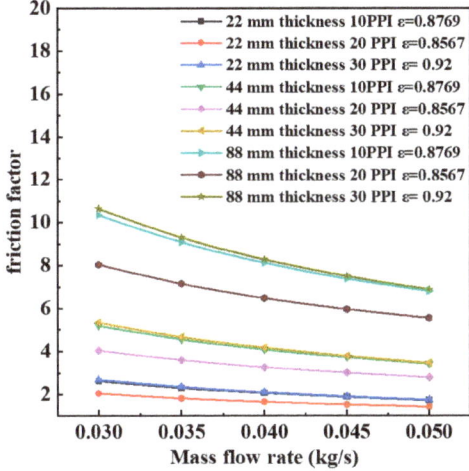

Figure 13. Variation of friction factor for different mass flow rates varying PPI and thickness of metal foam.

5.11. Effect of Ratio of Porous Nusselt Number to Empty Channel Nusselt Number

The heat transfer enhancement ratio is shown in Figure 14. As the PPI and thickness of the metal foam increases, the heat transfer enhancement ratio also increases. The figure represents that with an increase in mass flow rate, the heat transfer enhancement ratio reduces for all cases of PPI and thicknesses. The heat transfer enhancement ratio is higher for lesser velocity than higher velocity because more time is taken for the fluid to flow through the metal foam in lower mass flow rate compared to higher mass flow rate. For 22 mm, 44 mm, and 88 mm, discrete metal foam arrangement in channel shows 30 PPI 0.92 porosity has 11.56%, 3.01%, and 2.41%, respectively, higher than 10 PPI 0.8769 porosity metal foam. With the same 30 PPI 0.92 porosity for an increase in thickness of the metal foam, the heat transfer enhancement increases for 88 mm and 44 mm metal foam thickness which is 12.67% and 10.49%, respectively, more compared to the 22 mm metal foam thickness.

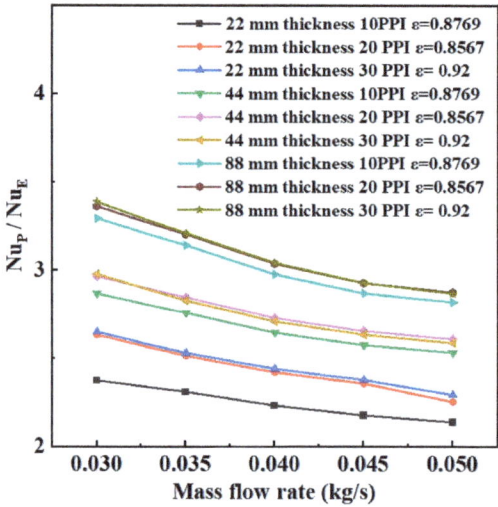

Figure 14. The change in the ratio of Nu_P/Nu_E for different mass flow rates for different PPI and different thicknesses.

5.12. Effect of Performance Factor

Figure 15 shows the performance factor distribution with an increase in mass flow rate. The figure shows that the 22 mm thickness of 20 PPI 0.8567 porosity metal foam is higher than other 44 mm and 88 mm metal foam thickness for 10 PPI 0.8769 porosity and 30 PPI 0.92 porosity. The maximum performance factor for the 20 PPI 0.8567 porosity is 0.0055, 0.0050, and 0.0044 at 22 mm, 44 mm, and 88 mm metal foam thickness, respectively. It is noticed that the performance factor has the maximum value near to 0.0055 at a lower mass flow rate and reduces as the mass flow rate increases.

5.13. Effect of Different Material Metal Foam

Figure 16 shows the difference between different material performance factors. The material considered for comparison is copper, aluminum, and nickel. The figure shows the performance factor for nickel is minimal compared to aluminum and copper. Since copper has high thermal conductivity, the absorber plate and outlet temperature difference are less than nickel. It is observed that the temperature difference reduces as the thermal conductivity increases. The copper metal foam has 3.13% and 9.63% lesser mean temperature difference of the absorber plate temperature and bulk mean fluid temperature than aluminum and nickel, respectively, for 88 mm thick metal foam.

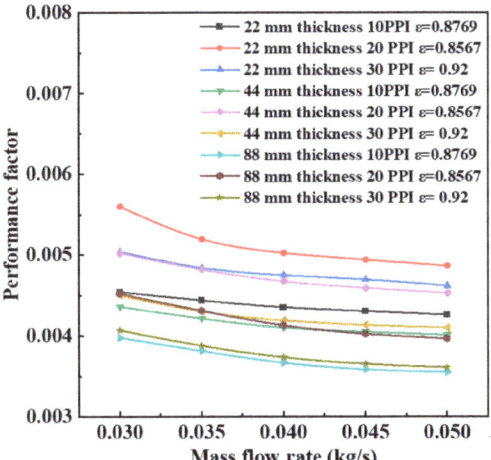

Figure 15. The variation of performance factors for different mass flow rates for different PPI and different thicknesses.

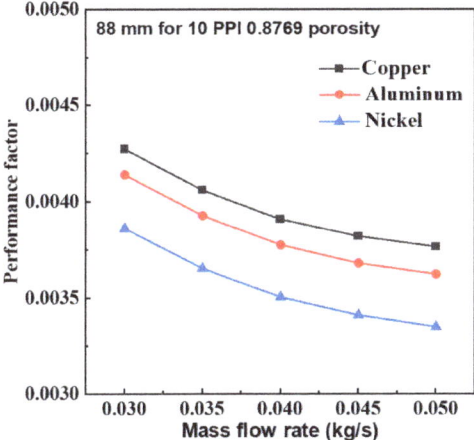

Figure 16. Variation of performance factors for different mass flow rates for different material.

6. Conclusions

The two-dimensional rectangular channel was modelled to evaluate the effect of partial filling of different porosity of copper metal foam in SAH. The complete length of the rectangular channel was 2.35 m and the height was 0.120 m. The computational analysis was performed for three different thicknesses with variation in PPI and porosity of the copper metal foam. Based on the current investigation, the following points are observed:

- With increasing mass flow rate, the outlet temperature decreases for the empty channel as well as for the partially filled porous channel in all cases of PPI and porosity. The same is achieved for different thickness of metal foam. The average temperature difference between the absorber plate and bulk mean fluid temperature is lowest for 88 mm thick metal foam than 22 mm and 44 mm thick metal foam.
- The Nusselt number is higher at higher mass flow rate and rises with increasing PPI and thickness of metal foam. The Nusselt number is highest for 88 mm metal foam, rather than 22 mm and 44 mm thick metal foam The Nusselt number for 22 mm,

- 44 mm, and 88 mm thicknesses is 157.64%, 183.31%, and 218.60%, respectively, higher than the empty channel.
- The pressure drop increases with higher thickness and it increases with increase in mass flow rate. Amongst the 10 PPI, 20 PPI and 30 PPI copper metal foam, the 20 PPI gives a lesser pressure drop than 10 PPI and 30 PPI metal foam for 22 mm, 44 mm and 88 mm thickness. The highest pressure drop belongs to 30 PPI, having 28% and 2% more average pressure drop than 20 PPI and 10 PPI, respectively, for 22 mm, 44 mm, and 88 mm thickness.
- The performance factor is higher for lower velocity, irrespective of PPI and porosity. The 20 PPI 0.8567 porosity with 22 mm thick metal foam has highest performance factor compared to all 10 PPI and 30 PPI metal foam. For mass flow rate of 0.03 kg/s, the maximum performance factor for the 20 PPI 0.8567 porosity is 0.0055, 0.0050, and 0.0044 at 22 mm, 44 mm, and 88 mm metal foam thickness, respectively.
- The temperature difference of the absorber plate and the bulk mean fluid temperature depend on thermal conductivity of material. The copper has lowest temperature difference of the absorber plate and bulk mean fluid temperature compared to aluminum and nickel because of its thermal conductivity.
- With respect to performance factor, 22 mm 20 PPI 0.8567 porosity is best in terms of pressure drop and cost involved in manufacturing the solar air heater.

The effect of porous media in the heat and fluid flow equations can be included by accounting for permeability, inertia coefficient, and effective thermal conductivity for solid and fluid, effective viscosity and interfacial heat transfer coefficient as well as thermal dispersion. All these parameters depend on porosity, strut diameter and topology of the metal foam. The equations used in this study for determination of permeability, inertia coefficient and interfacial heat transfer coefficient as well as those for effective thermal conductivity for the solid and fluid are calculated based on the porosity and strut diameter for metal foams [11,40]. However, the topology effect should be included for more accurate results by performing a representative elementary analysis.

The current study could be improved by using different geometrical parameters and thermal properties of metal foam. Further, the optimum distance between two discrete metal foams relative to other partially filled scenarios could also be explored.

Author Contributions: Conceptualization, R.D. and N.G.; methodology, R.D., N.G. and M.M.; software, R.D. and N.G.; validation, R.D. and N.G.; formal analysis, R.D. and N.G.; investigation, R.D., N.G. and M.M.; resources, R.D. and N.G.; data curation, R.D.; writing—original draft preparation, R.D.; writing—review and editing, R.D., N.G. and M.M.; visualization, R.D.; supervision, N.G. All authors have read and agreed to the published version of the manuscript.

Funding: This research received no external funding.

Data Availability Statement: Not applicable.

Acknowledgments: The author would like to thank Dnyandeo P. Nathe, Government Polytechnic Nashik Maharashtra 422101 and Abhay Wagh, Technical Education Maharashtra, Higher and Technical Education, Government of Maharashtra for giving the opportunity to do research under the quality improvement programme (QIP) in Mechanical Engineering Department, National Institute of Technology Surathkal Karnataka India. The author is also thankful to AICTE India for providing the QIP scheme facility.

Conflicts of Interest: The authors declare no conflict of interest between the contents of this work and any reported studies.

Nomenclature

A_C	Collector area in (m^2)
A_p	Absorber plate area in (m^2)
a, b	Constants for monthly average daily global radiation
a_{sf}	Interfacial surface area (m^{-1})
CFD	Computational fluid dynamics
CI	Inertial form coefficient
Cp	Specific heat of fluid (J/kg K)
Ct	Constant for top loss coefficient
d	The spacing between the glass cover and absorber plate (m)
D_h	Hydraulic diameter
d_f	Fibre diameter (m)
d_p	Pore diameter (m)
ΔP	The pressure drops across the collector in Pa.
FPSC	Flat plate Solar collector
F´	The solar air heater efficiency factor
F_R	The solar air heater heat removal factor
f	Friction factor
f_c	Normalizing factor
f_t	Constant for top loss coefficient
HTC	Heat transfer coefficient (W/m^2 K)
$\overline{H_d}$	Monthly average of the daily diffuse radiation on a horizontal surface (kJ/m^2-day)
$\overline{H_g}$	Monthly average of the daily global radiation on a horizontal surface at a location (kJ/m^2-day)
H_0	The daily radiation falls on a horizontal surface at the location, kJ/m^2
$\overline{H_o}$	The mean value of global radiation for each day of the month, kJ/m^2
h	Heat transfer coefficient (W/m^2 K)
$\bar{h}$	Average heat transfer coefficient, W/m^2 K
h_e	The effective heat transfer coefficient W/m^2 K
hfp	The convective heat transfer coefficient between the absorber plate and the air stream, W/m^2 K
h_r	The radiative heat transfer coefficient (W/m^2 K)
h_{sf}	Interfacial heat transfer coefficient
h_w	The convective heat transfer coefficient at the top cover, W/m^2 K
I_b	Beam radiation, W/m^2
I_d	Diffuse radiation, W/m^2
$\overline{I_d}$	Monthly average of the hourly diffuse radiation on a horizontal surface (kJ/m^2-h)
$\overline{I_g}$	Monthly average of the hourly global radiation on a horizontal surface (kJ/m^2-h)
I_g	Global radiation, W/m^2
$\overline{I_o}$	Monthly average of the hourly extraterrestrial radiation on a horizontal surface (kJ/m^2-h)
I_{SC}	Spectral distribution of extraterrestrial solar radiation flux at mean sun-earth distance (W/m^2)
I_T	The total flux falling on a tilted surface at any instant (W/m^2)
j	Colburn j factor
K	Permeability (m^2)
k_{air}	Thermal conductivity of air (W/m K)
k_i	The thermal conductivity of insulation material (W/m K)
L	Length of the solar air heater (m)
LTE	Local thermal equilibrium
LTNE	Local thermal nonequilibrium model
M	Number of glass covers
$\dot{m}$	Mass flow rate kg/s
Nu	Nusselt number
Nu_P	Nusselt number of porous media
Nu_E	Nusselt number of empty channel
n	The day of the year
PPI	Pores per inch

Pr	Prandtl number
q	Heat flux (W/m^2)
q$_u$	Useful heat gain (W/m^2)
Re	Reynolds number
Re_{d_f}	Reynolds number by fiber diameter
Re$_p$	Reynolds number for porous media
RNG	Renormalization Group
r_b	The tilt factor for beam radiation
r_d	The tilt factor for diffuse radiation
r_r	The tilt factor for reflector radiation
SAH	Solar air heater
SWH	Solar water heater
S	The flux absorbed in the absorber plate (W/m^2)
$\overline{S}$	Monthly average of the sunshine hours per day at the location, hr
$\overline{S}_{max}$	The daily sunlight or sunshine hours per day
ω_s	The hour angle at sunrise or sunset on the horizontal surface in degree
ω_{st}	The hour angle at sunrise or sunset
T$_{bm}$	Mean bottom plate temperature in K
T$_{fi}$	The inlet temperature of the fluid in K
T$_{fo}$	The outlet temperature of a fluid in K
T$_{pm}$	Mean plate temperature in K
U$_b$	The bottom loss coefficient of solar air heater (W/m^2 K)
U$_L$	The overall loss coefficient (W/m^2 K)
U$_s$	The side loss coefficient (W/m^2 K)
U$_t$	The top loss coefficient fir solar air heater (W/m^2- K)
u	The velocity of fluid (m/s)
V$_\infty$	Wind velocity m/s
V	The average air velocity in m/s
W	The width of absorber plate in m
η$_P$	Performance factor
η$_i$	The instantaneous efficiency of the solar air heater
δ	Declination, in degree i.e., the angle made by the line joining the centers of the sun and the earth with the projection of this line on the equatorial plane
δ$_b$	The thickness of insulation material in m
Greek symbols	
ε	Porosity
ε$_C$	Glass cover emissivity
ε$_P$	Absorber plate emissivity
ε$_b$	Bottom plate emissivity
θ	The angle between an incident solar beam flux and the normal to a plane surface
β	The slope of the solar air heater with the horizontal surface.
Φ	Latitude of a location
λ_f	Thermal conductivity of the fluid W/m K
k	Thermal conductivity (W/m K)
ν	Kinematic viscosity (m^2/s)
ρ	Density of the fluid (kg/m^3)
σ	Stefan Boltzmann constant
τ	The transmissivity of the cover system of solar air heater
τ$_a$	The transmissivity obtained by considering only absorption
τ$_r$	The transmissivity obtained by considering only reflection and refraction
μ	Dynamic viscosity (kg/ms)
Subscript	
Abs	Absorber
b	Bulk mean fluid
f	fluid
i	Inlet
max	maximum

o   Outlet
s   Solid

References

1. Kansara, R.; Pathak, M.; Patel, V.K. Performance Assessment of Flat-Plate Solar Collector with Internal Fins and Porous Media through an Integrated Approach of CFD and Experimentation. *Int. J. Therm. Sci.* **2021**, *165*, 106932. [CrossRef]
2. Sukhatme, S.P. *Solar Energy*, 4th ed.; McGraw Hill: New York, NY, USA, 2018.
3. Garg, H.P. *Solar Energy Fundamentals and Applications*, 1st ed.; Mc Graw Hill: New York, NY, USA, 2016.
4. Ismael, M.A. Forced Convection in Partially Compliant Channel with Two Alternated Baffles. *Int. J. Heat Mass Transf.* **2019**, *142*, 118455. [CrossRef]
5. Jadhav, P.H.; Gnanasekaran, N.; Mobedi, M. Analysis of Functionally Graded Metal Foams for the Accomplishment of Heat Transfer Enhancement under Partially Filled Condition in a Heat Exchanger. *Energy* **2023**, *263*, 125691. [CrossRef]
6. Abd Al-Hassan, A.Q.; Ismael, M.A. Numerical Study of Double Diffusive Mixed Convection in Horizontal Channel with Composite Open Porous Cavity. *Spec. Top. Rev. Porous Media Int. J.* **2019**, *10*, 401–419. [CrossRef]
7. Ali Abd Al-Hassan, M.I. Effect of Triangular Porous Layer on the Transfer of Heat and Species in a Channel-open Cavity. In *Heat Transfer*; Wiley: New York, NY, USA, 2022. [CrossRef]
8. García-Moreno, F. Commercial Applications of Metal Foams: Their Properties and Production. *Materials* **2016**, *9*, 85. [CrossRef]
9. Trilok, G.; Gnanasekaran, N.; Mobedi, M. Various Trade-Off Scenarios in Thermo-Hydrodynamic Performance of Metal Foams Due to Variations in Their Thickness and Structural Conditions. *Energies* **2021**, *14*, 8343. [CrossRef]
10. Jadhav, P.H.; Trilok, G.; Gnanasekaran, N.; Mobedi, M. Performance Score Based Multi-Objective Optimization for Thermal Design of Partially Filled High Porosity Metal Foam Pipes under Forced Convection. *Int. J. Heat Mass Transf.* **2022**, *182*, 121911. [CrossRef]
11. Jadhav, P.H.; Gnanasekaran, N.; Perumal, D.A. Numerical Consideration of LTNE and Darcy Extended Forchheimer Models for the Analysis of Forced Convection in a Horizontal Pipe in the Presence of Metal Foam. *J. Heat Transf.* **2021**, *143*, 012702. [CrossRef]
12. Sharma, S.P.; Saini, J.S.; Varma, H.K. Thermal Performance of Packed-Bed Solar Air Heaters. *Sol. Energy* **1991**, *47*, 59–67. [CrossRef]
13. Rajarajeswari, K.; Alok, P.; Sreekumar, A. Simulation and Experimental Investigation of Fluid Flow in Porous and Non-Porous Solar Air Heaters. *Sol. Energy* **2018**, *171*, 258–270. [CrossRef]
14. Singh, S.; Dhruw, L.; Chander, S. Experimental Investigation of a Double Pass Converging Finned Wire Mesh Packed Bed Solar Air Heater. *J. Energy Storage* **2019**, *21*, 713–723. [CrossRef]
15. Kesavan, S.; Arjunan, T.V.; Vijayan, S. Thermodynamic Analysis of a Triple-Pass Solar Dryer for Drying Potato Slices. *J. Therm. Anal. Calorim.* **2019**, *136*, 159–171. [CrossRef]
16. Sözen, A.; Şirin, C.; Khanlari, A.; Tuncer, A.D.; Gürbüz, E.Y. Thermal Performance Enhancement of Tube-Type Alternative Indirect Solar Dryer with Iron Mesh Modification. *Sol. Energy* **2020**, *207*, 1269–1281. [CrossRef]
17. Singh, S. Experimental and Numerical Investigations of a Single and Double Pass Porous Serpentine Wavy Wiremesh Packed Bed Solar Air Heater. *Renew. Energy* **2020**, *145*, 1361–1387. [CrossRef]
18. Srinivas, K.E.S.; Harikrishnan, D.; Mobedi, M. Correlations and Numerical Modeling of Stacked Woven Wire-Mesh Porous Media for Heat Exchange Applications. *Energies* **2022**, *15*, 2371.
19. Li, X.; Yuan, F.; Tian, W.; Dai, C.; Yang, X.; Wang, D.; Du, J.; Yu, W.; Yuan, H. Heat Transfer Enhancement of Nanofluids with Non-Spherical Nanoparticles: A Review. *Appl. Sci.* **2022**, *12*, 4767. [CrossRef]
20. Mortazavifar, S.L.; Salehi, M.R.; Shahraki, M.; Abiri, E. Optimization of Light Absorption in Ultrathin Elliptical Silicon Nanowire Arrays for Solar Cell Applications. *J. Mod. Opt.* **2022**, *69*, 368–380. [CrossRef]
21. Sezer, N.; Khan, S.A.; Biçer, Y.; Koç, M. Enhanced Nucleate Boiling Heat Transfer on Bubble-Induced Assembly of 3D Porous Interconnected Graphene Oxide/Silver Nanowire Hybrid Network. *Case Stud. Therm. Eng.* **2022**, *38*, 102334. [CrossRef]
22. Kamath, P.M.; Balaji, C.; Venkateshan, S.P. Convection Heat Transfer from Aluminium and Copper Foams in a Vertical Channel—An Experimental Study. *Int. J. Therm. Sci.* **2013**, *64*, 1–10. [CrossRef]
23. Bayrak, F.; Oztop, H.F. Experimental Analysis of Thermal Performance of Solar Air Collectors with Aluminum Foam Obstacles Kapali Hücreli Alüminyum Köpük Engeller Sahip Hava Isitmali Güneş Kollektörlerinin Isil Performansinin Deneysel Analizi. *J. Therm. Sci. Technol.* **2015**, *35*, 11–20.
24. Dukhan, N.; Quinones, P.D. Convective Heat Transfer Analysis of Open Cell Metal Foam for Solar Air Heaters. *Int. Sol. Energy Conf.* **2003**, *36762*, 287–293.
25. Mancin, S.; Zilio, C.; Diani, A.; Rossetto, L. Experimental Air Heat Transfer and Pressure Drop through Copper Foams. *Exp. Therm. Fluid Sci.* **2012**, *36*, 224–232. [CrossRef]
26. Chen, C.C.; Huang, P.C. Numerical Study of Heat Transfer Enhancement for a Novel Flat-Plate Solar Water Collector Using Metal-Foam Blocks. *Int. J. Heat Mass Transf.* **2012**, *55*, 6734–6745. [CrossRef]
27. Jouybari, H.J.; Saedodin, S.; Zamzamian, A.; Nimvari, M.E.; Wongwises, S. Effects of Porous Material and Nanoparticles on the Thermal Performance of a Flat Plate Solar Collector: An Experimental Study. *Renew. Energy* **2017**, *114*, 1407–1418. [CrossRef]
28. Saedodin, S.; Zamzamian, S.A.H.; Nimvari, M.E.; Wongwises, S.; Jouybari, H.J. Performance Evaluation of a Flat-Plate Solar Collector Filled with Porous Metal Foam: Experimental and Numerical Analysis. *Energy Convers. Manag.* **2017**, *153*, 278–287. [CrossRef]

29. Hussien, S.Q.; Farhan, A.A.; Hussien, S.Q.; Farhan, A.A. Research The Effect of Metal Foam Fins on the Thermo-Hydraulic Performance of a Solar Air Heater. *Int. J. Renew. Energy* **2019**, *9*, 840–847.
30. Baig, W.; Ali, H.M. An Experimental Investigation of Performance of a Double Pass Solar Air Heater with Foam Aluminum Thermal Storage Medium. *Case Stud. Therm. Eng.* **2019**, *14*, 100440. [CrossRef]
31. Anirudh, K.; Dhinakaran, S. Performance Improvement of a Flat-Plate Solar Collector by Inserting Intermittent Porous Blocks. *Renew. Energy* **2020**, *145*, 428–441. [CrossRef]
32. Farhan, A.A.; Obaid, Z.A.H.; Hussien, S.Q. Analysis of Exergetic Performance for a Solar Air Heater with Metal Foam Fins. *Heat Transf.-Asian Res.* **2020**, *49*, 3190–3204. [CrossRef]
33. Anirudh, K.; Dhinakaran, S. Numerical Analysis of the Performance Improvement of a Flat-Plate Solar Collector Using Conjugated Porous Blocks. *Renew. Energy* **2021**, *172*, 382–391. [CrossRef]
34. Jadhav, P.H.; Nagarajan, G.; Perumal, D.A. Conjugate Heat Transfer Study Comprising the Effect of Thermal Conductivity and Irreversibility in a Pipe Filled with Metallic Foams. *Heat Mass Transf. Und Stoffuebertragung* **2021**, *57*, 911–930. [CrossRef]
35. Yadav, S.; Saini, R.P. Numerical Investigation on the Performance of a Solar Air Heater Using Jet Impingement with Absorber Plate. *Sol. Energy* **2020**, *208*, 236–248. [CrossRef]
36. Chen, Z.; Gu, M.; Peng, D. Heat Transfer Performance Analysis of a Solar Flat-Plate Collector with an Integrated Metal Foam Porous Structure Filled with Paraffin. *Appl. Therm. Eng.* **2010**, *30*, 1967–1973. [CrossRef]
37. Yadav, A.S.; Bhagoria, J.L. Heat Transfer and Fluid Flow Analysis of Solar Air Heater: A Review of CFD Approach. *Renew. Sustain. Energy Rev.* **2013**, *23*, 60–79. [CrossRef]
38. *Ansys Fluent R2 Student Version*; Ansys, Inc.: Canonsburg, PA, USA, 2022.
39. Vafai, K. *Handbook of Porous Media*, 2nd ed.; CRC Press: Boca Raton, FL, USA, 2005.
40. Calmidi, V.V.; Mahajan, R.L. Forced Convection in High Porosity Metal Foams. *J. Heat Transf.* **2000**, *122*, 557–565. [CrossRef]
41. Sharma, S.; Talukdar, P. Thermo-Mechanical Analysis of a Porous Volumetric Solar Receiver Subjected to Concentrated Solar Radiation. *Sol. Energy* **2022**, *247*, 41–54. [CrossRef]
42. Zhang, S.; Yao, Y.; Jin, Y.; Shang, Z.; Yan, Y. Heat Transfer Characteristics of Ceramic Foam/Molten Salt Composite Phase Change Material (CPCM) for Medium-Temperature Thermal Energy Storage. *Int. J. Heat Mass Transf.* **2022**, *196*, 123262. [CrossRef]
43. Cengel, Y.A. *Heat Transfer a Practical Approach*; McGraw-Hill: New York, NY, USA, 2003.

MDPI
St. Alban-Anlage 66
4052 Basel
Switzerland
Tel. +41 61 683 77 34
Fax +41 61 302 89 18
www.mdpi.com

Energies Editorial Office
E-mail: energies@mdpi.com
www.mdpi.com/journal/energies

www.ingramcontent.com/pod-product-compliance
Lightning Source LLC
LaVergne TN
LVHW070403100526
838202LV00014B/1379